Practical Solutions for Modern NLP Challenges

Mastering LLMs and SLMs for Real-World NLP in Cloud and Open-Source

Venkata Gunnu
Shubham Shah
Anvesh Minukuri
Jayanth Gopu

Apress®

Practical Solutions for Modern NLP Challenges: Mastering LLMs and SLMs for Real-World NLP in Cloud and Open-Source

Venkata Gunnu
malvern, PA, USA

Shubham Shah
Rajkot, Gujarat, India

Anvesh Minukuri
middletown, DE, USA

Jayanth Gopu
Hyderabad, Andhra Pradesh, India

ISBN-13 (pbk): 979-8-8688-2055-7
https://doi.org/10.1007/979-8-8688-2056-4

ISBN-13 (electronic): 979-8-8688-2056-4

Copyright © 2025 by Venkata Gunnu, Shubham Shah, Anvesh Minukuri and Jayanth Gopu

This work is subject to copyright. All rights are reserved by the Publisher, whether the whole or part of the material is concerned, specifically the rights of translation, reprinting, reuse of illustrations, recitation, broadcasting, reproduction on microfilms or in any other physical way, and transmission or information storage and retrieval, electronic adaptation, computer software, or by similar or dissimilar methodology now known or hereafter developed.

Trademarked names, logos, and images may appear in this book. Rather than use a trademark symbol with every occurrence of a trademarked name, logo, or image we use the names, logos, and images only in an editorial fashion and to the benefit of the trademark owner, with no intention of infringement of the trademark.

The use in this publication of trade names, trademarks, service marks, and similar terms, even if they are not identified as such, is not to be taken as an expression of opinion as to whether or not they are subject to proprietary rights.

While the advice and information in this book are believed to be true and accurate at the date of publication, neither the authors nor the editors nor the publisher can accept any legal responsibility for any errors or omissions that may be made. The publisher makes no warranty, express or implied, with respect to the material contained herein.

Managing Director, Apress Media LLC: Welmoed Spahr
Acquisitions Editor: Aditee Mirashi
Editorial Assistant: Gryffin Winkler
Copy Editor: Kezia Endsley

Cover designed by eStudioCalamar

Distributed to the book trade worldwide by Springer Science+Business Media New York, 1 New York Plaza, New York, NY 10004. Phone 1-800-SPRINGER, fax (201) 348-4505, e-mail orders-ny@springer-sbm.com, or visit www.springeronline.com. Apress Media, LLC is a Delaware LLC and the sole member (owner) is Springer Science + Business Media Finance Inc (SSBM Finance Inc). SSBM Finance Inc is a **Delaware** corporation.

For information on translations, please e-mail booktranslations@springernature.com; for reprint, paperback, or audio rights, please e-mail bookpermissions@springernature.com.

Apress titles may be purchased in bulk for academic, corporate, or promotional use. eBook versions and licenses are also available for most titles. For more information, reference our Print and eBook Bulk Sales web page at http://www.apress.com/bulk-sales.

Any source code or other supplementary material referenced by the author in this book is available to readers on GitHub. For more detailed information, please visit https://www.apress.com/gp/services/source-code.

If disposing of this product, please recycle the paper

Dedicated to the practitioners and engineers advancing NLP—from fine-tuning to inference, from open-source to cloud—turning language models into real-world solutions across modern NLP tasks

Table of Contents

About the Authors xvii

About the Technical Reviewers xix

Acknowledgments xxi

Introduction xxiii

Chapter 1: Introduction to LLMs, SLMs, and Modern NLP Challenges 1

   1.1 Introduction: Understanding LLMs and SLMs in the NLP Landscape 1

   1.2 What Is an LLM? 5

      1.2.1 Defining LLMs 5

      1.2.2 Key Features and Significance 6

      1.2.3 Evolution of LLMs 6

      1.2.4 Practical Applications of LLMs 7

      1.2.5 Challenges in Scaling LLMs 8

      1.2.6 Future Directions for LLMs 9

   1.3 What Is an SLM? 10

      1.3.1 Defining SLMs 10

      1.3.2 Key Features of SLMs 10

      1.3.3 Evolution of SLMs 11

      1.3.4 Practical Applications of SLMs 11

      1.3.5 Innovations Driving SLMs Forward 12

      1.3.6 Challenges and Limitations 12

      1.3.7 Case Studies and Real-World Scenarios 12

      1.3.8 Future Directions for SLMs 13

      1.3.9 Techniques Enhancing SLM Efficiency 13

TABLE OF CONTENTS

    1.4 Overview of Modern NLP Tasks 14
        1.4.1 Text Classification 14
        1.4.2 Sentiment Analysis 15
        1.4.3 Text Summarization 15
        1.4.4 Question Answering (QA) 16
    1.5 LLMs on AWS vs. Open-Source Frameworks 18
        1.5.1 AWS Suite Features for Modern NLP Tasks 18
        1.5.2 Open-Source Frameworks 21
        1.5.3 Benefits of Cloud Platforms vs. Open-Source Frameworks 22
        1.5.4 When to Choose Cloud Platforms vs. Open-Source Frameworks 23
        1.5.5 The Hybrid Approach 24
    1.6 LLM Lifecycle Activities 24
        1.6.1 Data Collection and Preparation 24
        1.6.2 Model Training and Fine-Tuning 26
        1.6.3 Model Evaluation and Testing 27
        1.6.4 Deployment and Monitoring 31
    1.7 LLM Selection Based on Cost, Compute, and Effectiveness and SLM Exploration ... 33
        1.7.1 The LLM vs. SLM Dilemma 34
        1.7.2 Effectiveness: Balancing Performance and Efficiency 36
        1.7.3 SLM Techniques and Innovations 38
        1.7.4 The Future of SLMs in NLP 39
    1.8 Summary 40

Chapter 2: Text Generation with LLMs and SLMs 41
    2.1 Text Generation with LLMs and SLMs 42
    2.2 Types of Text Generation Tasks 43
        2.2.1 Creative Writing 43
        2.2.2 Automated Content Generation 44
        2.2.3 Chatbots and Conversational Agents 46
        2.2.4 Generative vs. Extractive Tasks 47

2.3 Building and Deploying LLMs and SLMs for Text Generation 48
2.3.1 Choosing the Right LLM or SLM for Text Generation 52
2.3.2 Training and Fine-Tuning LLMs and SLMs 53
2.3.3 Efficient Inference and Deployment of LLMs and SLMs 54
2.4 Implementing Text Generation on AWS 56
2.4.1 Transition from Building SLMs to AWS Implementation 56
2.4.2 Deploying LLMs with AWS SageMaker 57
2.4.3 Real-Time Inference Using AWS Lambda (Continued) 59
2.4.4 Optimizing Models for Edge Deployment with Amazon SageMaker Neo 60
2.4.5 Integrating AWS Services for Comprehensive Text Generation Solutions 62
2.4.6 Use Case A: SLM for Technical Support Ticket Triage 63
2.4.7 Use Case B: Deploying an LLM for Marketing Content Generation 66
2.4.8 Best Practices and Architectural Considerations 70
2.4.9 Conclusion 73
2.5 Open-Source Implementation 73
2.5.1 Fine-Tuning GPT-2 LLM with Hugging Face 73
2.5.2 SLM Local Deployment 78
2.6 Industry Use Cases 79
2.6.1 Content Generation and Marketing 81
2.6.2 Industry-Level Best Practices 81
2.6.3 Chatbots and Customer Support 84
2.6.4 On-Device Text Generation 86
2.7 Summary 88

Chapter 3: Text Classification with LLMs and SLMs 89
3.1 Types of Classification Tasks 90
3.1.1 Binary Classification 90
3.1.2 Multi-Class Classification 92
3.1.3 Multi-Label Classification 93
3.1.4 Conclusion 97

TABLE OF CONTENTS

3.2 Implementation on AWS 97
3.2.1 Amazon Comprehend for Built-In Models 97
3.2.2 Custom Model Training with SageMaker 102
3.2.3 Inference and Deployment 104
3.3 Open-Source Implementation 105
3.3.1 Building Text Classifiers Using Hugging Face Transformers 105
3.3.2 Using SLMs for Faster Training and Inference 106
3.3.3 Data Preparation Techniques 112
3.3.4 Summary 113
3.4 Industry Use Cases 113
3.4.1 Spam Detection 113
3.4.2 Customer Feedback Categorization 114
3.4.3 Social Media Sentiment Analysis 115
3.4.4 Automated Customer Support 115
3.5 Summary 116

Chapter 4: Named Entity Recognition (NER) with LLMs and SLMs 117
4.1 Introduction to Named Entity Recognition (NER) 118
4.1.1 What Is Named Entity Recognition? 118
4.1.2 Why NER Matters: Real-World Applications 120
4.1.3 Techniques for NER: Traditional Models vs. LLMs 121
4.1.4 LLM Agents and Automated NER Pipelines 128
4.2 Implementation on AWS with SLMs and LLMs 130
4.2.1 Using Amazon Comprehend for NER Tasks 130
4.2.2 Training Custom NER Models on SageMaker 134
4.2.3 Deployment Considerations on AWS 139
4.2.4 Example Use Case: Real-World Deployment 144
4.2.5 Key Takeaways and Practical Insights 147
4.2.6 Summary of NER Implementation with AWS Services 148
4.3 Open-Source Implementation 149
4.3.1 From AWS Solutions to Open-Source Alternatives 149
4.3.2 Fine-Tuning with Hugging Face Transformers (BERT, RoBERTa, etc.) 149

4.3.3 Embracing Smaller Models: DistilBERT and TinyBERT 153

4.3.4 spaCy: Industrial-Strength NER Pipelines 154

4.3.5 Flair: Simple and Flexible NER Training 156

4.3.6 Model Evaluation and Optimization 158

4.3.7 Quantizing the Model for Efficient Inference 161

4.4 Low-level LLM Exploration: Small LLM Cost-Effective Alternatives 162

4.4.1 Cost-Effectiveness of SLMs in NER 162

4.4.2 Use Case: Small LLMs for Real-Time NER 169

4.4.3 Fine-Tuning SLMs for NER Tasks 169

4.5 Industry Use Cases 170

4.5.1 Resume Parsing 170

4.5.2 Legal Document Analysis 171

4.5.3 Customer Data Extraction 172

4.6 Summary 173

Chapter 5: Sentiment Analysis with LLMs and SLMs 175

5.1 Understanding Sentiment Analysis 176

5.1.1 Sentiment Analysis: Definition and Significance 176

5.1.2 Challenges in Analyzing Sentiment 178

5.1.3 LLMs in Sentiment Analysis 180

5.1.4 SLMs in Sentiment Analysis 182

5.1.5 LLMs vs. SLMs: A Comparative View 183

5.1.6 Example: Classifying Customer Reviews with Transformers 186

5.1.7 Next Steps: Implementing Sentiment Analysis on AWS 189

5.2 Implementing Financial Sentiment Analysis with AWS Services 189

5.2.1 Overview and Architecture 190

5.2.2 Using Amazon Comprehend for Financial Sentiment Analysis 191

5.2.3 Deploying SLMs on AWS Lambda for Real-Time Sentiment 193

5.2.4 Custom Model Training and Deployment Using Amazon SageMaker 199

5.2.5 Comparing Approaches: Comprehend vs. Custom Models vs. Lambda-deployed SLMs 205

5.2.6 Choosing the Best Approach 208

5.2.7 Conclusion and Next Steps 210

5.3 Open-Source Implementation ... 210
5.3.1 Implementing Sentiment Analysis with Hugging Face Transformers ... 210
5.3.2 MLOps Considerations for LLMs and SLMs in Sentiment Analysis ... 214
5.3.3 Techniques for Deployment of LLMs and SLMs ... 215
5.3.4 Data Labeling Techniques and Challenges for Sentiment Classification ... 218

5.4 Industry Use Cases ... 220
5.4.1 Brand Monitoring ... 220
5.4.2 Social Media Analysis ... 221
5.4.3 Customer Satisfaction Insights ... 221

5.5 Summary ... 224

Chapter 6: Question Answering (QA) ... 225

6.1 Introduction to Question Answering ... 226
6.1.1 Extractive Question Answering ... 226
6.1.2 Abstractive Question Answering ... 230
6.1.3 Large vs. Small Language Models in QA ... 234
6.1.4 LLM Agents and Multi-step Reasoning in QA ... 238

6.2 Implementation of QA Systems on AWS ... 243
6.2.1 Extractive vs. Abstractive QA on AWS ... 243
6.2.2 Model Selection Strategies for QA ... 245
6.2.3 Cost and Performance Benchmarks ... 253
6.2.4 Conclusion ... 257

6.3 Open-Source Implementation ... 258
6.3.1 Implementing QA with Hugging Face Transformers ... 258
6.3.2 Fine-Tuning T5 for Generative QA ... 260
6.3.3 Fine-Tuning DistilBERT for Extractive QA ... 261
6.3.4 Next-Gen Fine-Tuning Strategies ... 262
6.3.5 Deployment of QA Models ... 263
6.3.6 Inference Optimization Strategies for QA Models ... 264
6.3.7 Evaluation Metrics for QA Performance ... 267

TABLE OF CONTENTS

6.4 Industry Use Cases 272
- 6.4.1 Customer Support Chatbot 272
- 6.4.2 Knowledge Base Querying (Legal, HR) 274
- 6.4.3 Automated Assistants 276

6.5 Summary 279

Chapter 7: Text Summarization 281

7.1 Overview of Text Summarization 282
- 7.1.1 Extractive vs. Abstractive Summarization Techniques 282
- 7.1.2 Extractive Summarization Approaches (BERT and Other Models) 285
- 7.1.3 Abstractive Summarization Approaches (PEGASUS, BART, GPT-3, T5) 287
- 7.1.4 Recent Trends: Efficient SLMs, Retrieval-Augmented Summarization, and LLM Agents 290

7.2 Implementation on AWS: Using SageMaker and Lambda to Deploy Summarization at Scale 294
- 7.2.1 Architectural Overview: Deploying Summarization Pipelines at Scale on AWS 295
- 7.2.2 Deploying Extractive Summarization with DistilBERT on AWS Lambda 298
- 7.2.3 Deploying Abstractive Summarization with T5 on AWS SageMaker 300

7.3 Open-Source Summarization Systems 303
- 7.3.1 Implementing Summarization with Hugging Face Transformers 303
- 7.3.2 Fine-Tuning T5 for Abstractive Summarization 304
- 7.3.3 Fine-Tuning DistilBERT for Extractive Summarization 305
- 7.3.4 Next-Gen Fine-Tuning Strategies 305
- 7.3.5 Deployment of Summarization Models 306
- 7.3.6 Inference Optimization Strategies for Summarization Models 307
- 7.3.7 Evaluation Metrics for Summarization Quality 307
- 7.3.8 Conclusion: Building Future-Ready Summarization Systems 308

7.4 Industry Uses Cases 309
- 7.4.1 News Aggregation Use Case 309
- 7.4.2 Business Document Summarization Use Case 310
- 7.4.3 Automated Email Summarization Use Case 318

7.5 Summary 323

xi

TABLE OF CONTENTS

Chapter 8: Language Translation 325

8.1 Introduction to Language Translation 326
8.1.1 Translation Challenges 326
8.1.2 Importance of NLP in Translation 328
8.1.3 LLMs for Translation 330
8.1.4 SLMs for Translation 332

8.2 Implementation on AWS 335
8.2.1 Amazon Translate 336
8.2.2 Using LLMs on AWS (Custom Models) 341
8.2.3 Using SLMs on AWS (DistilBART, TinyBERT for Fast Translation) 346
8.2.4 Custom Training on SageMaker (T5, mBART, and so on) 351

8.3 Open-Source Translation Systems 356
8.3.1 Implementing Translation with Hugging Face Transformers 357
8.3.2 Fine-Tuning MarianMT for Custom Language Pairs 358
8.3.3 Fine-Tuning DistilBART for Lightweight Translation 358
8.3.4 Parameter-Efficient Translation with PEFT 358
8.3.5 Deploying Translation Models 359
8.3.6 Inference Optimization Strategies for Translation Models 360
8.3.7 Evaluation Metrics for Translation Quality 360

8.4 Industry Use Cases 361
8.4.1 Use Case: Global Customer Support Chatbots 361
8.4.2 Industry Use Case: Multilingual Content Delivery 363

8.5 Summary 365

Chapter 9: Dialogue Systems 367

9.1 Understanding Dialogue Systems 368
9.1.1 Rule-based vs. Generative Dialogue Systems 369
9.1.2 The Role of LLMs and SLMs in Dialogue Systems 375

9.2 Implementation on AWS 379
9.2.1 Amazon Lex: Rule-Based Conversational Agents 380
9.2.2 Amazon SageMaker: Fine-Tuning LLMs for Dialogue 387

TABLE OF CONTENTS

9.2.3 SLM Implementation: Using DistilGPT-2 and DistilT5 for Lightweight Deployment 394

9.3 Open-Source Implementation 404

    9.3.1 Rasa and Hugging Face Transformers 404

    9.3.2 Fine-Tuning a Generative SLM for Open-Source 407

9.4 Industrial Use Cases 407

    9.4.1 Virtual Customer Assistants 407

    9.4.2 Interactive Voice Response Systems (IVR) Use Case 410

    9.4.3 Healthcare Virtual Assistants 412

9.5 Summary 416

Chapter 10: Text Correction and Language Modeling 417

10.1 Introduction to Text Correction 418

    10.1.1 The Importance of Grammar and Spelling Correction 418

    10.1.2 Common Applications of Text Correction 419

    10.1.3 Traditional Approaches to Text Correction (Rules and Heuristics) 421

    10.1.4 AI-Powered Approaches (Statistical, Neural, and LLM-based) 423

10.2 Implementation on AWS: Building Text Correction Models with SageMaker 426

    10.2.1 Why Use AWS SageMaker for Text Correction? 426

    10.2.2 Data Preparation: Grammar Correction Dataset on S3 428

    10.2.3 Fine-Tuning an LLM on SageMaker 431

    10.2.4 Fine-Tuning a SLM and the Tradeoffs 435

    10.2.5 Deploying the Model on AWS SageMaker for Scalable Inference 438

10.3 Open-Source Implementation: Training with Hugging Face 443

    10.3.1 Fine-Tuning an LLM or SLM for Grammar Correction 444

    10.3.2 Evaluation and User Feedback 447

10.4 Industrial Use Cases 449

    10.4.1 Writing Assistance Tools 449

    10.4.2 Proofreading Applications 452

10.5 Summary 455

xiii

Chapter 11: Coreference Resolution and Text Entailment 457

11.1 Overview of Coreference Resolution 458
11.1.1 Importance of Understanding Context and Semantics 459
11.1.2 Coreference Resolution 461
11.1.3 Text Entailment 464

11.2 Implementation on AWS 467
11.2.1 Custom Model Development on SageMaker 467
11.2.2 LLMs for Coreference and Entailment 474
11.2.3 The SLM Approach 479

11.3 Open-Source Implementation for Coreference Resolution and Text Entailment 484
11.3.1 Coreference Resolution 485
11.3.2 Textual Entailment 486
11.3.3 Improving Model Accuracy 486

11.4 Industry Use Cases 488
11.4.1 Enhancing AI-driven Customer Support 488
11.4.2 Industry Use Cases: LegalTech and Contract Review 490

11.5 Summary 493

Chapter 12: Emerging Trends and Future Directions in NLP 495

12.1 LLM Evolution: Multimodality, Zero/Few-Shot, and Reasoning 496
12.1.1 From BERT/T5 Pretraining to Prompting Paradigms 496
12.1.2 In-Context Learning and Instruction-Tuning 498
12.1.3 RLHF, Multimodal Integration, and Tool Use 498
12.1.4 Long-Context Reasoning and Chain-of-Thought Advances 499

12.2 New Training Techniques and Responsible Alignment 499
12.2.1 Instruction-Tuning, RLHF, and Red Teaming 499
12.2.2 Bias Mitigation, Fairness-Aware Learning, and Governance 500
12.2.3 Empathy-Informed Reinforcement Learning and Human-in-the-Loop Protocols 500

12.3 SLM Innovation: Efficiency and Accessibility 500
12.3.1 Edge AI, Distilled/Quantized Models, and IoT Deployments 501
12.3.2 Benchmarking Efficiency: Latency, Memory, and Context Tradeoffs 501

12.3.3 Transfer Learning with SLMs: Knowledge Distillation, Task Adapters, and LoRA 501

12.3.4 Hybrid Collaboration: SLM–LLM Router Systems and Cascade Pipelines 502

12.4 Scaling Models: Tradeoffs Between Size, Accuracy, and Efficiency 502

12.4.1 The Three Axes of Scaling 503

12.4.2 From Dense to Sparse: Mixture-of-Experts (MoE) 506

12.4.3 SLMs: Efficiency First 510

12.5 Future Applications 515

12.6 Ethical Considerations 516

12.6.1 Bias Mitigation 517

12.6.2 Responsible AI 518

12.6.3 Environmental Sustainability 519

Index **521**

About the Authors

Venkata Gunnu is a senior executive director of knowledge management and innovation at JPM Chase. He is an executive with a successful background crafting enterprise-wide data and data science solutions, GenAI, process improvements, and data- and data science-centric products. He is a concept-to-implementation strategist with demonstrated success controlling multiple projects that elevate organizational efficiency while optimizing resources. Venkat is data-focused and analytical, with a track record of automating functions, standardizing data management protocol, and introducing new business intelligence solutions.

Shubham Shah is a software engineer with expertise in machine learning, cloud technologies, and AI-powered solutions. He has experience developing and optimizing Retrieval Augmented Generation (RAG) models, as well as integrating AI technologies like ChatGPT and Mistral for smarter, real-time information retrieval. Skilled in building scalable microservices and cloud-based architectures, he is passionate about solving complex challenges, improving system performance, and driving technological innovation. Shubham is always eager to learn, collaborate, and stay ahead in the fast-evolving tech landscape.

ABOUT THE AUTHORS

Anvesh Minukuri currently serves as a VP, senior lead ML engineer (LLM) at JP Morgan Chase, specializing in NLP applications. With a fervent interest in data science and artificial intelligence, he boasts more than 12 years in IT and ten years of experience in the analytics field, executing predictive, prescriptive, and GenAI solutions. Holding a master's degree from Oklahoma State University, he majored in data mining, following his bachelor's in computer science from JNTU University in India. Originating from South India, he began his career as a software engineer, catering to esteemed Fortune 500 clients such as GE, Cisco, and Tech Mahindra. Additionally, he aided stakeholders in capitalizing on the true value of AI and ML by using actionable data insights and was responsible for overseeing the design of ML. His primary focus lies in crafting data science solutions and machine learning pipeline frameworks to address predictive, prescriptive, and NLP challenges. Proficient in LLM model development, assessment, and optimization, he excels in NLP preprocessing, data storage, infrastructure management, and scaling endeavors. With a wealth of experience in driving adoption across diverse NLP applications, he has contributed significantly to the field through various published papers and posts in the realm of AI/ML.

Jayanth Gopu is a seasoned machine learning engineer with 12 years of experience, specializing in Python programming, LLMs, ModelOps, and automation technologies. With a strong background in deploying and optimizing machine learning models, he excels in creating efficient workflows that streamline the model lifecycle, from development to production. By leveraging AWS services, he designs and implements robust cloud-based solutions that enhance scalability and reliability. At an architect level, he has expertise in automation tools like Alteryx and UiPath, enabling the development of sophisticated automated processes that improve operational efficiency. He is known for guiding cross-functional teams and fostering a collaborative environment that encourages innovation and continuous improvement. Passionate about solving complex challenges and driving technological innovation, he is committed to mentoring emerging talent and staying at the forefront of advancements in the rapidly evolving field of machine learning engineering and automation.

About the Technical Reviewers

Kafui Nukunya is an experienced leader who has successfully led teams at Comcast and Walgreens and is currently serving as the head of customer machine learning and AI at Altice. Kafui holds a PhD in economics from Claremont Graduate University and an MA in economics from Bowling Green State University. With a solid foundation in econometrics, he integrates analytical expertise with cutting-edge machine learning techniques to deliver impactful business results.

Throughout his career, Kafui has led the development and implementation of advanced ML models, streamlined business processes, and driven transformative initiatives that have greatly enhanced operational efficiency and customer experience. He is known for his ability to build and guide high-performing teams, fostering a collaborative environment that produces innovative solutions at the intersection of technology and business.

Combining technical expertise, leadership acumen, and a strong econometric background, Kafui is dedicated to shaping the future of AI and machine learning, helping organizations harness the full potential of data-driven strategies.

Sundar Krishnan is an accomplished AI and data science leader with over 13 years of experience driving innovation and impact across industries. He currently serves as director of AI/ML at Altice/Optimum, where he leads the data science and ML engineering teams to develop solutions that enhance customer experience and enable personalized targeting strategies.

Before Altice, Sundar held a leadership role at CVS Health, where he spearheaded initiatives focused on improving member health outcomes through data-driven insights.

ABOUT THE TECHNICAL REVIEWERS

He holds a bachelor's degree from Government College of Technology, Coimbatore, and a master's degree from Oklahoma State University, Stillwater. Sundar is also the author of *Applied Data Science Using PySpark* and regularly shares his expertise through writing on Medium.

An active member of the AI community, Sundar frequently participates as a speaker, judge, and mentor in various industry forums and events.

Acknowledgments

This book could not have been written without the participation and support of many incredible individuals. I am profoundly grateful for the encouragement and assistance I received from my friends, family, and colleagues throughout this journey. Their unwavering belief in me, especially during the most challenging times, has been truly invaluable.

I would like to express my heartfelt thanks to Dr. Goutam Chakraborty, director of the master's business analytics program at Oklahoma State University, for introducing me to the field of data science over a decade ago. Serving as a pro-bono seasonal guest lecturer, I had the opportunity to design and deliver sessions on Apache Spark, anomaly detection, and introduction to LLMs, which helped shape the vision for this book. These experiences deepened my engagement with emerging technologies and inspired me to transform that content into a resource for a broader audience, ultimately leading to the creation of this book.

A special note of gratitude goes to my parents, Mr. Indrasena Reddy Minukuri and Mrs. Padma Thummala, whose constant motivation has always been a source of strength. I am equally thankful to my life partner, Mrs. Varshitha Yendapally, and my son, Avyaan Minukuri, for their unwavering support and love throughout this journey.

I am sincerely grateful to my co-authors, Mr. Shubham Shah and Mr. Jayanth Gopu, for their collaboration and insights. A very special thanks to Mr. Venkata Gunnu, not only for being a co-author and a mentor but also as the first person with whom I discussed the idea of writing this book, and his guidance and encouragement played a key role in bringing this project to life. My deepest appreciation also goes to the technical reviewers of this book, whose meticulous feedback and thoughtful suggestions significantly improved the final work.

—Anvesh Minukuri

Introduction

This comprehensive guide offers hands-on solutions to the most pressing challenges in modern *natural language processing* (NLP), combining cutting-edge theory and implementation with real-world applications. With a special focus on large language models (LLMs) and small language models (SLMs), this book is designed for data scientists, machine learning engineers, and NLP practitioners looking to understand modern NLP tasks and implementation. It can help practitioners scale and deploy modern NLP-GenAI intelligent systems across diverse platforms, from cloud infrastructure to edge devices.

LLMs and SLMs in Practice

The first three chapters establish the foundation by exploring the fundamentals of LLMs and SLMs, introducing modern NLP challenges and walking through task-specific architectures like BERT, GPT, Marian, and T5. Readers will gain a clear understanding of how to select the right model based on task requirements, compute budgets, latency constraints, and deployment environments. These chapters also provide comparative analyses of open-source frameworks (such as Hugging Face Transformers, SpaCy, and TensorFlow) alongside managed cloud platforms (including AWS SageMaker, Lambda, Transcribe, and Comprehend).

These chapters also examine tradeoffs between large-scale pretrained models versus compact, domain-specific alternatives. Special attention is given to SLM-first pipelines, including efficient transformer variants like BitNet and quantized models that enable low-latency inference on CPU, mobile, or edge devices.

INTRODUCTION

NLP Tasks: From Fine-Tuning to Deployment

Chapters 4 through 11 are structured as deep dives into specific NLP tasks, each following a consistent pattern:

- Data collection and preprocessing
- Model selection (LLM vs. SLM)
- Fine-tuning pipelines using Hugging Face and PyTorch
- Inference and evaluation
- Model compression and optimization (quantization, pruning, and distillation)
- Deployment strategies using Docker, FastAPI, Lambda, and SageMaker
- Monitoring and scaling best practices

Tasks covered include:

- Text generation (e.g., creative writing, autocomplete, knowledge completion)
- Text classification and sentiment analysis (including multi-label and zero-shot approaches)
- Summarization (extractive, abstractive, and hybrid methods)
- Translation (with multilingual models and inference benchmarking)
- Named entity recognition (NER)
- Question answering (QA) and retrieval augmented generation (RAG)
- Coreference resolution
- Grammar correction and entailment
- Dialogue and conversational agents

Each task is backed by end-to-end code in GitHub repository scripts and Docker containers.

Real-World Case Studies

Throughout the book, we explore domain-specific use cases, including:

- **Healthcare:** Automated triage and entity extraction from EMRs
- **Customer support:** Chatbots, sentiment routing, and intent classification
- **Legal tech:** Summarizing legal briefs and clause detection
- **Content marketing:** SEO optimization, copy generation, and grammar polishing

These case studies reflect real deployment pipelines and model choices based on resource availability, accuracy needs, and latency tolerance.

Advanced Topics and Future Trends

In the final chapter, we explore these emerging trends in NLP systems design:

- Agent-based LLM orchestration using tools like LangChain, CrewAI, and AutoGen
- Multimodal NLP and the integration of text, image, and speech
- Personalized NLP systems leveraging user data responsibly
- Sustainability and efficiency, focusing on distillation-first and quantization-first approaches
- Ethical AI, including bias mitigation, transparency, explainability, and compliance frameworks

We also walk through how to monitor production models, manage versioning, and optimize for cost-performance tradeoffs over time.

Companion GitHub Repository

All the NLP tasks, use cases, fine-tuning scripts, inference pipelines, and deployment templates are available from the GitHub companion repo at:

https://github.com/apress/ModernNLP-LLMSLM

CHAPTER 1

Introduction to LLMs, SLMs, and Modern NLP Challenges

This chapter explores the definitions, significance, and evolution of large language models (LLMs) and small language models (SLMs), offering an essential foundation for understanding how they function and how they can be applied to various NLP (natural language processing) tasks. This exploration will not only provide clarity on what makes each model unique but also help you grasp the considerations necessary when choosing between them for specific tasks or deployment scenarios.

1.1 Introduction: Understanding LLMs and SLMs in the NLP Landscape

LLMs and SLMs have significant real-world impacts across various industries. SLMs are increasingly being adopted for their efficiency and specialized capabilities. They are used in customer service automation, language translation, sentiment analysis, and market trend analysis. In the digital workplace, SLMs are enhancing productivity through applications like automated proofreading, email drafting, and data management.

While powerful, LLMs face challenges related to their size and resource requirements. The corpus for these models has been growing exponentially. GPT-3, released in 2020, was trained on 300 billion words. However, high-quality language data for training is expected to be exhausted by 2024 or 2025, with low-quality data potentially lasting until 2032.

Language models are based on the transformer architecture (see Figure 1-1), which uses self-attention mechanisms to process and generate sequences of text efficiently.

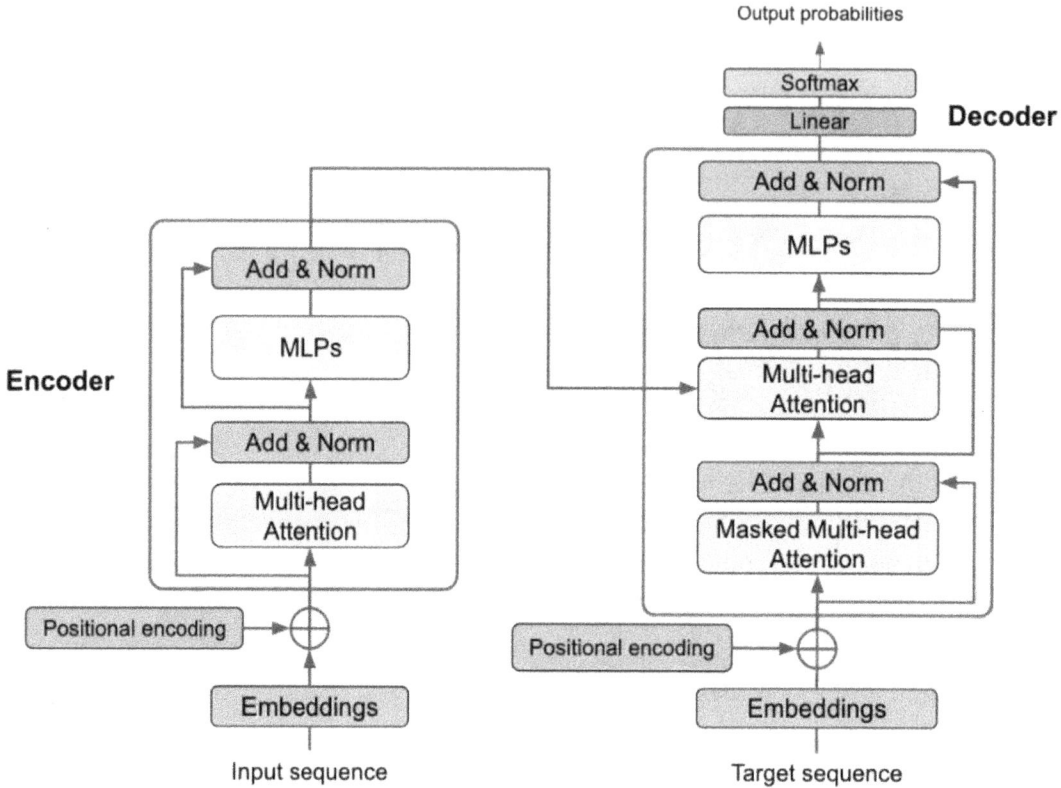

Figure 1-1. *Transformer architecture (image sourced from Google images)*

The Transformer architecture (introduced in 2017) has become the backbone of modern NLP due to its versatility and efficiency. The architecture consists of an encoder, a decoder, or a combination of both. Depending on the task, these components are used in different configurations. Table 1-1 shows a mapping of Transformer types to their sequence processing tasks, along with inputs and outputs.

Table 1-1. *Types of Transformer Models and Their Applications*

Type	Architecture	Input	Output	Common Use Cases
Encoder-decoder	Full transformer	Sequence	Sequence	Machine translation, text summarization, question answering
Encoder-only	Encoder stack	Sequence	Vector	Text classification, sentiment analysis, named entity recognition
Decoder-only	Decoder stack	Vector/prompt	Sequence	Text generation, language modeling

Mapping to sequence tasks:

- **Seq2Seq (Sequence-to-Sequence):**

    Encoder-Decoder architecture: The input sequence is processed by the encoder into a fixed-size vector representation, which is then decoded into the output sequence.

    Use cases: Machine translation, summarization.

- **Seq2Vec (Sequence-to-Vector):**

    Encoder-Only architecture: The input sequence is transformed into a fixed-size vector (embedding), commonly used for tasks where the output is a single vector or label.

    Use cases: Text classification, sentiment analysis.

- **Vec2Seq (Vector-to-Sequence):**

    Decoder-Only architecture: The input is typically a vector or a prompt, which the decoder uses to generate a sequence.

    Use cases: Text generation, language modeling.

CHAPTER 1 INTRODUCTION TO LLMS, SLMS, AND MODERN NLP CHALLENGES

The field of natural language processing (NLP) has undergone a remarkable transformation, driven largely by advances in LLMs and SLMs. These models have fundamentally altered how we interact with machines, enabling them to generate human-like text, understand complex language, and perform sophisticated language tasks. But what makes an LLM and an SLM different? And how have these models evolved over time to become the cornerstones of modern NLP?

SLMs and LLMs are expected to coexist, with SLMs filling in niche applications and LLMs handling more complex, general-purpose tasks. The future will likely see a convergence of techniques, with hybrid models combining the strengths of both approaches. See Tables 1-2 and 1-3.

Table 1-2. Common SLMs

Model Name	Parameters	Key Features
Qwen2	0.5B, 1B, 7B	Scalable, suitable for various tasks
Mistral Nemo	12B	Complex NLP tasks, local deployment
Llama 3.1	8B	Balanced power and efficiency
Phi-3.5	3.8B	Long context length (128K tokens), multilingual
TinyLlama	1.1B	Efficient for mobile and edge devices
MobileLLaMA	1.4B	Optimized for mobile and low-power devices
StableLM-zephyr	3B	Fast inference, efficient for edge systems
LaMini-GPT	774M - 1.5B	Multilingual, instruction-following tasks
Gemma2	9B, 27B	Local deployment, real-time applications
MiniCPM	1B - 4B	Balanced performance, English and Chinese optimized
GPT-4o Mini	Not specified	Superior textual intelligence, multimodal reasoning
Gemini Flash	Not specified	Low-latency, high-speed language processing

Table 1-3. Common LLMs

Model	Parameters	Estimated Cost	Key Features
GPT-4	2 trillion	$100 million	Advanced reasoning, multimodal capabilities
GPT-5	20 trillion	$1 billion	Estimated future model, not yet released
Gemini 1.5	Unknown	Not available	1 million token context window, multimodal
Llama 3.1	405 billion	Not available	128,000 token context, trained on 15 trillion tokens
Mixtral 8x22B	141 billion	Not available	Sparse mixture-of-experts, 39B active parameters
DBRX	132 billion	Not available	Mixture-of-experts architecture, 36B active parameters
Falcon 180B	180 billion	Not available	Open-source, competitive performance
GPT-3	175 billion	$4.6 million	Widely used, versatile text generation

1.2 What Is an LLM?

1.2.1 Defining LLMs

LLMs are advanced neural network architectures trained on extensive datasets to understand, generate, and interact with human language. Characterized by their massive parameter count (billions to trillions), LLMs enable complex language tasks such as text generation, translation, summarization, and more. These models grasp contextual nuances by leveraging the transformer architecture, which facilitates contextual understanding over extensive text sequences.

LLMs derive their power from their ability to process vast amounts of text data and learn statistical patterns that allow for highly coherent language output. By analyzing patterns in data, they can generate human-like responses, making them invaluable for applications ranging from conversational AI to content generation and beyond.

1.2.2 Key Features and Significance

1. **Contextual language representation:** LLMs are adept at modeling the relationships between words within broad contexts, making them versatile for various NLP tasks. Their contextual comprehension is critical for applications requiring nuanced language understanding, such as legal document analysis or medical report summarization.

2. **Revolutionizing NLP tasks:**
   - **GPT-3:** Excels in creative text generation, coding assistance, and open-ended question answering with applications like writing assistants or automated storytelling.
   - **BERT:** Pioneers contextual embeddings for tasks like named entity recognition (NER) and question answering, which are widely used in search engines and customer support systems.
   - **T5:** Adaptable to diverse tasks via a text-to-text format, including summarization, translation, and classification, empowering academic research and enterprise-level applications.

3. **Scalability and power:** Scaling up datasets and model parameters has enabled LLMs to surpass traditional models in performance, reducing the need for extensive task-specific fine-tuning. This scalability ensures that domain-specific tasks, like financial forecasting or scientific literature analysis, can leverage pretrained LLMs with minimal customization.

1.2.3 Evolution of LLMs

1. **Pre-2010s: Traditional NLP approaches:** Early NLP systems relied heavily on rule-based techniques and statistical models like n-grams and Hidden Markov Models (HMMs). While functional for simple tasks like spell-checking or basic sentiment analysis, these systems struggled with understanding context or handling long-range dependencies between words.

2. **2013-2017: Word embeddings and deep learning:** The introduction of word embeddings, such as Word2Vec and GloVe, marked a significant improvement. These models provided dense vector representations of words, capturing semantic similarities and relationships. However, they were limited in their word-level understanding and lacked the ability to process sentence or document-level context effectively.

3. **2017: The transformer breakthrough:** The transformer architecture, introduced by Vaswani et al., revolutionized NLP by enabling models to process sequences using attention mechanisms. Unlike recurrent neural networks (RNNs), transformers could efficiently capture dependencies over long text sequences. BERT (bidirectional encoder representations from transformers) and its derivatives brought bidirectional contextual understanding, reshaping tasks like text classification and question answering.

4. **2020-present: Scaling with GPT-3 and beyond:** OpenAI's GPT-3 demonstrated the power of scaling models to billions of parameters. Its ability to perform zero-shot and few-shot learning allowed it to handle diverse tasks without task-specific fine-tuning. GPT-4 introduced multimodal capabilities, enabling the processing of both text and images. These advancements have broadened the applicability of LLMs to areas such as healthcare diagnostics, educational tools, and creative industries.

1.2.4 Practical Applications of LLMs

LLMs have diverse applications across various industries. They are used in conversational AI for virtual assistants and chatbots, content creation for generating marketing copy and blog posts, healthcare for summarizing patient records and extracting medical information, programming assistance for generating code snippets, and legal document analysis for extracting key clauses and flagging potential risks. Table 1-4 lists and explain some LLM uses cases.

Table 1-4. *LLM Use Cases and Applications*

Use Case	Description
Conversational AI	Companies like OpenAI and Google use LLMs to power virtual assistants and chatbots capable of maintaining coherent and contextually relevant dialogues. For example, customer support systems powered by LLMs can handle complex queries while providing detailed, context-sensitive responses.
Content creation	Tools like Jasper and Copy.ai utilize LLMs to generate marketing copy, blog posts, and product descriptions, saving time for content creators and marketers.
Healthcare applications	LLMs assist in summarizing patient records, extracting relevant information from medical literature, and even generating preliminary diagnostic reports.
Programming assistance	GitHub Copilot, built on OpenAI's Codex (a derivative of GPT), enables developers to generate code snippets and receive contextual suggestions, significantly enhancing productivity.
Legal document analysis	LLMs like GPT-4 are used to analyze lengthy legal contracts, extract key clauses, and flag potential risks, enabling faster decision-making in legal and corporate environments.

1.2.5 Challenges in Scaling LLMs

On February 12, 2025, a company using an AI LLM chatbot for customer engagement faced increased inference times and rising scaling costs during peak traffic. This scenario highlights the challenges of optimizing LLMs for low latency and cost-efficiency in real-time applications. LLMs have revolutionized natural language processing, but their massive scale poses significant challenges. Here's a summary of the five key challenges:

1. **Resource demands:** Training and inference for LLMs require substantial computational resources, including clusters of GPUs or TPUs. For instance, training GPT-3 reportedly required several petaflop-days of compute power, making it inaccessible to smaller organizations.

2. **Data quality and bias:** The effectiveness of LLMs heavily depends on the quality of their training data. Poorly curated datasets can introduce biases, resulting in models propagating harmful stereotypes or misinformation. Continuous evaluation and dataset improvement are critical to address these challenges.

3. **Environmental impact:** The energy consumption in training large models raises concerns about sustainability. Research into energy-efficient training methods, such as sparsity techniques and model distillation, is gaining momentum to mitigate these effects.

4. **Interpretability:** Understanding how LLMs arrive at their outputs remains a challenge. This "black box" nature makes it difficult to trust LLMs in high-stakes applications like healthcare or law, where transparency is crucial.

1.2.6 Future Directions for LLMs

1. **Specialized LLMs:** Industry-specific LLMs are being developed to cater to niche requirements, such as BioGPT for biomedical research or FinBERT for financial analysis.

2. **Energy efficiency:** Innovations like sparse attention mechanisms and optimized hardware accelerators aim to reduce the environmental footprint of training and deploying LLMs.

3. **Ethical AI development:** Efforts to improve fairness, accountability, and transparency are gaining traction, ensuring that LLMs can be deployed responsibly.

By understanding the defining features, evolution, and practical applications of LLMs, you can appreciate their transformative impact on NLP and their potential to drive innovation across industries.

1.3 What Is an SLM?

1.3.1 Defining SLMs

SLMs are efficient, compact versions of LLMs, optimized for resource-constrained environments. While LLMs boast billions of parameters, SLMs typically range between tens of millions to a few billion, making them suitable for on-device or edge deployments. Unlike their larger counterparts, SLMs focus on task-specific efficiency rather than handling a broad range of general-purpose tasks.

1.3.2 Key Features of SLMs

1. **Efficiency and speed:**
   - SLMs require significantly less computational power and memory, enabling them to run on smartphones, IoT devices, and embedded systems.
   - Their optimized architectures ensure real-time responses, which is crucial for applications such as voice assistants and predictive text systems.

2. **Cost-effective solutions:**
   - Training and deploying SLMs is economically viable, especially for startups and small-scale enterprises.
   - Lower hardware and maintenance costs make SLMs accessible to various industries.

3. **Task-specific optimization:**
   - SLMs excel at narrow-domain tasks such as sentiment analysis, spam detection, and on-device text prediction.
   - Techniques like distillation and pruning enhance task efficiency without compromising accuracy.

1.3.3 Evolution of SLMs

1. **From rule-based models to compact transformers:** The journey of SLMs began with rule-based systems and statistical approaches, evolving into lightweight neural architectures inspired by transformers like BERT. Models such as DistilBERT and MobileBERT epitomize this transition by maintaining the performance of larger models while reducing complexity.

2. **Advancements in compression techniques:** Techniques like model pruning, quantization, and knowledge distillation have enabled the creation of high-performing yet compact models. DistilBERT, for instance, retains 97% of BERT's performance while being 40% smaller and 60% faster.

1.3.4 Practical Applications of SLMs

1. **Real-time language processing:**

   - **Predictive text and autocorrect:** Many mobile keyboards leverage SLMs to offer context-aware suggestions and corrections.

   - **Voice assistants:** SLMs power on-device voice recognition and command processing for assistants like Siri and Google Assistant.

2. **Low-latency applications:**

   - **Chatbots:** Businesses use SLM-driven chatbots for instant customer query handling, especially in industries like e-commerce and banking.

   - **IoT integration:** Smart devices such as thermostats and security systems rely on SLMs for natural language command interpretation.

3. **Privacy-focused deployments:**

   - SLMs enable on-device processing of sensitive data, such as healthcare records or financial transactions, reducing the need for cloud-based computation and ensuring user privacy.

1.3.5 Innovations Driving SLMs Forward

1. **Knowledge distillation:** SLMs inherit knowledge from larger models, achieving comparable performance while being significantly smaller. DistilBERT exemplifies this approach, balancing size and efficiency.

2. **Pruning and quantization:** These techniques remove redundant parameters and reduce precision in computations, creating smaller, faster models without substantial performance loss.

3. **Sparse attention mechanisms:** Sparse transformers, like Longformer, enable SLMs to handle longer sequences effectively, expanding their applicability to document summarization and legal text analysis.

1.3.6 Challenges and Limitations

1. **Task-specific training:** SLMs often require extensive fine-tuning to achieve optimal performance on specific tasks, demanding high-quality, domain-specific datasets.

2. **Scalability constraints:** While SLMs are efficient, their limited parameter size can restrict their ability to generalize across diverse tasks compared to LLMs.

3. **Balancing performance and size:** Reducing model size often involves tradeoffs in accuracy, necessitating careful optimization to avoid significant performance degradation.

1.3.7 Case Studies and Real-World Scenarios

1. **Mobile applications:** A popular keyboard app uses SLMs for multilingual text prediction, offering seamless typing experiences without draining battery life or requiring Internet connectivity.

2. **Healthcare:** An on-device SLM powers a diagnostic tool that processes patient symptoms and provides preliminary assessments, ensuring data privacy in remote regions.

3. **E-commerce:** SLMs analyze customer reviews in real-time, categorizing feedback to improve product recommendations and customer satisfaction.

1.3.8 Future Directions for SLMs

1. **Federated learning:** Enabling distributed training across devices to create personalized SLMs without compromising user privacy.

2. **Adaptive models:** Developing SLMs capable of dynamically adjusting complexity based on input size and computational resources.

3. **Energy-efficient deployments:** Research into hardware accelerators and energy-efficient algorithms will make SLMs even more accessible for edge and IoT applications.

SLMs are reshaping the NLP landscape by bringing powerful language capabilities to constrained environments. Their ongoing evolution promises even greater integration into everyday applications, from personalized assistants to secure, on-device processing solutions.

1.3.9 Techniques Enhancing SLM Efficiency

1. **Model pruning:** Reduces unnecessary parameters while preserving model performance.

2. **Knowledge distillation:** Transfers knowledge from larger models to smaller ones, balancing performance and efficiency.

3. **Quantization:** Reduces precision in model parameters for smaller memory footprints with minimal performance tradeoffs.

> **COMPARING SLMS AND LLMS**

Task suitability:

- LLMs dominate in creative or knowledge-intensive tasks requiring extensive context.
- SLMs excel in focused, high-throughput tasks where efficiency is key.

Deployment scenarios:

- LLMs are best for cloud-based systems with scalable infrastructure.
- SLMs are better for local devices or edge deployments that prioritize speed and resource conservation.

In summary, SLMs provide a pragmatic solution for applications demanding efficiency and specificity, complementing LLMs' expansive capabilities. By leveraging advancements in techniques like distillation and pruning, SLMs are poised to address an increasingly diverse range of NLP challenges.

1.4 Overview of Modern NLP Tasks

When you interact with digital systems—whether it's asking a virtual assistant for the weather, translating text between languages, or automatically summarizing a lengthy article—you are engaging with tasks that fall under the domain of NLP. At its core, NLP seeks to bridge the gap between human language and machine understanding, enabling computers to process, interpret, and generate language in ways that mimic human communication.

This section explores some of the most common NLP tasks. Each task represents a unique challenge in language understanding. We illustrate that different types of models—LLMs and SLMs—have specific strengths in tackling these challenges.

1.4.1 Text Classification

Text classification automates categorizing text, like sorting product reviews as positive, negative, or neutral. LLMs like GPT-3 and fine-tuned BERT excel at handling large datasets, while smaller models like DistilBERT and ALBERT are used in mobile apps for efficiency.

Key examples:

- **Spam detection:** Automatically classifies emails as spam or not spam.

- **Topic labeling:** Categorizes news articles into predefined topics such as politics, sports, or entertainment.

Model usage: LLMs such as GPT-3 and fine-tuned versions of BERT are commonly used for general text classification, while smaller models like DistilBERT or ALBERT are more suitable for on-device classification due to their efficiency.

1.4.2 Sentiment Analysis

Sentiment analysis determines the emotional tone of a text. LLMs like RoBERTa and GPT-4 offer high accuracy, while smaller models are used for real-time, on-device analysis.

LLMs like RoBERTa and GPT-4 have made sentiment analysis more accurate and reliable thanks to their ability to capture context and subtleties in language. Meanwhile, SLMs optimized for real-time applications are invaluable when quick, on-device analysis is needed.

Levels of sentiment analysis:

- **Document level:** Analyzes the overall sentiment of a document.

- **Sentence level:** Identifies the sentiment of individual sentences.

- **Aspect level:** Detects the sentiment about specific aspects or entities within the text.

Model usage: Models like BERT, RoBERTa, and GPT-3 are widely used for sentiment analysis due to their strong contextual understanding. SLMs, when optimized for mobile deployment, are ideal for real-time sentiment detection in low-resource environments.

1.4.3 Text Summarization

In today's information-driven world, we are constantly bombarded with an overwhelming amount of text—news articles, reports, emails, and more. Reading through every piece of information in detail can be impractical. Text summarization offers a way to quickly grasp the essence of a lengthy document by generating a shorter version.

Text summarization techniques can be categorized into two main types: extractive summarization and abstractive summarization:

- **Extractive summarization** involves selecting key sentences directly from the original text.

- **Abstractive summarization**, on the other hand, is more sophisticated. It generates new sentences that capture the core ideas, much like how a human would write a summary.

Early models relied on extractive approaches due to their simplicity. However, with the advent of transformer-based models, abstractive summarization has become increasingly effective. Models like T5 and GPT-4 have been fine-tuned to produce human-like summaries, making them invaluable tools for content aggregation platforms, news summarizers, and even academic research tools.

While LLMs excel at abstractive summarization, extractive summarization can often be handled effectively by smaller models like BERTSum, which are optimized for faster, more resource-efficient processing. This makes SLMs a preferred choice for real-time applications, such as summarizing news articles on a mobile app.

1.4.4 Question Answering (QA)

Imagine reading a lengthy document and needing to find a specific piece of information. Instead of scanning the entire text manually, what if you could simply ask a question and receive an instant answer? This is the essence of question-answering (QA) systems—a capability that has revolutionized how people retrieve information.

QA systems are broadly categorized into two types:

- **Closed-domain QA** focuses on answering questions within a specific domain or dataset.

- **Open-domain QA**, on the other hand, retrieves information from vast, diverse sources, such as the entire web.

The introduction of models like BERT transformed closed-domain QA by enabling a highly accurate extraction of answers from a given context. With open-domain QA, however, the challenge is more complex—requiring not just extraction but also the generation of coherent answers. This is where LLMs like GPT-3 and GPT-4 shine. These models can answer a wide variety of questions, often without needing fine-tuning, by leveraging the vast knowledge encoded in their parameters.

Nevertheless, QA systems are not without challenges. Open-domain QA requires significant computational resources, making LLMs less practical for scenarios where quick, on-device answers are needed. In such cases, fine-tuned SLMs can provide an efficient alternative.

Text generation might be the most captivating NLP task of all. The ability of machines to generate human-like text—whether it's a creative story, an article, or even code—has transformed how we think about automation in content creation.

The concept of text generation isn't new. Early models used simple statistical approaches, like *n-grams,* to predict the next word in a sequence. However, these models struggled with coherence and context. The introduction of neural networks, particularly LLMs like GPT-3, revolutionized text generation by enabling models to produce coherent, contextually relevant, and stylistically diverse text at scale.

Applications of text generation are vast, including:

- **Chatbots and virtual assistants:** Enabling natural, human-like conversations.

- **Creative writing:** Assisting authors in generating story ideas or drafting content.

- **Code generation:** Automatically generating code snippets from natural language descriptions.

While LLMs dominate text generation due to their ability to handle complex prompts, SLMs are often used in domain-specific scenarios where computational efficiency is paramount.

Finally, at the heart of many NLP tasks lies *language modeling*—the ability of a system to predict the next word in a sequence, given a context. Language models are the foundation of tasks like autocomplete, speech recognition, and machine translation.

There are two main types of language models:

- **Unidirectional models**, such as GPT, predict the next word based only on the previous context.

- **Bidirectional models**, such as BERT, consider both previous and subsequent contexts to understand the meaning of the text better.

While unidirectional models are more suitable for generative tasks like text completion, bidirectional models excel in understanding tasks like classification and question answering.

You've now explored the most common NLP tasks and learned about their significance, evolution, and the models that power them. With this foundation in place, the next section dives deeper into how these tasks can be tackled using both LLMs and SLMs, and how choosing the right model depends on the specific requirements of each task.

1.5 LLMs on AWS vs. Open-Source Frameworks

When it comes to deploying LLMs and SLMs, the choice of platform plays a crucial role in determining the efficiency, scalability, and cost-effectiveness of the solution. Cloud-based platforms such as AWS (Amazon Web Services) and open-source frameworks like Hugging Face, SpaCy, and TensorFlow each offer unique benefits depending on the nature of the task and infrastructure requirements. This section explores the advantages of both cloud-based tools and open-source frameworks and provides guidance on when to choose one over the other.

1.5.1 AWS Suite Features for Modern NLP Tasks

AWS provides a robust suite of services designed to meet the diverse needs of LLMs and SLMs. These services are equipped to handle the high computational demands of LLMs, such as model training, inference, and deployment while offering flexible, cost-effective solutions for SLMs that require low-latency responses and efficient resource management. AWS's combination of scalable infrastructure, machine learning tools, and optimized environments allow users to develop, train, and deploy models at scale, making it a go-to choice for both small and large NLP applications. For LLMs, it provides scalability and powerful hardware for resource-intensive models. For SLMs, it offers managed services reducing operational overhead.

Cloud platforms like AWS provide essential benefits for managing the resource demands of LLMs, enabling efficient training, deployment, and scaling without requiring extensive on-premise infrastructure investments. The key advantages include:

1. **Easy scalability:**
   - **Dynamic resource allocation:** Cloud platforms allow real-time scaling of resources based on needs, such as increasing compute power during model training or adjusting resources during inference stages.

- **Elastic infrastructure:** AWS EC2 offers instant access to both CPU and GPU instances, enabling organizations to scale infrastructure as required for LLMs.

- **Global reach:** AWS's global data centers ensure model deployment in multiple regions, optimizing for high availability and low latency for users.

2. **Managed services:**

   - **Simplified setup and maintenance:** AWS SageMaker provides pre-configured environments, built-in algorithms, and automated pipelines, easing model training and deployment without manual infrastructure management.

   - **Model versioning and monitoring:** Built-in version control and monitoring tools (like SageMaker Model Monitor) help track model performance, detect issues like model drift, and maintain continuous improvements.

   - **Integrated security and compliance:** Services like AWS IAM ensure secure data handling with encryption, identity management, and compliance with regulations like GDPR and HIPAA.

3. **Powerful hardware:**

   - **High-performance GPUs/TPUs:** AWS provides access to powerful hardware such as NVIDIA GPUs and Google TPUs, optimized for the parallel processing needed in LLM training.

   - **Optimized ML instances:** AWS EC2 P3 and P4 instances are specifically designed to handle ML workloads, offering the computational power required by LLMs.

   - **Custom chips (e.g., AWS Inferentia):** AWS Inferentia chips offer cost-effective, high-performance inference tailored for large-scale LLM deployments, especially for NLP tasks.

By leveraging these capabilities, cloud platforms like AWS make it easier to manage large-scale LLMs and reduce infrastructure complexity and costs. Table 1-5 outlines the applicable AWS services.

CHAPTER 1   INTRODUCTION TO LLMS, SLMS, AND MODERN NLP CHALLENGES

Table 1-5. *AWS Services and Their Use Cases*

AWS Service	Key Features	Use Cases
AWS SageMaker	– Pre-built Jupyter notebooks for experimentation – Support for CPU/GPU instances – Auto-scaling model deployment – Distributed training for LLMs	Model training, fine-tuning, and deployment for complex NLP tasks (e.g., text generation, question answering)
AWS Lambda	– Event-driven compute service – Pay-per-use pricing – Integrates with API Gateway – Ideal for real-time inference	Real-time inference for NLP tasks like sentiment analysis, entity recognition, and real-time question answering
AWS Elastic Inference	– Cost-effective GPU-powered instances – Scales inference environment – Supports TensorFlow, PyTorch, and MXNet	Running large transformer models (e.g., GPT, BERT) at scale with cost-efficient GPU instances for production environments
AWS Deep Learning AMIs	– Pre-configured environments with deep learning frameworks – Optimized for GPU-powered instances – Easy integration with SageMaker	Training and deploying large transformer models like GPT, BERT, and T5, or fine-tuning models for specific applications like translation or summarization
AWS Inferentia	– Custom-designed chip for machine learning inference – Low-latency, high-throughput inference – Compatible with major ML frameworks like TensorFlow	Fast and cost-effective inference for LLMs at scale, especially for real-time text generation or question-answering tasks
AWS Elastic Kubernetes Service (EKS)	– Automated container deployment and management – Supports GPU instances – Integrates with SageMaker for end-to-end pipelines	Large-scale deployment of LLMs in microservices architecture (e.g., conversational AI, multi-task models)

1.5.2 Open-Source Frameworks

1. Hugging Face:

    - **Use cases:** Offers pretrained models for SLMs (e.g., BERT, DistilBERT) and LLMs (e.g., GPT-3, T5). Hugging Face provides a flexible platform for training, fine-tuning, and deploying models for various NLP tasks.

    - **Benefits:** For SLMs, it provides cost-effective pretrained models for specific tasks like text classification and sentiment analysis. For LLMs, it offers flexibility in customizing large models and running them on local or cloud infrastructure.

2. LangChain and LlamaIndex:

    - **Use cases:** These frameworks are designed to help build applications with SLMs and LLMs by facilitating tasks such as model chaining, retrieval, and indexing. They support integration with various data sources and LLMs for complex NLP pipelines.

    - **Benefits:** Both SLMs and LLMs provide powerful tools to manage workflows, making it easier to build applications that combine various models or data processing steps.

3. Ollama:

    - **Use cases:** Ollama allows local execution and customization of both SLMs and LLMs, making it suitable for environments where data privacy and control are priorities.

    - **Benefits:** Ideal for organizations that need to run models locally without relying on cloud services. This enhances data privacy and offers more control over model execution.

4. Fairseq: Facebook's Research-Focused Library

    - **Use cases:** Fairseq is an open-source sequence-to-sequence learning toolkit from Facebook AI Research (FAIR). It supports training SLMs and LLMs using various neural network architectures, including transformers.

- **Benefits:** Implements top-performing models like BART, T5, and RoBERTa. Extensive flexibility in model configuration. Great community support from Facebook AI

5. OpenNLP: Apache's NLP Suite

    - **Use cases:** Apache OpenNLP is a toolkit for classic NLP tasks, including tokenization, sentence splitting, part-of-speech tagging, named entity recognition (NER), and parsing. It is ideal for small- to medium-sized SLMs used in tasks like text classification and entity recognition. OpenNLP is typically not used for training large models but can be combined with other libraries for specific NLP tasks.

    - **Benefits:**

      For SLMs: OpenNLP is lightweight and fast, making it well-suited for smaller models and traditional NLP tasks. It provides an efficient solution for deploying and integrating models focused on tasks such as named entity recognition and part-of-speech tagging.

      For LLMs: OpenNLP can be combined with larger models but is not typically used for training LLMs. It is better suited for deploying smaller models or serving specific NLP functions in production environments. It also integrates well with frameworks like Hugging Face for more complex tasks.

1.5.3 Benefits of Cloud Platforms vs. Open-Source Frameworks

For LLMs:

- **Cloud platforms:** Provide scalability, powerful hardware, and easy workflow integration. Cloud platforms like AWS handle the immense computational requirements of LLMs, making them the best choice for large models that need to scale.

- **Open-source:** Offer flexibility for LLMs, allowing for customization and fine-tuning. Open-source frameworks like Hugging Face enable developers to deploy models across different environments, including private servers or cloud infrastructure, while maintaining complete control over the model.

For SLMs:

- **Cloud platforms:** AWS Lambda, SageMaker, and Comprehend are ideal for lightweight models that require real-time processing and easy deployment. The managed services reduce operational complexity.

- **Open-source:** For SLMs, open-source tools like Hugging Face or LangChain provide cost-effective alternatives, with models available for various NLP tasks, offering flexibility and privacy.

1.5.4 When to Choose Cloud Platforms vs. Open-Source Frameworks

For LLMs:

- **Cloud platforms:** Best suited when dealing with huge models that demand high computational power, such as GPT-3, or when seamless scalability is required for production environments.

- **Open-source:** Ideal when focusing on custom model design, fine-tuning, or prioritizing control over data privacy (especially for sensitive applications).

For SLMs:

- **Cloud platforms:** Opt for AWS if you need managed services for deployment and infrastructure, reducing the need for hands-on management of smaller models.

- **Open-source:** Open-source frameworks are more cost-effective for lightweight, task-specific models that require high customization and flexibility.

1.5.5 The Hybrid Approach

Combining cloud services (for scalability and infrastructure) with open-source tools (for customization and flexibility) is often the best approach. For example, running Hugging Face models on AWS infrastructure allows for easy scaling and extensive customization, maximizing cost-efficiency and performance.

In summary, by choosing the right tool based on model size, resource requirements, and flexibility, organizations can strike the right balance between cloud platforms and open-source frameworks for both SLMs and LLMs.

1.6 LLM Lifecycle Activities

As explored in the previous section, choosing between cloud-based solutions (like AWS SageMaker) and open-source frameworks (such as Hugging Face Transformers) hinges on factors including cost, scalability requirements, and the complexity of your NLP tasks. Once you've made that choice—or even if you're considering a hybrid approach—the next significant step is understanding the end-to-end lifecycle of working with LLMs and SLMs. This lifecycle includes data collection, model training, evaluation, and deployment to real-world applications.

In essence, the success of an NLP project doesn't just depend on selecting the right model or platform; it also relies heavily on how well each lifecycle activity is executed. For instance, a state-of-the-art LLM may fail to deliver real business value if it's not trained on the right type of data or if it's deployed without proper monitoring and optimization strategies. Conversely, even a smaller model—like an SLM optimized for mobile environments—can produce surprisingly robust results if meticulously fine-tuned with domain-specific data and carefully monitored over its lifecycle.

1.6.1 Data Collection and Preparation

At the heart of any NLP initiative is the data that fuels your model. Whether you're aiming to build a sentiment analysis tool for social media posts or an advanced conversational chatbot, the quality and relevance of your training data can make or break your project. For large-scale models, the sheer volume of data is often a differentiating factor; GPT-3 and GPT-4 were trained on massive text corpora from diverse sources like books, websites, and code repositories. However, quality is just as crucial as quantity. Poorly filtered or non-representative datasets can introduce biases and compromise model accuracy in production.

1.6.1.1 Cupcakes and Customer Feedback: A Data-Driven Story

Imagine you own a small bakery and collect online reviews to understand which cupcake flavors people love the most. Before training your language model on these comments, you remove spam, fix obvious spelling mistakes, and translate any non-English text. This process, sometimes called "data cleaning and augmentation," helps prevent messy information from confusing your model. With a neat dataset, your model can more accurately spot trends and guide you on which flavors deserve a spot on your bakery's menu.

1.6.1.2 Key Considerations for Data Collection and Preparation

1. **Data sourcing:** Identify where your text data is coming from. Familiar sources include web crawls, internal databases (e.g., customer support logs, product descriptions), and open datasets (e.g., Wikipedia).

2. **Cleaning and filtering:** Remove irrelevant or low-quality text to reduce noise. This may involve eliminating duplicate records, filtering out profanity, or discarding incomplete data.

3. **Data augmentation:** Techniques like back-translation (translating text to another language and back again) can help expand your dataset, especially for smaller, domain-specific tasks.

4. **Ethical and legal compliance:** Always ensure you have the right to use the data and consider anonymizing sensitive information. For instance, healthcare-related NLP tasks often require stringent HIPAA or GDPR compliance.

By setting a strong data foundation, you reduce the potential for bias in your models and maximize the likelihood of achieving meaningful results in downstream tasks like classification, entity recognition, or text generation. As the following sections explore, every phase of the LLM and SLM lifecycle—model training and fine-tuning, evaluation and testing, and deployment and monitoring—builds on this crucial first step of data preparation.

1.6.2 Model Training and Fine-Tuning

Once you've collected and prepared your data, the next step is to train your model or fine-tune an existing one. Training an LLM from scratch can be extremely resource-intensive—requiring vast amounts of text data, powerful hardware (like GPUs or TPUs), and days or even weeks of compute time. This is why many practitioners opt for fine-tuning a pretrained model (e.g., GPT, BERT, or T5). Fine-tuning allows you to take a model that already "knows" language patterns and adapt it to your specific task using a more modest dataset and fewer computational resources.

1.6.2.1 Scenario in Action: Fine-Tuning a Cupcake Chatbot

After preparing your customer feedback data, the next step is to train or fine-tune your chatbot model. Instead of building a language model from scratch, you start with a pretrained model like GPT-3, which already understands general language patterns. You then feed it examples of typical customer interactions specific to your bakery—such as inquiries about flavors, order statuses, and customization options. This fine-tuning process allows the chatbot to grasp the unique aspects of your business, enabling it to respond accurately and helpfully to customer queries within just a few hours or days rather than the weeks it would take to train a model from the ground up.

1.6.2.2 Key Considerations for Training and Fine-Tuning

1. **Hardware and compute resources:**
   - **LLMs:** Training (or re-training) a large model from scratch generally demands specialized hardware clusters and deep learning frameworks like TensorFlow or PyTorch.
   - **SLMs:** Smaller models can often be trained on a single GPU or CPU, making them more accessible for teams with limited resources.

2. **Hyperparameter tuning:**
   - Adjusting parameters like learning rate, batch size, and number of epochs can dramatically impact your model's performance.

- Tools such as Weights & Biases or Amazon SageMaker Experiments help track these hyperparameters and automate tuning processes.

3. **Data size vs. quality:**
   - A massive dataset is not always better. For specialized tasks, focusing on domain-relevant, high-quality examples (e.g., legal contracts for a legal assistant or medical notes for a healthcare system) can yield better results than sheer volume.
   - Balanced data is crucial—if your dataset is heavily skewed, your model might learn biases that impair real-world performance.

4. **Regular updates:**
   - LLMs and SLMs may need periodic re-training or fine-tuning to stay current—especially in fast-changing domains like news, finance, or tech.
   - Scheduling updates (e.g., monthly or quarterly) helps keep your model's responses relevant and accurate over time.

By tailoring a pretrained model to the nuances of your dataset and domain, you strike a balance between performance and practicality. This approach leverages the "language understanding" your model already has, saving you both time and money, while delivering solutions that can quickly adapt to changing user needs or evolving industry conditions.

1.6.3 Model Evaluation and Testing

After fine-tuning your language model, measuring its performance and validating its outputs is crucial. Even the most powerful LLM can produce unsatisfactory results if it's not rigorously tested on relevant data. Evaluation helps you identify potential weaknesses—like poor handling of out-of-distribution queries or biased outputs—and informs decisions about further tuning or data adjustments. See Table 1-6.

Table 1-6. Key Metrics for Evaluating NLP Models

Metric	Description	Formula	Threshold for Interpretation
Accuracy	Measures the proportion of correct predictions made by the model.	Accuracy = (TP + TN) / (TP + TN + FP + FN)	Higher values indicate better performance (e.g., 85% is good).
F1-Score	Balances precision and recall, especially useful for imbalanced datasets.	F1 = 2 * (Precision * Recall) / (Precision + Recall)	Closer to 1 is ideal; a value of 0.5 suggests poor performance.
Perplexity	Measures how well a language model predicts a sequence of words.	Perplexity = 2^(H(p))	Lower values indicate better performance (e.g., 10 > 100).
BLEU	Measures the similarity between generated and reference texts via n-grams.	BLEU = BP * exp(Σ(p_n)) Range: 0–1. Higher values (e.g., 0.8) indicate strong alignment.	Range: 0 to 1. Higher values (e.g., 0.8) indicate good alignment.
ROUGE	Evaluates overlap and content capture of reference text using n-grams.	ROUGE = (Recall of n-grams) / (Total n-grams in reference). Higher scores (e.g., 0.7) indicate better content capture.	Range: 0 to 1. Higher scores (e.g., 0.7) indicate good content capture.
Answer Correctness	Measures if the response is factually correct and addresses the query.	Manual Evaluation (Binary/Graded Scale)	Scale: 0 to 5. Higher scores indicate better factual accuracy.
Relevance	Assesses how well the response aligns with the user's intent and context.	Manual Evaluation (Scale 1 to 5). Higher scores indicate better relevance.	Higher scores indicate better contextual relevance (e.g., 4 or 5).

(continued)

Table 1-6. (*continued*)

Metric	Description	Formula	Threshold for Interpretation
Hallucination	Evaluates the frequency of fabricated or misleading responses.	Hallucination Rate = (Erroneous responses / Total responses) * 100. Percentage of erroneous responses.	Lower values indicate a more reliable model (e.g., <5% hallucinations).
ROUGE-L	Measures the longest common subsequence overlap between generated and reference text.	ROUGE-L = LCS_Length / Reference_Length	Higher scores (closer to 1) indicate better content capture.
Average Sentence Length	Evaluates the fluency and coherence of the text generated by the model based on sentence length.	Avg Sentence Length = (Σ sentence_length_i) / N	Lower values suggest more concise and focused text; very long sentences might indicate poor coherence.
Word Error Rate (WER)	Measures the discrepancy between the generated text and the reference text by counting insertion, deletion, and substitution errors.	WER = (S + D + I) / N × 100 where SSS is substitution, DDD is deletion, and III is insertion errors. • **S** = number of substitutions • **D** = number of deletions • **I** = number of insertions • **N** = total number of words in the reference transcript	Lower WER indicates better model accuracy, with values close to 0 being ideal.

(*continued*)

Table 1-6. (*continued*)

Metric	Description	Formula	Threshold for Interpretation
Execution Time	The time taken by the model to process a given input usually measured in seconds.	Time taken for model processing.	Lower values indicate faster responses;, critical for real-time tasks.
Model Fairness	Assesses if the model's predictions are biased based on sensitive attributes (e.g., gender, race).	Various fairness metrics (e.g., demographic parity, equalized odds).	Equal performance across different groups indicates no bias.
Model Robustness	Evaluate the model's stability when exposed to noisy, adversarial, or out-of-distribution data.	Measured by performance drop on perturbed input data.	Smaller performance drops indicate better robustness.
Memory Usage	The amount of memory (RAM) consumed by the model during inference.	Measured in GB or MB.	Lower values are ideal for deployment in resource-constrained environments.
Throughput	The number of processed inputs per unit of time (e.g., sentences per second).	Throughput = Number_of_samples / Time_taken	Higher throughput indicates faster processing speed.
Latency	The time taken to generate a response after an input is provided.	Measured in milliseconds or seconds.	Lower latency is crucial for real-time applications.

1.6.3.1 Scenario in Action: Validating a Cupcake Chatbot

After fine-tuning your cupcake chatbot, you evaluate its performance using quantitative metrics (like intent recognition accuracy) and human feedback. While automated metrics assess accuracy, human evaluations help identify issues with response clarity and helpfulness. This combined approach ensures that the chatbot communicates effectively and meets quality standards.

1.6.3.2 Common Pitfalls in Model Testing

- **Overfitting:** If the model scores extremely high on the training data but performs poorly on unseen questions, it might be memorizing rather than understanding.

- **Data leakage:** Unintentionally including test examples in your training set can inflate scores and give a false sense of confidence.

- **Bias and fairness issues:** Even well-intentioned models can produce biased outputs if trained on skewed datasets. Regular audits—including checking for demographic or linguistic biases—are essential.

By combining quantitative metrics (like accuracy and perplexity) with qualitative assessments (like user feedback), you thoroughly understand model's strengths and weaknesses. This holistic approach ensures that your LLM or SLM is truly production-ready and equipped to handle the variety of inputs it will encounter in real-world scenarios.

1.6.4 Deployment and Monitoring

With your model fine-tuned and evaluated, deployment ensures it is accessible and operational for real-world use. The deployment strategy should balance scalability, efficiency, and cost while ensuring performance.

1.6.4.1 Deployment Strategies for LLMs

1. **Cloud-based deployments (e.g., AWS, Azure, and Hugging Face Inference Endpoints):**

   Cloud platforms like AWS SageMaker and Hugging Face simplify deployment, offering scalable GPU and TPU resources for high-performance inference. For instance, Hugging Face's Inference Endpoints allow seamless hosting and integration with minimal setup, which is ideal for small- to medium-scale operations.
   Use case: Deploying a cupcake chatbot via Hugging Face can automate customer interactions, allowing businesses to respond quickly to queries about flavors, delivery times, and custom orders.

2. **Distributed deployments:**

   For applications requiring high throughput, distributed strategies—such as load balancing between GPUs or torrent-style decentralized inference (e.g., Petals)—ensure efficient resource use. These methods split model computations across multiple servers, reducing latency and increasing scalability.

3. **Edge and browser-based deployments (e.g., WebGPU):**

   Models like WebLLM run directly in users' browsers using WebGPU, eliminating server dependencies. This approach improves privacy by keeping computations on the client side and reduces latency for real-time applications like chatbots.
   Use case: Implementing the cupcake chatbot directly in browsers enables faster user responses, particularly during high-traffic periods like holidays.

4. **Quantized and optimized models:**

   Techniques like quantization reduce the model's memory footprint, making deployment feasible on resource-constrained devices like mobile phones or edge devices. Local LLM offers a solution for running smaller, quantized models locally without GPUs, catering to offline or privacy-sensitive applications.

1.6.4.2 Monitoring and Versioning

Once deployed, ongoing monitoring is vital to maintain model reliability. Key practices include:

- **Performance tracking:** Measure metrics like latency, throughput, and accuracy to ensure consistent performance.

- **Error logging and feedback loops:** Analyze logs to identify patterns in failures or poor responses and incorporate user feedback to refine the model.

- **Version control:** Maintain a clear record of updates and changes, allowing rollbacks if a new version underperforms.

1.6.4.3 Scenario in Action: Deploying and Monitoring a Cupcake Chatbot

Suppose your cupcake chatbot is deployed using Hugging Face Inference Endpoints. You configure the endpoint to handle up to 1,000 requests per minute during busy periods like holidays. Regular monitoring reveals a slight delay during peak hours, prompting you to optimize the deployment by leveraging distributed GPU inference. Additionally, customer feedback highlights repeated misunderstandings about custom orders, leading you to fine-tune the model with updated training data. These proactive measures ensure that the chatbot continues to provide accurate and efficient responses.

A well-executed deployment strategy with diligent monitoring ensures your model performs reliably and evolves to meet user needs. The next section explores choosing between LLMs and SLMs, balancing costs, resources, and task requirements while leveraging the latest innovations in model optimization.

1.7 LLM Selection Based on Cost, Compute, and Effectiveness and SLM Exploration

In the rapidly evolving landscape of NLP, the choice between LLMs and SLMs has become a critical decision point for developers, researchers, and organizations. This section delves into the intricacies of selecting the right model based on cost, computational requirements, and effectiveness, while also exploring the latest innovations in SLM technology.

1.7.1 The LLM vs. SLM Dilemma

As we navigate the complex world of language models, we find ourselves at a crossroads. On one side, we have the behemoths of the AI world—LLMs like GPT-3, BERT, and T5—capable of understanding and generating human-like text across a wide range of topics and tasks. On the other side, we have the more nimble and resource-efficient SLMs, designed to perform specific tasks with remarkable speed and efficiency.

The choice between these two classes of models is not always straightforward. It requires a careful consideration of various factors, including the specific requirements of the task at hand, the available computational resources, and the overall goals of the project. This section explores these considerations in detail.

1.7.1.1 Cost Considerations

When it comes to deploying language models, cost is often a primary concern. LLMs, with their vast number of parameters and complex architectures, come with a hefty price tag. Training an LLM from scratch can cost millions of dollars, not to mention the ongoing expenses associated with inference and fine-tuning.

Consider the case of GPT-3, one of the largest language models to date. With 175 billion parameters, the computational resources required to train and run this model are staggering. Organizations looking to leverage such models often need to rely on cloud-based solutions or invest in expensive hardware infrastructure.

On the other hand, SLMs offer a more cost-effective alternative. These models, designed to be lightweight and efficient, can often run on commodity hardware or even mobile devices. This significantly reduces both the initial investment and ongoing operational costs.

For example, a startup developing a chatbot for customer service might find the cost of deploying a full-scale LLM prohibitive. In this case, a carefully chosen and fine-tuned SLM could provide satisfactory performance at a fraction of the cost.

1.7.1.2 Computational Requirements

Closely related to cost considerations are the computational requirements of language models. LLMs, by their very nature, demand substantial computational power. They require high-end GPUs or TPUs for efficient training and inference, and they often necessitate distributed computing setups for optimal performance.

The computational demands of LLMs can be a significant barrier to entry for many organizations. Small- to medium-sized businesses, in particular, may find it challenging to allocate the necessary resources for running these models. Moreover, the high energy consumption associated with LLMs has raised concerns about their environmental impact.

SLMs, in contrast, are designed to be computationally efficient. They can often run on CPUs or low-end GPUs, making them suitable for deployment on edge devices or in resource-constrained environments. This efficiency translates to lower energy consumption and a smaller carbon footprint.

Consider a scenario where real-time language processing is required on a mobile device. An LLM would be impractical in this context due to its size and computational demands. An SLM, however, could be easily integrated into a mobile app, providing quick and efficient language processing capabilities without draining the device's battery or requiring constant Internet connectivity.

1.7.1.3 Task Suitability

While cost and computational requirements are important factors, the suitability of a model for the specific task at hand is paramount. LLMs excel at tasks that require broad knowledge and context understanding. Their ability to generate coherent and contextually relevant text makes them ideal for applications such as:

1. Open-ended text generation
2. Complex question-answering systems
3. Multi-task language understanding
4. Creative writing assistance

For instance, an AI-powered writing assistant designed to help authors generate ideas and overcome writer's block would benefit greatly from the vast knowledge and creative capabilities of an LLM.

SLMs, on the other hand, shine in more focused, domain-specific tasks. They are particularly well-suited for applications that require:

1. Real-time processing
2. Low-latency responses
3. Deployment on edge devices
4. Specific, narrow-domain expertise

A sentiment analysis tool for social media monitoring, for example, might perform better with a well-tuned SLM that can quickly process short text snippets in real-time.

1.7.2 Effectiveness: Balancing Performance and Efficiency

When evaluating the effectiveness of language models, it's crucial to consider both performance and efficiency. LLMs generally outperform SLMs in tasks that require extensive knowledge and context understanding. Their ability to maintain coherence over long passages of text and to draw connections between diverse concepts gives them an edge in complex language tasks.

However, this superior performance comes at a cost. LLMs are often slower to generate responses, especially when deployed on less powerful hardware. They may also struggle with tasks that require quick, real-time processing of short text snippets.

SLMs, while potentially less powerful in absolute terms, can deliver faster responses with lower resource consumption. In many real-world applications, the ability to provide quick, reasonably accurate results is more valuable than slower, marginally more accurate ones.

Consider the following example: A large e-commerce platform wants to implement a product recommendation system based on customer reviews. An LLM might be able to provide more nuanced and contextually relevant recommendations by deeply analyzing the review text. However, if these recommendations take several seconds to generate for each user, it could lead to a poor user experience. An SLM, on the other hand, might provide slightly less sophisticated recommendations, but do so almost instantaneously, leading to a smoother browsing experience for customers.

This example illustrates a common tradeoff in NLP applications: the balance between performance and efficiency. In many cases, the "best" model is not necessarily the most powerful one, but the one that best meets the specific needs of the application while operating within the given constraints.

1.7.2.1 Infrastructure Requirements

LLMs typically require robust cloud infrastructure or high-performance on-premises hardware. They are well-suited for deployment in environments where:

1. Powerful GPU clusters are available
2. High-bandwidth network connections can handle large data transfers
3. Scalability is a primary concern

For example, a large-scale language translation service processing millions of requests daily would likely benefit from the power and scalability offered by cloud-deployed LLMs.

SLMs, conversely, can often be deployed on more modest hardware. They are ideal for scenarios where:

1. On-device processing is necessary (e.g., for privacy reasons)
2. Internet connectivity is limited or unreliable
3. Low-latency responses are crucial

A voice-activated smart home device, for instance, might use an SLM for basic command recognition to ensure quick response times even without Internet connectivity.

1.7.2.2 Scalability and Maintenance

When it comes to scalability, LLMs have a clear advantage. Their ability to handle a wide range of tasks makes them highly adaptable as system requirements evolve. However, this flexibility comes with increased complexity in terms of maintenance and updates.

SLMs, being more specialized, may require less frequent updates and are often easier to maintain. However, they may need to be replaced or significantly modified if the task requirements change substantially.

Consider a startup developing an AI writing assistant. Initially, they might deploy an SLM focused on grammar correction and basic style suggestions. As the company grows and user demands become more sophisticated, they might transition to an LLM capable of more advanced features like content generation and contextual recommendations.

1.7.3 SLM Techniques and Innovations

The field of SLM research is rapidly evolving, with new techniques and innovations constantly emerging. These advancements are making SLMs increasingly attractive for a wide range of NLP tasks. This section explores some of the key areas of innovation.

1.7.3.1 Model Pruning

Model pruning is a technique used to reduce the size of neural networks by removing unnecessary connections or neurons. In the context of SLMs, pruning can significantly reduce model size while maintaining most of the original performance.

For example, researchers have demonstrated that it's possible to prune BERT, a popular language model, to less than 5% of its original size while retaining 95% of its performance on certain tasks. This dramatic reduction in size makes it feasible to deploy BERT-like models on mobile devices or in other resource-constrained environments.

1.7.3.2 Knowledge Distillation

Knowledge distillation is a process where a smaller model (the "student") is trained to mimic the behavior of a larger model (the "teacher"). This technique allows SLMs to benefit from the knowledge captured by larger models without incurring the full computational cost.

DistilBERT, for instance, is a smaller version of BERT created through knowledge distillation. It retains 97% of BERT's language understanding capabilities while being 40% smaller and 60% faster. This makes DistilBERT a popular choice for applications requiring BERT-like performance but with lower computational requirements.

1.7.3.3 Quantization

Quantization involves reducing the precision of the model's parameters, typically from 32-bit floating-point numbers to 8-bit integers. This technique can significantly reduce the memory footprint and computational requirements of a model with minimal impact on performance.

For example, quantized versions of models like MobileBERT have been successfully deployed on mobile devices, enabling on-device natural language understanding for applications like virtual assistants and real-time translation.

1.7.3.4 Sparse Attention Mechanisms

Traditional transformer models use dense attention mechanisms, which can be computationally expensive, especially for long sequences. Sparse attention mechanisms, such as those used in models like Longformer and Big Bird, allow SLMs to process longer sequences more efficiently.

These innovations enable SLMs to handle tasks that were previously the domain of LLMs, such as document-level sentiment analysis or long-form question answering, while maintaining the efficiency advantages of smaller models.

1.7.4 The Future of SLMs in NLP

As research in SLMs continues to advance, you can expect these models to play an increasingly important role in the NLP landscape. Some potential future developments include:

1. **Task-specific SLMs:** Highly optimized models designed for specific NLP tasks, outperforming general-purpose LLMs in terms of efficiency and sometimes even accuracy.

2. **Adaptive SLMs:** Models that can dynamically adjust their size and complexity based on the input, optimizing the tradeoff between performance and efficiency in real-time.

3. **Federated learning for SLMs:** Techniques that allow SLMs to learn from distributed data sources without compromising privacy, enabling more personalized and secure NLP applications.

4. **Hardware-optimized SLMs:** Models designed to take full advantage of specialized hardware accelerators, pushing the boundaries of what's possible with on-device NLP.

As these innovations continue to emerge, the line between LLMs and SLMs may become increasingly blurred. We may see a future where the choice is not between large and small models, but rather a spectrum of models optimized for different tasks, deployment scenarios, and resource constraints.

1.8 Summary

- You learned the basics of LLMs and SLMs, their evolution, and their role in NLP tasks like text generation, translation, and summarization.

- You reviewed cloud platforms (AWS) and open-source tools for training, fine-tuning, and deploying models.

- You examined the LLM and SLM deployment lifecycle, from data collection to testing and monitoring.

- You learned how to choose between LLMs and SLMs based on cost, compute, and task complexity.

Great job! The next chapter dives into practical implementation strategies for LLMs and SLMs. Stay tuned!

CHAPTER 2

Text Generation with LLMs and SLMs

This chapter explains text generation techniques using LLMs and SLMs. It explores their applications in creative writing, automated reporting, and chatbot development, with practical guidance on implementing and deploying these models using AWS and other open-source frameworks.

The chapter begins by covering several types of text generation tasks, then delves into how LLMs and SLMs can be used effectively. The chapter also includes hands-on tips for deploying and fine-tuning models for real-time applications and industry-specific use cases like content generation, customer support, and on-device text generation.

This chapter covers:

- **Types of text generation tasks**: Creative writing, automated content, and chatbots.

- **Building and deploying SLMs**: Choosing, training, and deploying small models for efficient text generation.

- **Implementing on AWS**: Deploying models using AWS services for scalable text generation.

- **Open-source implementation**: Fine-tuning and deploying models with open-source tools.

- **Industry use cases**: Text generation in content marketing, customer support, and more.

By the end of this chapter, you will clearly understand how to apply both LLMs and SLMs to text generation tasks and how to implement them effectively.

CHAPTER 2   TEXT GENERATION WITH LLMS AND SLMS

2.1 Text Generation with LLMs and SLMs

Text generation is a transformative AI technology with diverse applications, from creative writing to automated content creation and sophisticated chatbots. This field leverages LLMs and SLMs, each offering unique strengths (see Table 2-1).

Table 2-1. Popular LLMs and SLMs for Text Generation

Model Name	Type	Tokens	Parameters	Training Features	Fine-Tuning Features	Inference Features
GPT-4	LLM	Very Large	Very Large (Estimated to be trillions)	Massive dataset, web scrape data, books, etc.	Task-specific data for customization, reinforcement learning	High-quality output, context-awareness, good for complex and creative tasks
GPT-3	LLM	Very Large	175 Billion	Massive dataset, web scrape data, books, etc.	Task-specific data for customization	High-quality output, context-aware
T5	LLM	Large	220 Million - 11 Billion	Colossal Clean Crawled Corpus	Task-specific data for customization	Good for translation and summarization
BERT	SLM	Medium	110 Million - 340 Million	BooksCorpus and English Wikipedia	Task-specific data for customization	Good for text classification, question-answering
DistilGPT-2	SLM	Small	82 Million	WebText dataset	Task-specific data for customization	Efficient, fast inference
TinyBERT	SLM	Small	14.5 Million	General domain data	Task-specific data for customization	Extremely efficient, low-latency

2.2 Types of Text Generation Tasks

Text generation is a transformative technology with applications across industries, from creative writing and content automation to sophisticated conversational agents. To leverage its full potential, you must distinguish between the different tasks and understand how they align with various model capabilities.

This section explores the foundational aspects of text generation tasks, the roles of LLMs and SLMs, and the distinction between generative and extractive tasks.

2.2.1 Creative Writing

One of the hallmark applications of LLMs is creative writing. These models excel at generating:

- **Stories and novels:** LLMs such as GPT-4 can produce intricate narratives, maintaining character development, plot arcs, and thematic consistency across multiple chapters.

- **Poetry:** With appropriately crafted prompts, models can mimic poetic styles ranging from Shakespearean sonnets to free verse.

- **Scripts:** Screenwriters and playwrights can use LLMs to draft dialogues or scene outlines, iterating based on feedback.

2.2.1.1 How LLMs Drive Creativity

LLMs are trained on vast corpora encompassing diverse genres and styles, enabling them to:

- Understand and emulate nuanced language patterns.

- Incorporate thematic elements from user-defined prompts.

- Suggest variations or alternative continuations to stimulate ideation.

For example, a prompt such as:

"Write a suspenseful introduction for a thriller novel involving a cyber heist."

Can yield:

> "The room was bathed in a cold, flickering glow from the monitors, their screens displaying cascading streams of encrypted data. Alyssa's fingers danced across the keyboard, her pulse synchronized with the ticking clock on the wall. Every keystroke brought her closer to unlocking the digital vault—and to the point of no return."

2.2.1.2 Technical Considerations

- **LLM selection:** Use GPT-3 or GPT-4 models for high linguistic fluency and coherence.

- **Prompt engineering:** Provide detailed and context-rich prompts to steer the model's creative output.

- **Fine-tuning:** Adapt the model on domain-specific datasets, such as mystery novels or science fiction, for genre-focused writing.

2.2.2 Automated Content Generation

Beyond creativity, text generation plays a pivotal role in automating structured and semi-structured content, including:

- **News summaries:** LLMs generate concise reports from live feeds or lengthy articles.

- **Product descriptions:** For e-commerce platforms, models like T5 can automate descriptions based on product metadata.

- **Reports:** SLMs such as DistilGPT-2 can draft financial or business reports, transforming raw data into readable summaries.

2.2.2.1 Differentiating Between LLMs and SLMs

- **LLMs:** Handle high complexity, contextual interdependencies, and abstract reasoning. Ideal for drafting comprehensive and nuanced reports.

- **SLMs:** Optimized for efficiency, excelling in template-driven or repetitive tasks like generating consistent product descriptions.

Example of LLM in reporting:

A company's quarterly earnings report requires interpreting financial data and crafting a narrative for stakeholders. A prompt like:

> *"Summarize the financial performance of Q4 2024 for XYZ Corp, highlighting revenue growth, challenges, and future strategies."*

Can yield:

> "XYZ Corp achieved a 15% revenue growth in Q4 2024, driven by strong sales in the North American market. However, supply chain disruptions impacted operating margins. Moving forward, the company plans to diversify suppliers and invest in automation to enhance resilience."

An example of SLM in product descriptions: Given metadata like "Wireless headphones, 40-hour battery life, noise-cancelling," an SLM can generate:

> "Experience superior audio with these wireless headphones. Enjoy up to 40 hours of uninterrupted listening, complemented by advanced noise-cancelling technology."

2.2.2.2 Technical Considerations

- **Dataset preparation:** Curate structured datasets (e.g., product attributes or financial summaries) for effective fine-tuning.
- **Model compression:** Employ model distillation to adapt LLM capabilities into efficient SLM variants for repetitive tasks.

2.2.3 Chatbots and Conversational Agents

The proliferation of chatbots and virtual assistants underscores the importance of text generation in real-time, interactive settings. Applications include:

- **Customer support:** Automating responses to FAQs or troubleshooting steps.
- **Personal assistants:** Enabling conversational interfaces for scheduling, reminders, or recommendations.
- **Enterprise bots:** Assisting employees with HR queries or internal documentation.

SLMs, such as TinyBERT or DistilGPT-2, are particularly suited for:

- **Low-latency requirements:** Ensuring rapid responses in high-traffic scenarios.
- **Domain-specific tuning:** Adapting to specialized vocabulary, such as legal or medical jargon.

2.2.3.1 Chatbot Interaction Example

User: *"How do I reset my password?"*

SLM response: "To reset your password, click Forgot Password at the login screen. You'll receive an email with a reset link. Follow the instructions to set a new password."

2.2.3.2 Enhancing Conversational Quality with LLMs

LLMs contribute to chatbot systems by using:

- **Context awareness:** Maintaining conversation continuity across multiple turns.
- **Personalization:** Adapting responses based on user profiles and interaction history.
- **Handling ambiguity:** Generating clarifying questions when user intent is unclear.

2.2.3.3 Technical Considerations

- **Pipeline configuration:** Use SLMs for initial intent classification and LLMs for generating complex or nuanced responses.
- **Hybrid deployment:** Integrate LLMs and SLMs in a serverless architecture (e.g., AWS Lambda) to balance cost and performance.
- **Edge optimization:** Deploy SLMs on edge devices for offline capabilities in customer-facing applications.

2.2.4 Generative vs. Extractive Tasks

Text generation tasks broadly fall into two categories: generative and extractive. Understanding these paradigms helps in selecting the right model and approach.

Generative tasks involve creating new content from scratch, guided by input prompts. Examples include:

- **Essay writing:** Crafting argumentative essays from minimal topic outlines.
- **Dialogue generation:** Simulating realistic conversations for training chatbots.

LLMs like GPT-4 are ideal for generative tasks, leveraging their extensive training data and ability to synthesize information creatively.

Extractive tasks, on the other hand, focus on extracting relevant information from existing data. Examples include:

- **Summarization:** Condensing long articles into key points.
- **Question-answering:** Retrieving specific answers from a document or knowledge base.

SLMs excel with extractive tasks, offering:

- Faster inference speeds.
- Reduced memory requirements.

2.2.4.1 Practical Comparison

Generative task: *"Summarize the importance of renewable energy in combating climate change."*

Output: "Renewable energy reduces greenhouse gas emissions by replacing fossil fuels. It's a critical step in achieving global sustainability goals."

Extractive task: Given a document, *"Highlight the main reasons for using solar energy."*

Output: "Solar energy is renewable, reduces carbon footprint, and lowers electricity costs."

2.2.4.2 Technical Considerations

- Choose generative approaches for tasks requiring creativity and fluency.
- Opt for extractive methods when accuracy and precision are paramount.

By understanding the diversity of text generation tasks and aligning them with appropriate LLMs and SLMs, organizations can effectively harness AI to meet their unique requirements. Whether crafting compelling narratives or enabling real-time conversational agents, the key lies in leveraging the strengths of these models in the right context.

2.3 Building and Deploying LLMs and SLMs for Text Generation

SLMs are crucial for scenarios requiring efficient, real-time performance and resource-conscious deployments. This section explores the key aspects of building and deploying SLMs, which are outlined in Table 2-2.

Table 2-2. *Training and Deployment Steps for SLMs and LLMs*

Step	Description	Code Snippet
| 1. Data collection | Collect raw textual data from multiple sources (datasets, web scraping, etc.). | ```python
import datasets
# Load a publicly available dataset
dataset = datasets.load_dataset("wikitext", "wikitext-103-raw-v1")
print(dataset["train"][0])
``` |
| 2. Data preprocessing | Clean, tokenize, and format the data suitable for training. | ```python
import re
from transformers import AutoTokenizer
tokenizer = AutoTokenizer.from_pretrained("bert-base-uncased")
def preprocess_text(text):
    text = re.sub(r'\s+', ' ', text.strip()) # Remove extra spaces
    return tokenizer(text, truncation=True, padding="max_length", max_length=512)
sample_text = "This is an example sentence."
tokenized_data = preprocess_text(sample_text)
print(tokenized_data)
``` |

(*continued*)

Table 2-2. (*continued*)

Step	Description	Code Snippet
3. Dataset preparation	Convert data into batches, tokenize it, and create input-output mappings.	`from torch.utils.data import DataLoader`
		`def tokenize_function(examples): return tokenizer(examples["text"], truncation=True, padding="max_length", max_length=512)`
		`tokenized_datasets = dataset.map(tokenize_function, batched=True) train_dataloader = DataLoader(tokenized_datasets["train"], batch_size=8, shuffle=True)`
4. Model selection	Choose between an SLM (e.g., DistilBERT) and an LLM (e.g., GPT-3, LLaMA).	`from transformers import AutoModelForCausalLM`
		`model = AutoModelForCausalLM.from_pretrained("gpt2") # Choose a small or large model`
5. Model configuration	Define hyperparameters (batch size, learning rate, layers, etc.).	`from transformers import TrainingArguments`
		`training_args = TrainingArguments( output_dir="./results", num_train_epochs=3, per_device_train_batch_size=8, learning_rate=5e-5 )`

(*continued*)

Table 2-2. (*continued*)

Step	Description	Code Snippet

6. Training	Train the model using GPUs/TPUs with loss computation and optimization.	`from transformers import Trainer`
		`trainer = Trainer(`
		`model=model,`
		`args=training_args,`
		`train_dataset=tokenized_`
		`datasets["train"]`
		`)`
		`trainer.train()`

7. Evaluation and validation	Test the model on validation data and adjust any hyperparameters.	`results = trainer.evaluate()` `print(results)`

8. Fine-tuning (optional)	Further, train a pretrained model on domain-specific data.	`trainer.train(resume_from_`
		`checkpoint=True) # Resume training`
		`with new dataset`

9. Model optimization	Quantize, prune, or distill to reduce size and improve inference.	`from optimum.onnxruntime import`
		`ORTModelForSequenceClassification`
		`optimized_model =`
		`ORTModelForSequenceClassification.`
		`from_pretrained("bert-base-uncased")`
		`optimized_model.save_pretrained("./`
		`optimized_model")`

10. Model deployment	Export and deploy the model using APIs, cloud platforms, or local inference.	`from transformers import pipeline`
		`nlp_pipeline = pipeline("text-`
		`generation", model="gpt2")`
		`output = nlp_pipeline("The AI`
		`revolution is", max_length=50)`
		`print(output)`

(*continued*)

Table 2-2. (*continued*)

Step	Description	Code Snippet

11. Inference and monitoring	Serve the model for real-time or batch inference and monitor its performance.	`import torch`
		
		`def generate_text(prompt):`
		`    inputs = tokenizer(prompt, return_tensors="pt")`
		`    outputs = model.generate(**inputs, max_length=50)`
		`    return tokenizer.decode(outputs[0], skip_special_tokens=True)`
		
		`print(generate_text("The future of AI is"))`

2.3.1 Choosing the Right LLM or SLM for Text Generation

Selecting an appropriate LLM or SLM depends on task requirements, computational constraints, and deployment scenarios. While LLMs like GPT-4 and PaLM-2 are suited for complex, high-resource applications, SLMs like DistilGPT-2 and TinyBERT cater to efficiency-driven use cases. Table 2-3 illustrates these comparisons.

Table 2-3. *Comparing LLMs and SLMs for Text Generation*

Criteria	LLM (e.g., GPT-4, PaLM-2)	SLM (e.g., DistilGPT-2, TinyBERT)
Task complexity	Ideal for open-ended, high-context text generation	Suitable for structured content or classification tasks
Computational requirements	Requires high-end GPUs/TPUs	Optimized for edge devices with minimal latency
Inference speed	Higher latency requires optimization	Faster inference, suitable for real-time applications
Training data scale	Trained on massive datasets	Trained on smaller, domain-specific data
Deployment suitability	Cloud-based or high-resource environments	Mobile, IoT, or on-device applications

Consider these factors when choosing between LLMs and SLMs:

1. **Task complexity:**
   - For open-ended text generation (e.g., creative writing, long-form content), opt for LLMs like GPT-4.
   - For structured or repetitive tasks (e.g., chatbots, product descriptions), an SLM such as DistilGPT-2 is sufficient.

2. **Resource constraints:**
   - SLMs like TinyBERT are preferable for mobile and IoT devices due to lower computational demands.
   - LLMs require cloud-based deployments with high-performance GPUs or TPUs for optimal operation.

3. **Latency requirements:**
   - LLMs may require optimization techniques (e.g., model parallelism, distillation) to meet real-time processing demands.
   - SLMs are inherently faster and better suited for low-latency applications like real-time chatbots.

2.3.2 Training and Fine-Tuning LLMs and SLMs

LLMs and SLMs can be fine-tuned on domain-specific datasets to enhance their performance for specialized applications. Fine-tuning an LLM requires substantial computational resources, while fine-tuning an SLM can be done efficiently on smaller hardware setupsf.

2.3.2.1 Steps for Fine-Tuning

1. **Dataset preparation:**
   - Gather and preprocess a high-quality dataset aligned with the target application.
   - Ensure labelled datasets for supervised fine-tuning (e.g., sentiment analysis, entity recognition).

2. **Choosing the right fine-tuning framework:**

   - **For LLMs**: Use DeepSpeed, FSDP, or LoRA to optimize memory usage during fine-tuning.

   - **For SLMs**: Use Hugging Face Transformers for efficient training on constrained hardware.

3. **Evaluation metrics:**

   - **Text generation**: BLEU, ROUGE, and perplexity.

   - **Classification tasks:** Accuracy, F1-score, and AUC-ROC.

4. **Hyperparameter optimization:**

   - Adjust learning rate, batch size, number of epochs, and weight decay to maximize performance while preventing overfitting.

5. **Transfer learning:**

   - Use pretrained models and apply transfer learning techniques to adapt LLMs and SLMs to specific domains (e.g., legal, finance, healthcare).

2.3.3 Efficient Inference and Deployment of LLMs and SLMs

Optimizing inference is crucial for deploying LLMs and SLMs at scale. LLMs typically require significant computational resources, whereas SLMs can be deployed efficiently in resource-constrained environments.

2.3.3.1 Optimization Techniques

1. **Quantization:**

   - Reduce model precision from FP32 to INT8 or BF16 for faster inference.

   - Tools: ONNX Runtime, TensorRT, TensorFlow Lite.

2. **Pruning:**
   - Remove less significant model weights to reduce memory footprint while maintaining accuracy.

3. **Distillation:**
   - Train a smaller student model using an LLM as a teacher, preserving knowledge while reducing complexity.

4. **Efficient parallelism for LLMs:**
   - Implement model parallelism (TPU/GPU splitting) to speed up LLM inference.
   - Use DeepSpeed ZeRO or FSDP (Fully Sharded Data Parallelism) for distributed training.

2.3.3.2 Deployment Strategies

1. **Cloud-based inference (For LLMs):**
   - LLMs are best deployed in cloud environments like AWS, Azure, or Google Cloud to leverage high-performance GPUs.
   - Use model serving frameworks such as TorchServe or NVIDIA Triton Inference Server.

2. **On-device deployment (For SLMs):**
   - SLMs can be deployed on mobile, edge, or IoT devices using TensorFlow Lite or ONNX Runtime.

2.3.3.3 Real-World Use Cases

1. **LLM-powered AI assistant in Enterprise SaaS:**
   - Deploy GPT-4 or PaLM-2 as an AI-powered knowledge assistant for customer support automation.
   - Optimize inference using DeepSpeed ZeRO for faster response times.

2. **SLM-based chatbot for mobile banking:**

    - Fine-tune TinyBERT for financial queries.

    - Quantize the model to minimize latency and memory usage for seamless mobile deployment.

Organizations can achieve state-of-the-art text generation while balancing computational efficiency by carefully selecting, fine-tuning, and optimizing LLMs and SLMs. The following section explores real-world cloud deployment strategies, focusing on scalable and cost-effective implementations using AWS, Azure, and GCP.

2.4 Implementing Text Generation on AWS

Previous sections of this book explored the foundations of text generation using LLMs and SLMs, along with the essential steps involved in building and deploying SLMs for various enterprise use cases. This section focuses on Amazon Web Services (AWS)—a powerful and highly scalable cloud platform that enables robust hosting, real-time inference, and optimization of text generation models.

By leveraging AWS services such as Amazon SageMaker, AWS Lambda, and Amazon SageMaker Neo, organizations can efficiently manage the entire machine learning lifecycle of their text generation applications—from development and training to deployment and optimization. This section walks through the crucial AWS offerings, illustrating how they integrate with text generation workflows and presenting practical use cases that demonstrate how to implement real-world solutions.

2.4.1 Transition from Building SLMs to AWS Implementation

After training and fine-tuning your language model (whether an SLM for a niche domain or a general-purpose LLM), the next challenge is making it available to users in a secure, fast, and scalable manner. While local deployments or self-managed servers can work for small-scale prototypes, cloud platforms are the gold standard for enterprise-grade solutions. AWS specifically provides:

- **High availability and scalability** to handle unpredictable traffic.

- **Managed infrastructure** to reduce operational overhead.

- **Built-in monitoring and logging** for performance insights and debugging.
- **Integration with a broad ecosystem** of services (e.g., AWS Lambda, Amazon API Gateway, AWS IoT, Amazon SQS) for a comprehensive end-to-end workflow.

With these capabilities, AWS offers a seamless transition from model development to production deployment, empowering teams to focus on iteration and innovation rather than wrestling with infrastructure complexities.

2.4.2 Deploying LLMs with AWS SageMaker

AWS SageMaker is a fully managed service that covers the entire machine learning workflow—data labelling, model training, tuning, deployment, and monitoring. For text generation tasks, SageMaker supports various deep learning frameworks (e.g., TensorFlow, PyTorch) and enables one-click deployment of large models like GPT-2, T5, or custom SLMs. The upcoming pages explore:

1. **Creating a SageMaker notebook instance** for model training and fine-tuning.
2. **Packaging and containerizing** your text generation model.
3. **Deploying to a SageMaker endpoint** for real-time inference.
4. **Monitoring and scaling** to handle production loads.

The following snippet illustrates how to establish a simple SageMaker session using the AWS SDK for Python (Boto3). We expand on these concepts in subsequent pages:

https://github.com/anvcse562/ModernNLP-LLMSLM/blob/main/Ch02/ch2.3/sample-sagemaker-init-2-3-1.py

To illustrate how to set up a text generation model on SageMaker, we'll dive deeper into the practical steps of model training, containerization, and deployment.

2.4.2.1 Preparing Your Training Environment

Before deploying a model to a SageMaker endpoint, you typically need to train or fine-tune it within a SageMaker environment. You can use the following approaches:

- **Built-in containers:** AWS provides pre-built Docker images for deep learning frameworks like PyTorch, TensorFlow, and Hugging Face.
- **Custom Docker containers:** For unique libraries or dependencies, you can create your container and register it in Amazon ECR (Elastic Container Registry).

The following is a minimal example of using a Hugging Face Deep Learning Container for training a text generation model (e.g., GPT-2 or T5). This snippet shows how to initiate a training job using SageMaker's Estimator interface.

https://github.com/anvcse562/ModernNLP-LLMSLM/blob/main/Ch02/ch2.3/create-sagemaker-training-job-2-3-1-1.py

Key points:

1. `entry_point` specifies the Python script that contains the training logic (e.g., downloading the dataset, training the model, and saving checkpoints).
2. `source_dir` must include all required Python modules and scripts for the training process.
3. `transformers_version` and `pytorch_version` indicate which Hugging Face and PyTorch versions to use.
4. `hyperparameters` are passed into the training script, allowing for dynamically configured model details and training parameters.

2.4.2.2 Deploying a SageMaker Endpoint

Once the training job is completed, SageMaker stores the model artifacts (e.g., `.bin` or `.pt` files) in an S3 bucket. You can then create a real-time endpoint for inference:

https://github.com/anvcse562/ModernNLP-LLMSLM/blob/main/Ch02/ch2.3/create-sagemaker-endpoint-2-3-1-2.py

- `deploy()`: Automatically packages your fine-tuned model and creates an HTTPS endpoint you can invoke.
- `predict()`: Sends a payload (in this case, a prompt for text generation) and returns the generated response from the model.

Note Depending on the model size and traffic, consider using GPU-backed instances (e.g., `ml.g4dn.xlarge`) for faster inference.

2.4.3 Real-Time Inference Using AWS Lambda (Continued)

AWS Lambda offers a serverless approach for quick, on-demand inference. While it may not be optimal for huge models (due to memory and cold start limitations), it can be highly efficient for smaller SLMs or specialized text classification tasks.

2.4.3.1 Packaging Your Model for Lambda

To deploy an SLM to Lambda, you need to package the following:

1. Dependencies (e.g., PyTorch, Transformers library) in a Lambda Layer or within a custom container.
2. Model artifacts (or a smaller quantized version) in the function's code or an attached EFS (Elastic File System) volume for larger files.

A high-level workflow involves the following steps:

1. **Create a ZIP** file or Docker image containing your model and inference script.
2. **Upload** the package to AWS Lambda (directly or via ECR for Docker-based Lambda).
3. **Configure** function memory and timeouts to accommodate loading and running the model.

2.4.3.2 Sample AWS Lambda Function for Text Classification

Here is a simplified example of a Lambda function handler, showing how you might classify an input text into predefined categories. This snippet can be adapted to handle various SLM-based text generation or classification tasks.

https://github.com/anvcse562/ModernNLP-LLMSLM/blob/main/Ch02/ch2.3/lambda-with-slm-2-3-2-1.py

>   Key steps:
>
>   1. **Model loading:** Placed in a separate function (`load_model()`), it only executes once per Lambda container lifecycle, minimizing cold start overhead.
>   2. **Input handling:** Expects JSON input from an API call.
>   3. **Inference:** PyTorch is used for forward passes, and the predicted class is obtained from logits.
>   4. **Response:** Returns a JSON response.

To connect this Lambda function to a public-facing endpoint, you can use Amazon API Gateway, creating a RESTful API for your text classification or generation use case.

2.4.4 Optimizing Models for Edge Deployment with Amazon SageMaker Neo

As organizations increasingly require *on-device inference*—for example, in IoT devices, mobile apps, or remote locations—there is a growing need to run text generation models efficiently at the edge. Running language models locally can help reduce latency, enhance data privacy, and enable offline capabilities. However, typical LLMs (GPT-2, T5, etc.) can be large, demand substantial power, and are too complex to deploy on constrained hardware.

Amazon SageMaker Neo addresses these challenges by compiling and optimizing machine learning models to run on various hardware platforms with improved performance. Neo supports common frameworks (e.g., TensorFlow, PyTorch) and can target multiple CPU and GPU architectures. This is particularly useful for specialized language models operating on custom devices or embedded systems.

2.4.4.1 How SageMaker Neo Works

1. **Model compilation:** Neo analyzes the model's computational graph and generates an optimized version tailored to the target hardware. It uses a machine learning compiler called TVM to perform graph-level optimizations.

2. **Runtime execution:** Neo provides a lightweight runtime that can be deployed alongside the optimized model. This runtime executes the compiled artifacts for efficient inference.

3. **Multi-platform support:** You can compile once and run anywhere—x86, ARM, NVIDIA GPUs, and more. This ensures flexibility when transitioning models from the cloud to edge environments.

Here is a simplified code snippet showing how to compile a PyTorch or TensorFlow model using Neo for an edge target device (e.g., an ARM-based AWS Graviton processor).

```
https://github.com/anvcse562/ModernNLP-LLMSLM/blob/main/Ch02/ch2.3/create-sagemaker-training-job-2-3-1-1.py
```

Key parameters in the compilation process:

- `input_model`: Points to the model (or `Estimator`, `Model` object) you want to optimize.
- `target_instance_family`: Specifies the hardware environment (e.g., `ml_c5` for AWS C5 instances, `ml_inf` for Inferentia-based instances, or even `jetson` for NVIDIA Jetson devices).
- `output_path`: S3 location where the compiled model artifacts will be stored.
- `framework`: Indicates which deep learning framework you used (PyTorch, TensorFlow, etc.).

After the compilation is complete, you can either deploy the optimized model to an Amazon SageMaker endpoint or download it for use in a custom environment (e.g., on-premises server, Raspberry Pi, NVIDIA Jetson). This significantly expands the deployment possibilities for text generation applications that must operate outside the standard cloud infrastructure.

2.4.4.2 Considerations for Text Generation at the Edge

1. **Model size:** Even with optimization, large models can be challenging to run on edge devices. Techniques like model distillation, quantization, and pruning can reduce the footprint while maintaining acceptable accuracy.

2. **Latency:** Edge deployments can drastically reduce roundtrip times since inference doesn't rely on cloud connectivity. However, you must ensure the device has sufficient compute resources for real-time performance.

3. **Memory constraints:** Some SLMs might require gigabytes of memory. Validate your device's RAM and storage capacity before attempting on-device deployment.

4. **Data privacy:** Sensitive user data remains on the device, avoiding potential compliance issues related to sending data over the network.

2.4.5 Integrating AWS Services for Comprehensive Text Generation Solutions

While you can deploy your text generation models individually with SageMaker, Lambda, or SageMaker Neo, most enterprise solutions require multiple AWS services working in tandem. For example:

- **AWS IoT** for edge device connectivity.

- **Amazon API Gateway** to expose a RESTful or WebSocket API for your model endpoint.

- **AWS Step Functions** or **Amazon EventBridge** for orchestrating end-to-end workflows.

- **Amazon S3** and **Amazon DynamoDB** for storing logs, model metadata, or inference results.

High-level architecture:

1. **Model training and optimization:** Train your LLM or SLM on SageMaker and optionally optimize it with Neo.

2. **Deployment:** Launch the model as a SageMaker endpoint (for large models) or containerize it for Lambda (for smaller models).

3. **Exposure via API:** Use Amazon API Gateway to create a secure, scalable API.

4. **Orchestration:** Incorporate AWS Step Functions to chain multiple steps (e.g., data preprocessing, inference, post-processing, logging).

5. **Monitoring and logging:** Leverage Amazon CloudWatch for real-time metrics and centralized logging.

2.4.6 Use Case A: SLM for Technical Support Ticket Triage

One of the most common text-based challenges in enterprise environments is technical support ticket triage. Support teams often receive large volumes of incoming tickets, each requiring manual review to determine the correct category (e.g., "Billing," "Technical," "Account Management") before routing to the appropriate team. This manual process can be time-consuming and error-prone.

By deploying an SLM on AWS, you can automate classification and significantly reduce the workload on support staff. AWS Lambda offers a cost-effective, serverless solution for this use case—which is ideal when you need near real-time responses and do not need a large GPU instance. The following section walks through the problem, solution architecture, and implementation steps with code snippets.

2.4.6.1 Problem Statement

- **High ticket volume:** Hundreds or thousands of support tickets may arrive daily, overwhelming staff.

- **Delayed response:** Customers experience slower resolution times because tickets are stuck in triage queues.

- **Inconsistent categorization:** Humans may label tickets differently, leading to confusion and inefficiency.

2.4.6.2 Proposed Solution: SLM Deployed on AWS Lambda

1. **Automated classification:** Use a trained SLM (or a moderately sized transformer model) to classify tickets into predefined categories based on the text content.

2. **Serverless architecture:** Deploy the model on AWS Lambda, triggered via an API Gateway or an EventBridge rule (e.g., new ticket event).

3. **Real-time inference:** Lambda handles incoming requests on-demand, categorizing each ticket text in milliseconds to seconds, depending on model size and complexity.

4. **Scalability:** Lambda automatically scales up to handle traffic bursts and scales down when idle, ensuring you only pay for compute time consumed.

2.4.6.3 Architecture Overview

1. **Support ticket source:** Tickets may originate from a web form, chatbot, or email system.

2. **Amazon API Gateway:** Exposes a secure REST endpoint that accepts ticket data (subject, description, metadata).

3. **AWS Lambda function:** Executes the classification logic using the SLM.

4. **Routing logic:** After classification, the Lambda function can tag the ticket and store the result in Amazon DynamoDB or another database, or it can be published in an Amazon SQS queue for processing by the relevant support team.

A simple flow could look like this:

User or System ➤ API Gateway ➤ Lambda Classifier ➤ DynamoDB / SQS / Notification

2.4.6.4 Implementation Steps and Code Snippet

1. **Train and fine-tune the model:**

    - Fine-tune a transformer-based classifier (e.g., DistilBERT, BERT, RoBERTa) on historical support tickets.

    - Ensure the final model artifact is uploaded to Amazon S3.

2. **Set up a Lambda function:**

    - In the AWS Lambda console, create a new function.

    - Configure the function's memory size and timeout based on your model's loading and inference requirements.

3. **Add model and dependencies:**

    - Use a Lambda layer (or a container image) to include the PyTorch/Transformers libraries.

    - Store your model artifact in the Lambda package or load it from S3 during cold start.

    Here is a minimal example of a Lambda function (`lambda_function.py`) that loads a PyTorch SLM to classify incoming tickets. This code assumes you've stored the model artifact and tokenizer in S3 or packaged them with the function.

    https://github.com/anvcse562/ModernNLP-LLMSLM/blob/main/Ch02/ch2.3/slm-usecase-lambda-2-3-5-4.py

    Notes:

    - **Global model loading:** By placing the `model` and `tokenizer` as global objects, you ensure they are loaded only once per Lambda container's lifecycle, reducing cold start overhead for subsequent invocations.

    - **Memory and timeout**: Models like DistilBERT are relatively small, making them suitable for a 512–2048 MB Lambda function. Larger models may require more memory or a different approach (e.g., SageMaker Endpoint).

- **Categories:** In this example, we assume three categories. You can expand or alter these to match your support workflows.

4. **Configure API Gateway:**
   - Create and integrate a new API in API Gateway with your Lambda function.
   - Map the incoming request JSON to the `event["body"]` in your Lambda function so that it can extract the `ticket_text`.

5. **Test and monitor:**
   - Submit a test request with sample ticket text through API Gateway or the Lambda console.
   - Check Amazon CloudWatch Logs to see your function's execution details (e.g., latency, any errors).

Once deployed, each incoming support ticket can be automatically classified within seconds, enabling near real-time routing to the appropriate support channel.

2.4.7 Use Case B: Deploying an LLM for Marketing Content Generation

Modern marketing teams often struggle to keep up with the demand for fresh, high-quality content—from ad copy and social media posts to email campaigns and blog articles. LLMs have shown remarkable potential in automating content generation, helping teams scale their creative output while maintaining consistency and brand voice.

In this use case, we explore how to deploy an LLM on Amazon SageMaker to generate engaging marketing material on demand. We walk through the problem, solution architecture, and implementation steps, complete with a sample code snippet.

2.4.7.1 Problem Statement

- **High volume of content:** Marketing departments need to produce ads, emails, blog posts, and social media content across multiple channels.

- **Resource constraints:** Copywriters and content strategists often get overwhelmed by tight deadlines and multiple simultaneous campaigns.

- **Maintaining quality and consistency:** Manual processes can lead to inconsistent messaging and brand tone across different platforms.

2.4.7.2 Proposed Solution: LLM Deployment on Amazon SageMaker

1. **Automated text generation:** Train or fine-tune an LLM (e.g., GPT-2, T5, or a Hugging Face model) to understand your brand voice and generate high-quality marketing copy.

2. **Scalable inference:** Deploy the model on SageMaker, where you can select instance types suitable for high-throughput or low-latency scenarios.

3. **Real-time or batch generation:** Use real-time endpoints to generate copy on demand or schedule batch jobs for bulk content creation.

4. **Integration with marketing platforms:** Connect the SageMaker endpoint to marketing automation tools (e.g., Amazon Pinpoint, Marketo, HubSpot) for seamless content delivery.

2.4.7.3 Architecture Overview

A typical workflow for marketing content generation might look like this:

1. **S3 storage:** Historical marketing content, such as emails and ad copy stored in Amazon S3 for training and fine-tuning.

2. **SageMaker training:** Fine-tune your chosen LLM with brand-relevant data, such as product descriptions, tone guidelines, and existing copy.

3. **SageMaker endpoint:** Deploy the fine-tuned model as a real-time inference endpoint.

4. **Triggering generation:**

    - **On-demand:** Marketers use a simple UI or API to request new content (e.g., "Generate a product description for our new smartwatch").

    - **Automated:** Marketing automation tools (via AWS Lambda or Step Functions) periodically request bulk content for upcoming campaigns.

5. **Review and edit:** Copywriters refine and approve AI-generated content.

6. **Publish**: Finalized content is pushed to relevant marketing channels (email newsletters, social media posts, blog platforms).

2.4.7.4 Implementation Steps

1. **Prepare the data:**

    - Gather relevant marketing data and store it in an S3 bucket.

    - Clean and format the text (e.g., removing personally identifiable information, standardizing brand guidelines).

2. **Fine-tune the model:**

    - Use a Hugging Face or custom PyTorch/TensorFlow container on SageMaker to fine-tune your LLM on the curated marketing dataset.

    - Adjust hyperparameters (epochs, batch size, learning rate) to balance training cost and model performance.

3. **Deploy the model:**

    - After training, use **estimator.deploy()** (or create a Model object) to spin up a SageMaker real-time endpoint.

    - Select an **instance type** that aligns with performance requirements, such as an **ml.g4dn.xlarge** for GPU-based inference or **ml.m5.xlarge** for CPU-based workloads.

4. **Generate marketing content:**

    - Interact with the deployed endpoint via a custom script or an API.

    - Pass prompt data that includes context, desired style, or brand-specific keywords.

2.4.7.5 Code Snippet: Generating Content on a SageMaker Endpoint

Here is a simplified Python example showing how you might call a SageMaker endpoint to generate marketing copy. This snippet uses a Hugging Face model and an endpoint named `"marketing-llm-endpoint"`.

`https://github.com/anvcse562/ModernNLP-LLMSLM/blob/main/Ch02/ch2.3/llm-usecase-with-sagemaker-2-3-6-5.py`

Key points to note:

- **Prompt engineering:** Including style, audience, and branding instructions in your prompt can helps shape the content's tone and relevance.

- **Inference parameters:** Adjust parameters like `temperature`, `top_k`, `top_p`, and `max_length` to control creativity and length.

- **Security and access control:** Use AWS Identity and Access Management (IAM) and Amazon VPC endpoints to restrict who can access your SageMaker endpoint.

2.4.7.6 Integration with Marketing Platforms

To seamlessly feed generated content into your marketing workflows, integrate the following:

- **AWS Lambda:** Create a Lambda function that polls the SageMaker endpoint for new content and automatically posts it to social media via third-party APIs (e.g., Twitter, LinkedIn).

- **AWS step functions:** Orchestrate a multi-step pipeline—generate, review, schedule—for larger campaigns.

- **Amazon Pinpoint:** Craft and send targeted email or push campaigns with AI-generated copy to segmented user lists.

With this architecture, you can scale your marketing efforts on demand without overwhelming your content team. The AI handles the heavy lifting of initial draft generation, while human experts refine and finalize the message.

2.4.8 Best Practices and Architectural Considerations

With the examples of technical support ticket triage (Use Case A) and marketing content generation (Use Case B), we have demonstrated how AWS services like SageMaker, Lambda, and SageMaker Neo can address diverse text generation requirements. However, achieving production-grade performance, cost-effectiveness, and reliability requires careful planning. This section delves into best practices and architectural considerations to ensure your text generation solutions on AWS are robust, scalable, and optimized.

2.4.8.1 Cost Optimization

1. **Instance selection:**
   - **GPU instances** (e.g., `ml.g4dn.xlarge`) may be necessary for acceptable inference latency for large models. However, if your workloads are intermittent or primarily CPU-friendly (e.g., smaller or quantized models), a CPU instance (`ml.m5.xlarge`) could be more cost-effective.
   - **Evaluate spot instances** for training jobs where interruptions are acceptable, significantly reducing training costs.

2. **Autoscaling strategies:**
   - **SageMaker automatic scaling:** Configure autoscaling policies to add or remove inference instances based on metrics such as CPU/GPU utilization or number of requests per minute.
   - **On-Demand Lambda invocations:** For smaller SLMs, Lambda inherently scales with demand, but be mindful of concurrency limits and cold start implications.

3. **Multi-model endpoints:**
   - Deploy multiple models behind a single SageMaker endpoint to consolidate resources. This is especially useful if you maintain several specialized models that are rarely used but must remain available.

4. **Model compression and distillation:**
   - Techniques like distillation, quantization, and pruning reduce model size, leading to lower inference costs and better performance on CPU-only instances.
   - Combine compression with Amazon SageMaker Neo to further optimize inference speed.

2.4.8.2 Monitoring, Logging, and Observability

1. **Amazon CloudWatch metrics and logs:**
   - **Monitor** endpoint latency, error rates, and CPU/GPU usage to detect bottlenecks proactively.
   - **Set up alarms** to notify your team when metrics deviate from normal ranges.

2. **AWS X-Ray tracing:**
   - For complex microservices architectures, use AWS X-Ray to trace requests from the API layer through the inference service, identifying performance bottlenecks in real-time.

3. **Logging inference requests:**
   - Use SageMaker Logging or Amazon CloudWatch Logs to record input prompts and model outputs.
   - Redact or anonymize sensitive data to comply with privacy regulations (e.g., GDPR, HIPAA).

4. **Performance profiling:**

   - Leverage tools like Amazon SageMaker Debugger to profile training jobs. Manual profiling or load testing can reveal inefficiencies in your code or data pipelines for inference.

2.4.8.3 Common Pitfalls and How to Avoid Them

1. **Underestimating resource requirements:**

   - Large LLMs require ample GPU memory. Test your model's inference requirements locally or in a development environment before finalizing your endpoint configurations.

2. **Ignoring cold starts in Lambda:**

   - Models loaded in Lambda can cause high latency during the first invocation in a new container. Mitigate by provisioning concurrency for critical workloads or using smaller, optimized models.

3. **Incorporating insufficient data security measures:**

   - Your model might process sensitive information. Use VPC endpoints, KMS encryption, and IAM roles to safeguard data in transit and at rest.
   - Ensure only authorized services/users can invoke your endpoints (e.g., via AWS PrivateLink).

4. **Using overly generic prompts:**

   - Poor prompt design can lead to irrelevant or low-quality outputs for text generation tasks. Incorporate prompt engineering best practices, specifying style, tone, and context.

5. **Having unclear model governance:**

   - Document training data sources, model lineage, and versioning. Use SageMaker Model Registry to track model versions and facilitate collaboration between data scientists and DevOps teams.

2.4.9 Conclusion

Throughout this subtopic, we:

- Introduced how key AWS services—Amazon SageMaker, AWS Lambda, and SageMaker Neo—enable the end-to-end lifecycle of text generation models.

- Detailed real-world use cases for SLMs and LLMs, highlighting the flexibility of AWS architectures in addressing different operational constraints (serverless vs. managed endpoints vs. edge deployment).

- Explored best practices for scaling, cost optimization, monitoring, and security, ensuring your text generation solutions are both efficient and robust.

By following these guidelines, teams can seamlessly integrate AI-driven text generation into their existing AWS environments—enhancing productivity, reducing manual workloads, and driving innovative user experiences.

2.5 Open-Source Implementation

This section explores the process of fine-tuning and deploying pretrained language models using open-source frameworks. Specifically, we focus on GPT-2, DistilGPT-2, T5-XXL, and ALBERT for text generation tasks, such as automated content generation, product descriptions, and chatbot responses. This section also explores how to optimize and deploy these models locally on edge devices using TensorFlow Lite and ONNX for real-time performance.

2.5.1 Fine-Tuning GPT-2 LLM with Hugging Face

Fine-tuning is essential in adapting a pretrained model to solve a specific task. Fine-tuning involves training a pretrained model (such as GPT-2 or DistilGPT-2) on a new dataset, enabling it to learn task-specific nuances while retaining the general knowledge from its pretraining phase.

2.5.1.1 Technical Requirements for Fine-Tuning

- **Hugging Face Transformers:** This library offers state-of-the-art machine learning models and tools for natural language processing (NLP) tasks. It supports easy integration of pretrained models, tokenizers, and custom datasets.

- **PyTorch or TensorFlow:** Both are deep learning frameworks for training and fine-tuning models. In this example, we focus on PyTorch due to its flexibility and widespread use in the research community.

- **Task-specific dataset:** The dataset should represent the problem you want the model to solve. Examples include product descriptions, customer support dialogues, and automated reports.

- **Computational resources:** Fine-tuning large models like GPT-2 requires significant hardware resources (e.g., GPUs). You should ideally use cloud services like AWS or Google Colab for high-performance training.

2.5.1.2 Model Fine-Tuning Parameters

You need to thoroughly understand and strategically apply the model fine-tuning parameters outlined in Table 2-4, including their definitions, typical thresholds, and optimal values based on the specific use case.

Table 2-4. *Model Fine-Tuning Parameters*

Parameter	Definition	Threshold/Range	Optimal Use
max_length	Maximum length of the generated sequence.	Integer (e.g., 50, 100, 512)	Control output size. Typically, 50-100 for shorter text and 200-512 for long text.
temperature	Controls the randomness of the model's predictions. Higher values lead to more random outputs.	0.0 to 1.0	Lower (0.7) for more deterministic outputs; higher (1.0) for creative, diverse text.

(continued)

Table 2-4. (*continued*)

Parameter	Definition	Threshold/Range	Optimal Use
top_k	Limits the model to sampling from the top K most likely words during generation.	Integer (e.g., 5, 50)	Typically, 40-50 for diverse output or 5-10 for more focused output.
top_p	Top-p sampling (nucleus sampling) limits the model to sampling from the smallest set of words with cumulative probability p.	0.0 to 1.0	Typically around 0.85-0.95 for a balance between diversity and focus.
num_beams	Beam search is used to improve the output quality by exploring multiple possibilities.	Integer (e.g., 1-5)	Typically, 3-5 beams for better quality, 1 for faster and more focused results.
no_repeat_ngram_size	Prevents repetition of n-grams of a given size during generation.	Integer (e.g., 2, 3)	Often set to 2 or 3 to reduce repetitive phrases.
repetition_penalty	Penalizes repeated phrases or tokens in the output, reducing their likelihood.	Float (e.g., 1.0 to 2.0)	A value >1.0 (e.g., 1.5) typically reduces repetition.
length_penalty	Controls the length of the output, where higher values encourage longer outputs.	Float (e.g., 1.0 to 2.0)	Use a value >1.0 to encourage longer output (useful for reports or detailed text).
early_stopping	Stops the generation when the model generates an end-of-sequence token.	Boolean (True/False)	Set to True for quicker responses and better control over output length.
no_repeat_ngram_size	Prevents repeating sequences of words or tokens.	Integer (2 or 3)	Set to 2-3 to prevent the model from generating repetitive sequences.

Step 1: Set Up the Environment

Ensure that you have all the necessary libraries installed.

```
pip install transformers datasets torch
```

Step 2: Load the Pretrained GPT-2 Model

You can load the pretrained model and tokenizer using Hugging Face's API. The model will serve as the starting point for fine-tuning:

```
from transformers import GPT2LMHeadModel, GPT2Tokenizer # Load GPT-2 and
                                                         its tokenizer
model_name = 'gpt2'  # Or use 'distilgpt2' for smaller models
tokenizer = GPT2Tokenizer.from_pretrained(model_name)
model = GPT2LMHeadModel.from_pretrained(model_name)
```

Step 3: Prepare the Dataset

We use a custom dataset for this example, but the process is the same for any text-based dataset. The Amazon Product Review dataset is used to generate automated product descriptions:

```
from datasets import load_dataset
# Load dataset - this can be replaced with your own task-specific dataset
dataset = load_dataset('amazon_polarity')  # A dataset for sentiment
analysis, replace as needed
```

Step 4: Tokenize the Dataset

Before training, you need to preprocess and tokenize the dataset. The tokenizer will convert the raw text into numerical tokens that the model can understand.

```
def preprocess_function(examples):    # Tokenize the text
    return tokenizer(examples['text'], truncation=True, padding='max_
    length', max_length=512)
tokenized_datasets = dataset.map(preprocess_function, batched=True)
```

Step 5: Set Up Fine-Tuning

Now that the dataset is ready, you can fine-tune the model using the Trainer API from Hugging Face. This API simplifies the training process by managing training loops, evaluation, and logging.

```
from transformers import Trainer, TrainingArguments
training_args = TrainingArguments(    # Define training arguments
    output_dir="./results",  # Output directory for model and logs
    evaluation_strategy="epoch",   # Evaluate after each epoch
    num_train_epochs=3,   # Number of training epochs
    per_device_train_batch_size=4,   # Batch size per device (adjust based
                                     on GPU availability)
    logging_dir="./logs",   # Log directory
    logging_steps=10,)
    trainer = Trainer(
    model=model,
    args=training_args,
    train_dataset=tokenized_datasets['train'],
    eval_dataset=tokenized_datasets['test'],)
trainer.train() # Train the model
```

Step 6: Save the Fine-Tuned Model

Once the training is complete, save the fine-tuned model for later use:

```
model.save_pretrained('./finetuned_model')
tokenizer.save_pretrained('./finetuned_model')
```

Step 7: Generate Text with the Fine-Tuned Model

To generate text, you load the fine-tuned model and tokenizer, then generate predictions based on input prompts:

```
model = GPT2LMHeadModel.from_pretrained('./finetuned_model') # Load the fine-
                                                              tuned model
tokenizer = GPT2Tokenizer.from_pretrained('./finetuned_model')
def generate_response(prompt, max_length=50): # Generate text
```

```
    input_ids = tokenizer.encode(prompt, return_tensors='pt')
    output = model.generate(input_ids, max_length=max_length, num_return_
    sequences=1, no_repeat_ngram_size=2, top_p=0.92, temperature=0.85)
    return tokenizer.decode(output[0], skip_special_tokens=True)
response = generate_response("Write a product description for a new
smartwatch")
print(response)
```

2.5.2 SLM Local Deployment

Once you've fine-tuned a language model for a specific task, the next challenge is to deploy it to allow for real-time inference on edge devices (e.g., smartphones, embedded systems, or IoT devices). DistilGPT-2, a smaller variant of GPT-2, is an ideal candidate for such tasks due to its efficiency in both computation and memory.

2.5.2.1 Deployment with TensorFlow Lite

TensorFlow Lite is a lightweight version of TensorFlow optimized for running machine learning models on mobile and embedded devices. Converting a model to TensorFlow Lite allows it to run efficiently on edge devices with limited computational power.

Step 1: Convert the Model to TensorFlow Format

Before converting to TensorFlow Lite, you must first convert the model to a TensorFlow format. The following code shows how to convert a DistilGPT-2 model from Hugging Face into TensorFlow format:

```
import tensorflow as tf
from transformers import TFAutoModelForCausalLM # Load the fine-tuned model
model = TFAutoModelForCausalLM.from_pretrained('./finetuned_model')
# Convert the model to TensorFlow Lite
converter = tf.lite.TFLiteConverter.from_keras_model(model)
tflite_model = converter.convert() # Save the TensorFlow Lite model
with open('distilgpt2_edge.tflite', 'wb') as f:
    f.write(tflite_model)
```

Step 2: Load and Inference on Edge Device

Now that the model is in TensorFlow Lite format, it can be deployed on mobile or edge devices. The following is an example of how to perform inference on a mobile device using TensorFlow Lite:

```
import tensorflow as tf
interpreter = tf.lite.Interpreter(model_path="distilgpt2_edge.tflite")
# Load the TensorFlow Lite model
interpreter.allocate_tensors()
# Set up input tensor
input_data = prepare_input_data("Generate a product description for a smartwatch...")
interpreter.set_tensor(interpreter.get_input_details()[0]['index'], input_data)
interpreter.invoke() # Perform inference
output_data = interpreter.get_tensor(interpreter.get_output_details()[0]['index']) # Get output
generated_text = decode_output(output_data)
print("Generated Text:", generated_text)
```

2.6 Industry Use Cases

In various industries, text generation powered by advanced language models transforms how businesses interact with customers, streamline operations, and create content. For customer support and chatbots, fine-tuning models like DialoGPT on domain-specific data enables intelligent, human-like conversations that improve customer satisfaction and reduce response times. On-device text generation optimizes technical support ticket categorization, utilizing SageMaker Neo to deploy models on devices, automating the process and enhancing efficiency in real-time. In content generation and marketing, models such as T5-base automate the creation of product descriptions and marketing materials, saving time and ensuring consistency across large volumes of content. These industries use cases highlight the versatility of text generation models in solving practical business challenges while driving innovation and operational efficiency.

Language models are powerful tools that can be used for various tasks, but they often require some degree of customization to achieve optimal performance. There are three primary methods for customizing LLMs or SLMs, listed in Table 2-5.

Table 2-5. *Three Methods for Customizing LLMs and SLMs*

Method	Description	Use Case	Pros	Cons
Training	Training a model from scratch with large amounts of data.	Complex tasks like creating a new chatbot or language generation model from scratch. In our illustration, we utilize DialoGPT-small. Note: The later chapters provide the demonstration of pretraining SLMs.	Full control over the model's behavior.	Requires a large dataset and computational resources.
Fine-tuning	Adapting a pretrained model to a specific domain/task using domain-specific datasets.	Fine-tuning DialoGPT-small on multi_woz_v22 for customer support dialogues.	Faster than training from scratch; uses less data.	May require specialized domain data for accuracy.
Prompt engineering	Customizing a model's response to a given prompt without modifying the underlying model.	Generating specific text outputs for product descriptions or customer support with T5-base using prompt customization.	Simple, low-cost, and quick.	Limited to prompt design, less control over behavior.

Each customization method—training, fine-tuning, and prompt engineering—has its strengths. Training is the most powerful but also the most resource-intensive. Fine-tuning balances power and efficiency, making it an excellent option for specific tasks like chatbots or content generation. Prompt engineering is the most efficient but may not offer the same depth of customization. Often, a combination of methods is used, such as fine-tuning an LLM and refining it with prompt engineering for optimal results. This chapter demonstrates practical implementation of text generation by illustrating diverse industry use cases.

2.6.1 Content Generation and Marketing

This involves automating content creation for product descriptions, blogs, and marketing campaigns.

Approach: T5-base LLM can generate product descriptions, advertisements, and other marketing content. Fine-tuning the model on industry-specific data helps generate high-quality, contextually relevant content.

Use case: Marketing teams can automate the generation of product descriptions, reducing time and ensuring consistency across product listings.

Best approach: Fine-tuning on domain-specific datasets, combined with the proper deployment infrastructure, enhances efficiency in content creation at scale.

2.6.2 Industry-Level Best Practices

- **Modularity:** Split code into distinct modules for better organization and maintainability (e.g., `model.py`, `api.py`, `config.py`).

- **Logging:** Track application events, errors, and operational insights. Helps with debugging and monitoring.

- **Configuration management:** Manage model parameters using configuration files or environment variables.

- **Experiment tracking:** Tools like MLFlow and Weights & Biases can be used to track models and experiments.

- **Error handling and monitoring:** Comprehensive error handling is crucial while monitoring tools like Prometheus or Grafana can track app performance in real-time.

Table 2-6 shows a step-by-step breakdown of the execution process. You can access the full code of this use case from this link:

https://github.com/anvcse562/ModernNLP-LLMSLM/blob/main/Ch02/ch2.5/text_generation_api_productdescription.ipynb

Table 2-6. Step-by-Step Breakdown of the Execution Process

Step	Code Snippet	Explanation
Model invocation	self.tokenizer=AutoTokenizer. from_pretrained(model_name) self. model=AutoModelForSeq2SeqLM.from_ pretrained(model_name).to("cuda" if torch. cuda.is_available() else "cpu")	Load the pretrained model and tokenizer from Hugging Face. The tokenizer converts text to the format the model expects, and the model is loaded on GPU if available.
Model fine-tuning	def generate_content(self, product_ description, target_audience, desired_tone, max_length=config['max_length']): prompt = f"Write a {desired_tone} marketing copy for {product_description} targeting {target_audience}." input_ids = self.tokenizer(prompt, return_ tensors="pt").input_ids.to(self.model.device) output = self.model.generate(input_ids, max_length=max_length, temperature=config['temperature'], top_k=config['top_k'], top_p=config['top_p'], num_beams=config['num_beams'], no_repeat_ngram_size=config['no_repeat_ ngram_size'], repetition_penalty=config['repetition_ penalty']) generated_text = self.tokenizer. decode(output[0], skip_special_tokens=True) return generated_text	Fine-tune the pretrained model for specific tasks (e.g., generating marketing content). The model generates content based on the input prompt, using parameters to control randomness, diversity, and quality.

(continued)

Table 2-6. (*continued*)

Step	Code Snippet	Explanation

Configuration parameters	`config = {'api_key': 'YOUR_SIMULATED_API_KEY',`
	`'model_name': 't5-base',`
	`'max_length': 100,`
	`'temperature': 0.8,`
	`'top_k': 50,`
	`'top_p': 0.95,`
	`'num_beams': 5,`
	`'no_repeat_ngram_size': 2,`
	`'repetition_penalty': 1.5,`
	`'host': '0.0.0.0',`
	`'port': 5001}`
		Define flexible configuration settings that control the model's behavior, such as length, randomness, and diversity. These can be adjusted based on the use case.

Inference	`@app.route("/generate", methods=["POST"])`
	`def generate():`
	`if not authenticate(request):`
	`return jsonify({"error": "Unauthorized"}), 401`
	`try:`
	`data = request.get_json()`
	`product_description = data.get("product_description")`
	`target_audience = data.get("target_audience")`
	`desired_tone = data.get("desired_tone")`
	`if not all([product_description, target_audience, desired_tone]):`
	`return jsonify({"error": "Missing required parameters"}), 400`
	`content = content_generator.generate_content(product_description, target_audience, desired_tone)`
	`return jsonify({"content": content})`
	`except Exception as e:`
	`return jsonify({"error": str(e)}), 500`
		This part of the Flask API handles user input, validates the request, calls the model to generate content, and returns the output. It also logs successful and failed generation attempts.

(*continued*)

Table 2-6. (*continued*)

Step	Code Snippet	Explanation
| Flask app spinning | ```
def run_app():
    app.run(debug=False,host=config['host'],
    port=config['port'])
thread = Thread(target=run_app)
thread.start()
``` | The Flask app is started in a separate thread to handle multiple requests concurrently. A test request is sent to ensure the server starts up properly. |

2.6.3 Chatbots and Customer Support

This use case addresses improving customer interaction through automated and efficient response generation

Approach: Using fine-tuning techniques, an SLM model like Microsoft's DialoGPT-small can be adapted for the Customer Support Chatbot domain. The multi_woz_v22 dataset is commonly used for fine-tuning dialogue systems, enabling the model to provide relevant, context-based responses.

Use case: A customer support chatbot can assist in handling queries, troubleshooting, and offering personalized solutions by generating human-like responses based on the fine-tuned dataset.

Best approach: Fine-tuning SLMs with specialized datasets helps the model better understand the context and nuances of customer interactions.

Table 2-7 explains the execution steps. You can access the full code of this use case from this link:

https://github.com/anvcse562/ModernNLP-LLMSLM/blob/main/Ch02/ch2.5/Customer_Support_Chatbot_SLM_.ipynb
```

Table 2-7. Execution Steps

Step	Code	Explanation	Optimization Suggestions

Imports	`import os`
	`from datasets import load_dataset`
	`from transformers import GPT2Tokenizer, GPT2LMHeadModel, Trainer`
	
		Import necessary libraries for dataset handling, modeling, and training.	Import only needed libraries to save memory.

Dataset loading	`dataset = load_dataset("multi_woz_v22")`
	`train_sample = dataset['train'].select(range(100))`
	`valid_sample = dataset['validation'].select(range(20))`
	
		Load and sample the MultiWOZ dataset for training and validation.	Use a smaller subset for testing to save time.

Tokenizer setup	`tokenizer = GPT2Tokenizer.from_pretrained('microsoft/DialoGPT-small')`	Load the pretrained GPT-2 tokenizer.	Choose a domain-specific tokenizer if available.

Preprocessing	`def preprocess_function(examples):`
	`...`
	
		Preprocess and tokenize the dialogue data into a conversational format.	Use batch processing for efficiency.

Model initialization	`model = GPT2LMHeadModel.from_pretrained('microsoft/DialoGPT-small')` 	Load the GPT-2 model for fine-tuning.	Use a more specialized pretrained model for better results.
Training setup	`trainer = Trainer(model=model, args=training_args, train_dataset=train_data, eval_dataset=valid_data)` 	Initialize and configure the Trainer object for fine-tuning.	Apply gradient accumulation for large batches.

(*continued*)

Table 2-7. (*continued*)

Step	Code	Explanation	Optimization Suggestions
Model training	trainer.train()	Fine-tune the model using the training dataset.	Use mixed-precision training for faster performance.
Save fine-tuned model	trainer.save_pretrained("./chatbot_model") tokenizer.save_pretrained("./chatbot_model") 	Save the fine-tuned model and tokenizer for future inference.	Save checkpoints during training for safety.
Generate responses	def generate_response(prompt): ... 	Generate chatbot responses using the fine-tuned model.	Filter out repetitive responses.

2.6.4 On-Device Text Generation

You can speed up the categorization of technical support tickets using edge computing. On-device text generation for ticket categorization enhances speed, security, and efficiency by processing data locally. Fifty-eight percent of organizations report improved security with edge computing. Adoption is growing across industries.

Approach: An LLM is compiled using Amazon SageMaker Neo to deploy it on devices, making it capable of performing on-device inference. The model quickly and accurately classifies tickets into various categories (Billing, Technical).

Use case: Automating the categorization process of technical support tickets reduces manual labor, improves response times, and optimizes operations.

Best approach: Training and deploying LLMs with SageMaker Neo ensures that models are optimized for on-device usage, reducing latency and resource consumption.

Table 2-8 explains the execution steps. You can access the full code of this use case from this link:

https://github.com/anvcse562/ModernNLP-LLMSLM/blob/main/Ch02/ch2.5/sagemaker_neo_usecase.ipynb

Table 2-8. Execution Steps

Step	Explanation	Optimization Suggestions
Setup	Set up environment variables and define AWS S3 paths for model artifacts.	— Use environment variables for credentials. — Add error handling for missing S3 paths.
Model compilation with SageMaker Neo	Compile model for edge devices (ml_c5) to optimize performance.	— Test different instance families (e.g., ml_m5). — Use batch compilation for speed.
Deploying compiled model	List and download the compiled model artifact from S3 for deployment.	— Automate model checks via AWS Lambda. — Use parallel downloads for faster deployment.
On-device inference simulation	Simulate inference to classify support tickets.	— Use ONNX Runtime for efficient edge inference. — Apply quantization to reduce model size. — Implement asynchronous inference for better throughput.

The previous use cases demonstrate the use of open-source tools like Transformers and Hugging Face for local deployment of sequence-to-sequence language models, in addition to AWS services for model optimization through SageMaker Neo. This combination allows efficient on-device inference, improving processes like customer support ticket triage with minimal cloud reliance. By leveraging local deployments, industries such as telecommunications, e-commerce, and customer support can ensure faster, real-time responses, cut down on latency, and enhance user experiences while maintaining control over deployment environments.

2.7 Summary

In this chapter, you explored various text generation techniques using LLMs and SLMs. You learned how to apply these models to creative writing, automated reporting, and chatbot development tasks.

- You now understand the key differences between LLMs (e.g., GPT-3) and SLMs (e.g., DistilGPT-2) and how each can be leveraged for text generation tasks.

- You gained insight into generative and extractive text generation tasks and when to use LLMs or SLMs depending on the complexity and speed requirements.

- You learned the process of pretraining, fine-tuning, and optimizing these models for real-world applications.

- You explored how to deploy and implement these models using AWS and open-source frameworks, ensuring you can deploy them effectively for both cloud and edge-based applications.

- Finally, you reviewed industry-specific use cases, such as content generation and customer support, where these models have a significant impact.

Good job! You've now acquired a firm grasp of both the foundational and practical concepts, which will support your journey toward mastering text generation, fine-tuning models, inference, and deploying them in local and cloud environments. The next chapter dives into the next modern NLP task: text classification.

CHAPTER 3

Text Classification with LLMs and SLMs

This chapter explores practical applications and key nuances of text classification using LLMs and SLMs. It compares how models like BERT and RoBERTa (LLMs) and DistilBERT and TinyBERT (SLMs) perform in different classification scenarios, including binary, multi-class, and multi-label tasks.

The focus includes real-world applications such as spam detection, sentiment analysis, and text categorization. The chapter also dives into implementation strategies—leveraging AWS services like Amazon Comprehend and SageMaker for scalable deployments, as well as using Hugging Face Transformers for open-source solutions.

By the end of this chapter, you will have a clear understanding of when to use LLMs for deep semantic understanding and when to opt for SLMs for efficiency and cost-effectiveness.

This chapter covers the following key topics:

- **Types of classification tasks:** Understanding binary, multi-class, and multi-label classification, along with model selection tradeoffs.

- **AWS implementation:** Using Amazon Comprehend for built-in models, training custom models on SageMaker, and deploying SLMs for cost-efficient, real-time inference.

- **Open-source implementation:** Fine-tuning models with Hugging Face Transformers, optimizing data for training, and handling imbalanced datasets.

- **Industry use cases:** Exploring text classification in spam detection, customer feedback categorization, social media sentiment analysis, and automated customer support.

This chapter provides theoretical insights and hands-on guidance, equipping you with the tools to implement effective text classification solutions in real-world applications.

3.1 Types of Classification Tasks

Text classification is a fundamental task in natural language processing (NLP) that involves assigning predefined categories to text data. It underpins many real-world applications, from spam detection and sentiment analysis to document categorization and intent recognition.

With the rise of deep learning, LLMs like BERT and RoBERTa have demonstrated exceptional performance in classification tasks by capturing contextual nuances. However, SLMs such as DistilBERT and TinyBERT offer efficient alternatives for real-time, resource-constrained environments. This section explores various classification paradigms, their challenges, and optimal model selection strategies.

3.1.1 Binary Classification

Binary classification tasks involve categorizing text into one of two possible labels. Examples include spam vs. non-spam detection, positive vs. negative sentiment classification, and fraudulent vs. non-fraudulent transaction detection.

3.1.1.1 LLMs for Binary Classification

LLMs like BERT (Bidirectional Encoder Representations from Transformers) and RoBERTa (Robustly Optimized BERT Pre-Training Approach) excel at binary classification due to their deep contextual understanding.

Why LLMs?

- **Context awareness:** LLMs can interpret the meaning of a sentence holistically by leveraging bidirectional attention mechanisms.

- **Transfer learning:** Pretrained LLMs can be fine-tuned on domain-specific datasets, enabling them to adapt to specialized classification tasks.

- **Handling ambiguity:** LLMs distinguish between subtle linguistic nuances that traditional models struggle with.

The following example implements spam detection with BERT using Hugging Face's Transformers library:

```
from transformers import BertTokenizer, BertForSequenceClassification
from torch.utils.data import DataLoader, Dataset
import torch

# Load tokenizer and model
model_name = "bert-base-uncased"
tokenizer = BertTokenizer.from_pretrained(model_name)
model = BertForSequenceClassification.from_pretrained(model_name, num_labels=2)

# Sample text
texts = ["Congratulations! You've won a free iPhone. Click here to claim.",
         "Meeting at 5 PM. See you there!"]
labels = [1, 0]  # 1: Spam, 0: Not Spam

# Tokenization
tokens = tokenizer(texts, padding=True, truncation=True, return_tensors='pt')

# Perform inference
with torch.no_grad():
    outputs = model(**tokens)
    predictions = torch.argmax(outputs.logits, dim=-1)
print(predictions)
```

This example showcases BERT's ability to detect spam by considering phrase structure, context, and spam-like indicators (e.g., "free iPhone").

3.1.1.2 SLMs for Binary Classification

SLMs such as DistilBERT and TinyBERT provide computationally efficient alternatives, making them ideal for real-time spam detection and mobile applications.

Why SLMs?

- **Lower latency:** DistilBERT achieves 60% of BERT's performance with just half the model size.

- **Reduced memory footprint:** TinyBERT is designed for edge devices, consuming significantly fewer resources.

- **Fine-tuning speed:** Training and inference are significantly faster, making them suitable for deployment in production environments.

The following example performs sentiment analysis with DistilBERT:

```
from transformers import DistilBertTokenizer, DistilBertForSequenceClassification

tokenizer = DistilBertTokenizer.from_pretrained("distilbert-base-uncased")
model = DistilBertForSequenceClassification.from_pretrained("distilbert-base-uncased", num_labels=2)
```

3.1.2 Multi-Class Classification

Multi-class classification assigns text to one label from multiple possible categories. Examples include news categorization (politics, sports, technology), customer intent detection (order status, refund, complaint, inquiry), and topic classification.

3.1.2.1 LLMs for Multi-Class Classification

RoBERTa and ALBERT outperform standard BERT for multi-class tasks due to their optimized pretraining strategies.

Implementation strategy:

1. **Pretraining and fine-tuning:** Fine-tune on domain-specific data using categorical cross-entropy loss.

2. **Augmentation techniques:** Use paraphrasing, back-translation, and synonym replacement to enhance dataset diversity.

3. **Hyperparameter optimization:** Experiment with batch size, learning rate, and weight decay.

This example categorizes news with RoBERTa:

```
from transformers import RobertaTokenizer, RobertaForSequenceClassification
tokenizer = RobertaTokenizer.from_pretrained("roberta-base")
model = RobertaForSequenceClassification.from_pretrained("roberta-base", num_labels=5)
```

3.1.2.2 SLMs for Multi-Class Classification

SLMs such as MobileBERT and TinyBERT provide efficient alternatives for multi-class classification when latency is critical.

Consider the following tradeoffs:

Model	Parameters	Accuracy	Latency
BERT	110M	High	Slow
RoBERTa	125M	Higher	Slow
DistilBERT	66M	Medium	Fast
TinyBERT	14M	Lower	Fastest

3.1.3 Multi-Label Classification

In multi-label classification, each text instance can belong to multiple categories simultaneously. Applications include:

- **Product tagging:** A single product can belong to "Electronics" and "Gaming."

- **Medical text classification:** Multiple ICD-10 codes can be assigned to a patient record.

- **Multi-intent chatbots:** A user query can involve multiple intents (e.g., "I want to check my account balance and apply for a loan").

3.1.3.1 LLMs for Multi-Label Classification

BERT and XLNet excel at multi-label classification by learning interdependencies between labels. These models use *sigmoid activation functions* instead of softmax, allowing independent probabilities for each category.

For example, consider this multi-label product classification:

```
from torch.nn import Sigmoid

# Output layer for multi-label classification
model.classifier = torch.nn.Linear(model.config.hidden_size, num_labels)
model.classifier_activation = Sigmoid()
```

3.1.3.2 SLMs for Multi-Label Classification

SLMs like DistilBERT provide faster inference, making them suitable for large-scale tagging applications (e.g., categorizing millions of customer reviews).

Table 3-1 provides optimization techniques applicable to LLMs and SLMs, ensuring improved generalization, calibration, and efficiency. While LLMs benefit from these methods in large-scale training, SLMs leverage them to overcome data.

Table 3-1. *LLM and SLM Optimation Techniques*

Technique	LLM	SLM
Label smoothing	Instead of assigning a hard 0 or 1 to class labels, label smoothing assigns a small probability (e.g., 0.1) to incorrect classes and slightly reduces the probability of the correct class. This prevents overconfidence and improves generalization.	SLMs also benefit from label smoothing, especially in limited-data settings. Reducing overconfidence in smaller-scale models prevents overfitting and ensures better calibration, especially when fine-tuning on small datasets.
Adaptive thresholding	Standard classifiers use a fixed threshold (e.g., 0.5) for binary classification, but real-world datasets have class imbalance and varying confidence distributions. Adaptive thresholding dynamically adjusts the threshold based on dataset distribution, class frequencies, and validation performance.	SLMs, particularly those trained on domain-specific data, can suffer from inconsistent confidence levels. Adaptive thresholding ensures that these models do not become biased toward dominant class distributions, improving performance in low-data scenarios.
Temperature scaling (logit calibration)	Temperature scaling is a post-processing method that adjusts logits before applying softmax, making the probability scores more reliable. It applies a temperature parameter T where higher values smooth predictions: $P(y) = \frac{e^{z/T}}{\sum e^{z/T}}$.	For SLMs, temperature scaling is essential to avoid overconfident predictions, especially when using them in resource-constrained environments. Calibrating probabilities improves interpretability and reliability in downstream applications.

(*continued*)

Table 3-1. (*continued*)

Technique	LLM	SLM
Ensemble learning (bagging, boosting, stacking)	Ensemble learning combines multiple models to improve accuracy and robustness. Methods include bagging (e.g., Random Forest), boosting (e.g., XGBoost), and stacking (combining multiple models using a meta-learner). It reduces variance and improves generalization.	SLMs gain significant performance improvements through ensemble learning, especially when combining specialized models trained on different subsets of data. Techniques such as bagging and boosting help balance bias and variance in smaller architectures.
Bayesian optimization (hyperparameter tuning)	Bayesian optimization models the function of hyperparameters and selects values based on past evaluations. It balances exploitation (selecting known good hyperparameters) and exploration (trying new values), finding optimal hyperparameters faster.	Bayesian optimization is crucial for SLMs due to their limited computational resources. Efficient hyperparameter tuning ensures that smaller models achieve optimal performance with fewer training iterations.
Class rebalancing (handling imbalanced datasets)	Many classification tasks suffer from class imbalance. Techniques include class weighting (higher loss for minority classes), oversampling (replicating minority samples), and undersampling (reducing majority class samples). These prevent model bias toward majority classes.	SLMs, often deployed in specific domains, need class rebalancing to address skewed distributions. Class weighting, oversampling, and undersampling enhance performance when training data is scarce, preventing biases from dominating small datasets.
Mixup data augmentation (regularization technique)	Mixup generates synthetic data by interpolating two training samples: $x' = \lambda x_1 + (1 - \lambda) x_2$, $y' = \lambda y_1 + (1 - \lambda) y_2$. This reduces model reliance on memorization and increases robustness to noise.	Mixup is highly effective for SLMs, particularly when training data is limited. Generating synthetic samples improves generalization, making smaller models more resilient to data noise and adversarial perturbations.

3.1.4 Conclusion

Each classification paradigm requires careful model selection based on tradeoffs between accuracy, latency, and resource constraints, as summarized in Table 3-2.

Table 3-2. *Classification Type Considerations and Tradeoffs*

Classification Type	Best LLM	Best SLM	Key Considerations
Binary	BERT, RoBERTa	DistilBERT, TinyBERT	LLMs for high accuracy, SLMs for speed
Multi-class	RoBERTa, ALBERT	MobileBERT, TinyBERT	LLMs for complex relationships, SLMs for real-time tasks
Multi-label	XLNet, BERT	DistilBERT	LLMs for dependencies, SLMs for efficiency

Selecting the right model is crucial for achieving optimal performance in real-world applications.

3.2 Implementation on AWS

As organizations seek to deploy text classification models in production environments, AWS provides a suite of tools for both pretrained and custom model solutions. This section explores the practical implementation of text classification using AWS services, covering built-in models, custom training, and scalable deployment strategies.

3.2.1 Amazon Comprehend for Built-In Models

Amazon Comprehend is AWS's managed NLP service that offers pretrained models for text classification tasks such as sentiment analysis, spam detection, and topic categorization. This is an excellent option for organizations looking for a quick, low-maintenance solution without deep ML expertise.

Key features:

- **Pretrained models:** Amazon Comprehend provides out-of-the-box models for sentiment analysis, entity recognition, and topic classification.

- **Real-time and batch processing:** Supports both synchronous and asynchronous inference for handling real-time queries and large datasets.

- **Integration with AWS services:** Easily integrates with S3, Lambda, and other AWS services for seamless automation.

This example performs sentiment analysis with Amazon Comprehend:

```python
import boto3
import json

# Initialize Comprehend client
comprehend = boto3.client('comprehend', region_name='us-east-1')

# Sample text for analysis
texts = [
    "I absolutely love this product! It's fantastic.",
    "The service was terrible and I will not return.",
    "It was okay, not great but not bad either."
]

# Perform sentiment analysis
response = comprehend.batch_detect_sentiment(
    TextList=texts, LanguageCode='en'
)

# Print formatted sentiment results
for i, result in enumerate(response['ResultList']):
    print(f"Text: {texts[i]}")
    print(f"Sentiment: {result['Sentiment']}")
    print(f"Confidence Scores: {json.dumps(result['SentimentScore'],
    indent=2)}\n")
```

SLM use case: Amazon Comprehend can be combined with AWS Lambda for serverless, real-time sentiment analysis or spam detection, reducing infrastructure costs and enabling rapid responses in production applications.

In this case, we are integrating into a serverless architecture using AWS Lambda and API Gateway for real-time sentiment analysis.

```python
import boto3
import json

def lambda_handler(event, context):
    comprehend = boto3.client('comprehend')
    text = event['body']
    response = comprehend.detect_sentiment(Text=text, LanguageCode='en')
    return {
        'statusCode': 200,
        'body': json.dumps(response['Sentiment'])
    }
```

3.2.1.1 Deploying the Lambda Function

- Package the script and create a Lambda function in AWS.
- Attach necessary IAM permissions.
- Connect it to an API Gateway for external access.

Note For organizations that require fine-tuned NLP models, Amazon SageMaker offers the ability to train and deploy custom sentiment analysis models using state-of-the-art transformer architectures like BERT or DistilBERT.

Step 1: Prepare the Dataset

Assume you have a CSV file containing labeled sentiment data:

```
text,sentiment
"I love this product!",positive
"This is terrible. Would not recommend.",negative
```

Upload the dataset to an S3 bucket:

```python
import boto3
s3 = boto3.client('s3')

bucket_name = 'your-sagemaker-bucket'
file_name = 'sentiment_data.csv'
s3.upload_file(file_name, bucket_name, file_name)
```

Step 2: Fine-Tune a Pretrained Model

You can use Amazon SageMaker's built-in Hugging Face containers to fine-tune a DistilBERT model:

```python
from sagemaker.huggingface import HuggingFace
import sagemaker

role = sagemaker.get_execution_role()

hyperparameters = {
    'epochs': 3,
    'train_batch_size': 16,
    'model_name': 'distilbert-base-uncased'
}

huggingface_estimator = HuggingFace(
    entry_point='train.py',  # Training script
    source_dir='./scripts',
    instance_type='ml.p3.2xlarge',
    instance_count=1,
    role=role,
    transformers_version='4.6',
    pytorch_version='1.7',
    py_version='py36',
    hyperparameters=hyperparameters
)
huggingface_estimator.fit({'train': f's3://{bucket_name}/{file_name}'})
```

CHAPTER 3 TEXT CLASSIFICATION WITH LLMS AND SLMS

Step 3: Deploy the Model to Amazon SageMaker Endpoint

Once training is complete, deploy the model:

```
predictor = huggingface_estimator.deploy(
    initial_instance_count=1,
    instance_type='ml.m5.large'
)
```

Inference:

```
response = predictor.predict("This product is fantastic!")
print(response)
```

3.2.1.2 Deploying with SageMaker Neo for Optimization

To further optimize the model for cost-effective inference:

```
optimized_model = huggingface_estimator.compile_model(
    target_instance_family='ml_c5',
    input_shape={'input_ids': [1, 128]},
    output_path=f's3://{bucket_name}/optimized_model'
)
```

3.2.1.3 Sentiment Analysis with Amazon Comprehend vs. SageMaker

Table 3-3 shows a cost and computation comparison of Amazon Comprehend versus the SageMaker Fine-Tuned Model.

Table 3-3. *Cost and Computational Comparison*

Feature	Amazon Comprehend	SageMaker Fine-Tuned Model
Setup	Fully managed, no training	Requires training and setup
Cost	Pay per request (e.g., $0.0001 per 100 characters)	Cost of training + hosting
Inference time	Near real-time	Slightly higher latency
Customization	Limited to pretrained models	Fully customizable
Best use	General sentiment analysis	Domain-specific NLP models

3.2.1.4 Conclusion

- **Use Amazon Comprehend** when you need a fully managed service for faster sentiment analysis solution.

- **Use Amazon SageMaker** if you require a custom NLP model with higher accuracy tailored to your specific business domain. You can then optimize with SageMaker Neo to reduce computational costs and improve inference efficiency.

Both approaches serve different business needs, and choosing the right one depends on factors like customization, cost, and scalability.

3.2.2 Custom Model Training with SageMaker

Fine-tuning helps adapt pretrained models (like BERT, RoBERTa, DistilBERT, TinyBERT) to specific tasks such as sentiment analysis, topic classification, or entity recognition. Follow this process to fine-tune SLMs and LLMs on SageMaker:

1. **Choose a pretrained model:**
   - Select an SLM (e.g., DistilBERT, TinyBERT) or LLM (e.g., BERT, RoBERTa) from the Hugging Face Model Hub or SageMaker JumpStart.

2. **Prepare the data:**
   - Format data in CSV, JSON, or Parquet.
   - Tokenize using the Transformers library.

3. **Set up the SageMaker environment:**
   - Define an Amazon SageMaker Training Job.
   - Use SageMaker's Hugging Face Estimator.

4. **Fine-tune the model:**
   - Define hyperparameters.
   - Use distributed training if necessary.

5. **Evaluate and deploy:**
   - Deploy using SageMaker Endpoint.
   - Evaluate on test data.

Check out this code snippet for fine-tuning SLM/LLM on SageMaker:

https://github.com/anvcse562/ModernNLP-LLMSLM/blob/main/Ch03/ch3.2/FineTuneSLM.ipynb

For organizations that require domain-specific models with greater customization, Amazon SageMaker provides an end-to-end machine learning environment to train and deploy both LLMs and SLMs.

LLM training (BERT, RoBERTa) for deep understanding:

- Suitable for nuanced classification tasks requiring contextual awareness, such as detecting sarcasm in sentiment analysis.
- Requires higher computational resources and can be fine-tuned using SageMaker's built-in Hugging Face integration.

SLM training (DistilBERT, TinyBERT) for efficiency:

- Optimized for lower latency and cost efficiency while maintaining performance for simpler classification tasks.
- Can be trained using SageMaker's managed training instances with autoscaling capabilities.

This example fine-tunes a BERT model in SageMaker:

```
from sagemaker.huggingface import HuggingFace
huggingface_estimator = HuggingFace(
    entry_point='train.py',
    source_dir='./scripts',
    role='SageMakerExecutionRole',
    instance_type='ml.p3.2xlarge',
    transformers_version='4.6',
    pytorch_version='1.7',
    py_version='py36'
)
huggingface_estimator.fit({'train': 's3://my-bucket/train-data'})
```

3.2.3 Inference and Deployment

Once trained, models need to be deployed for inference in scalable and cost-efficient ways. AWS provides multiple deployment options:

1. **Scalable LLM deployments with SageMaker:**
   - Use SageMaker Endpoints for hosting large models like RoBERTa for high-throughput inference.
   - Supports multi-instance scaling and auto-recovery for production workloads.

2. **Low-latency SLM deployments with Lambda:**
   - Deploy models like DistilBERT in AWS Lambda for real-time inference with minimal cost.
   - Ideal for latency-sensitive applications like spam detection in email filtering.

This example deploys a DistilBERT model in Lambda:

```
import torch
from transformers import DistilBertTokenizer, DistilBertForSequenceClassification

def lambda_handler(event, context):
    model_path = "/opt/model"
    tokenizer = DistilBertTokenizer.from_pretrained(model_path)
    model = DistilBertForSequenceClassification.from_pretrained(model_path)
    text = event['text']
    inputs = tokenizer(text, return_tensors='pt')
    outputs = model(**inputs)
    prediction = torch.argmax(outputs.logits, dim=-1).item()
    return {"prediction": prediction}
```

This approach enables a cost-efficient, serverless deployment, reducing overhead while ensuring high availability.

By leveraging AWS services like Amazon Comprehend, SageMaker, and Lambda, organizations can implement robust text classification solutions tailored to their specific needs. Whether using pretrained models for quick deployment or training custom models for domain-specific applications, AWS provides the flexibility and scalability required for production-grade NLP systems.

3.3 Open-Source Implementation

This section explores how to build text classification models using open-source frameworks, particularly Hugging Face Transformers. It covers fine-tuning BERT and RoBERTa for text classification tasks, as well as leveraging SLMs (DistilBERT, TinyBERT) for optimized training and inference. Additionally, we discuss essential data preparation techniques, including tokenization, padding, and handling imbalanced datasets.

3.3.1 Building Text Classifiers Using Hugging Face Transformers

Hugging Face's Transformers library provides pretrained models that can be fine-tuned for various classification tasks. The primary models used include:

- **BERT (Bidirectional Encoder Representations from Transformers):** Effective for capturing contextual word representations.

- **RoBERTa (Robustly Optimized BERT Pretraining Approach):** An improved variant of BERT with dynamic masking and larger training data.

- **DistilBERT and TinyBERT:** Smaller, faster models for real-time classification.

See this link for more information:

https://github.com/anvcse562/ModernNLP-LLMSLM/blob/main/Ch03/ch3.3/FineTuneBERTTextClassification.ipynb

3.3.2 Using SLMs for Faster Training and Inference

For scenarios where computational efficiency is crucial, SLMs such as DistilBERT and TinyBERT offer lightweight alternatives to full-sized transformer models.

Advantages of SLMs:

- **Lower latency:** Ideal for real-time applications like chatbot classification.
- **Reduced memory footprint:** Enable deployment on edge devices.
- **Comparable accuracy:** With careful fine-tuning, SLMs can achieve performance close to LLMs.

This example uses DistilBERT for classification:

```
from transformers import DistilBertTokenizer, DistilBertForSequenceClassification

# Load pretrained DistilBERT tokenizer and model
tokenizer = DistilBertTokenizer.from_pretrained("distilbert-base-uncased")
model = DistilBertForSequenceClassification.from_pretrained("distilbert-base-uncased", num_labels=2)
```

Fine-tuning follows a process similar to BERT but with significantly faster execution and lower memory requirements.

3.3.2.1 Training Arguments and Speedup Tips

This section describes training arguments and provides tips on how to speed up training and inference for a spam detection model.

For the full SLM training and inference code, refer to this GitHub link:

https://github.com/anvcse562/ModernNLP-LLMSLM/blob/main/Ch03/ch3.4/SLM_Spam_Detection.ipynb

```
# Define training arguments
training_args = TrainingArguments(
    output_dir='./results',
    num_train_epochs=3,
    per_device_train_batch_size=16,
```

```
    per_device_eval_batch_size=64,
    warmup_steps=500,
    weight_decay=0.01,
    logging_dir='./logs', )
trainer = Trainer( # Initialize Trainer
    model=model,
    args=training_args,
    train_dataset=tokenized_train,
    eval_dataset=tokenized_test)
```

Table 3-4 lists each training argument and explains how to speed it up.

Table 3-4. Tips for Speeding Up Training Arguments

Training Argument	Definition	How to Speed Up
output_dir	Directory to save model outputs, checkpoints, and logs.	Use a fast storage system (e.g., SSDs or cloud storage).
num_train_epochs	Number of times to train the model on the full dataset.	Reduce the number of epochs for quicker results. Use early stopping.
per_device_train_batch_size	Batch size for training per device (GPU/CPU).	Increase batch size (if memory allows) for faster convergence.
per_device_eval_batch_size	Batch size for evaluation per device.	Use larger batch size for inference to speed up evaluation.
warmup_steps	Number of steps for learning rate warmup before starting actual training.	Reduce warm up steps if you can, but ensure convergence isn't impacted.
weight_decay	Regularization strength to reduce overfitting.	Keep default value unless fine-tuning significantly changes performance.
logging_dir	Directory to save logs for debugging.	Use lightweight logging to avoid overhead.
logging_steps	Frequency of logging during training (steps between logs).	Increase the number of steps between logging to reduce logging overhead.

(continued)

Table 3-5. (*continued*)

Training Argument	Definition	How to Speed Up
`learning_rate`	Learning rate for optimization.	Use adaptive learning rates like AdamW for faster convergence.
`fp16`	Use mixed precision for faster training with lower memory use.	Set `fp16=True` for faster and more memory-efficient training.
`gradient_accumulation_steps`	Number of steps to accumulate gradients before updating weights.	Increase gradient accumulation steps to simulate larger batch sizes.
`save_steps`	Frequency of saving model checkpoints.	Reduce frequency but ensure model recovery if needed.
`max_grad_norm`	Clip gradients to prevent exploding gradients during training.	Tune for optimal values to avoid unnecessary computation.

To accelerate inference after training, use strategies and tools that optimize performance, reduce latency, and boost throughput. Table 3-5 summarizes the ways you can speed up inference, including model storage, cloud deployment (e.g., Neo), and other techniques.

Table 3-5. Tips for Speeding Up Inference Methods

Method	Description	How to Speed Up
Model quantization	Reduces model size and speeds up inference by converting weights from 32-bit to 8-bit integers.	Convert models to lower precision (e.g., INT8) using libraries like TensorRT, Hugging Face's optimum.
Model pruning	Removes less important weights and connections from the model to make it smaller and faster.	Use pruning techniques (e.g., magnitude-based pruning) to reduce unnecessary parameters.
Distillation	Compresses a large model into a smaller one (student model) that mimics its behavior.	Train a smaller "student" model to approximate a larger "teacher" model for faster inference.

(*continued*)

Table 3-5. (*continued*)

Method	Description	How to Speed Up
ONNX (Open Neural Network Exchange)	Converts models to the ONNX format for optimized inference across hardware.	Export models to ONNX format and use ONNX Runtime for hardware-agnostic deployment.
TensorRT optimization	Optimizes deep learning models specifically for NVIDIA GPUs.	Use NVIDIA's TensorRT to optimize the model, making it faster by using precision tuning (e.g., FP16, INT8).
Deep learning compiler (e.g., AWS Neo)	Optimizes and compiles models to deploy them across various hardware platforms, providing better performance.	Use AWS SageMaker Neo to optimize and compile models for edge devices and cloud inference.
Model offloading to specialized hardware (e.g., TPUs, GPUs)	Offloads inference tasks to specialized hardware (like TPUs, GPUs) to accelerate processing.	Use hardware accelerators for inference such as NVIDIA GPUs, Google TPUs, and other custom accelerators.
Batch inference	Involves running multiple inferences at once in a batch to take advantage of hardware parallelism.	Increase batch size during inference for better throughput. This is particularly useful for GPUs or other hardware.
Edge deployment (with optimized models)	Deploys optimized models on edge devices for low-latency inference.	Use efficient models with tools like TensorFlow Lite, ONNX Runtime, or AWS SageMaker Neo to run models on edge devices.
Model parallelism	Splits the model into multiple parts and runs them on different devices for faster processing.	Implement model parallelism to distribute the model's layers across multiple devices (e.g., multiple GPUs).
Caching results for common queries	Caches frequently used results to avoid recomputing inferences.	Implement caching mechanisms for repeated queries (e.g., result caching or database caching).
Lazy loading of weights	Loads only necessary model parts during inference to save memory and speed up loading times.	Use lazy loading techniques to load model weights only when needed, especially for large models.

Table 3-6 shows the code implementations for optimizing inference: The same optimizations can apply to LLMs.

Table 3-6. *Code Implementations for Optimizing Inference*

Optimization Technique	Description	Code Snippet

Model quantization	Quantizes the model to reduce its size and accelerate inference, especially on hardware with support for lower precision computations.	`import torch`
		`from torch.quantization import quantize_dynamic`
		`# Load pretrained model`
		`model = BertForSequenceClassification.from_pretrained('bert-base-uncased')`
		`# Quantize the model`
		`quantized_model = quantize_dynamic(model, {torch.nn.Linear}, dtype=torch.qint8)`
		`quantized_model.save_pretrained('quantized_bert_model')`

ONNX export for inference speedup	Converts the model to ONNX format for faster inference with ONNX Runtime, optimizing execution on various hardware and platforms.	`import torch.onnx`
		`# Convert to ONNX format`
		`torch.onnx.export(model, input_tensor, "model.onnx", opset_version=12)`

(*continued*)

Table 3-6. (*continued*)

Optimization Technique	Description	Code Snippet
| ONNX Runtime for inference | Uses ONNX Runtime for hardware-optimized inference. You can use this on any supported hardware (CPU, GPU). | ```python
import onnxruntime as ort
# Load the ONNX model
session = ort.InferenceSession("model.onnx")
# Run inference
inputs = {session.get_inputs()[0].name: input_tensor.numpy()}
outputs = session.run(None, inputs)
``` |
| Deploy with AWS SageMaker Neo | Deploys the model to an edge device with optimized performance using SageMaker Neo, enabling faster inference. | ```python
from sagemaker.neo import NeoCompilationJob
# Compile the model for edge deployment
compilation_job = NeoCompilationJob.compile(
input_model_s3_uri='s3://your-bucket/model',
target_instance_family='ml_c5',
output_model_s3_uri='s3://your-bucket/compiled-model',
role='YourSageMakerRole',
framework='PYTORCH', framework_version='1.8')
compilation_job.wait()
``` |
| Use TorchScript for deployment | TorchScript allows you to optimize models for deployment by serializing them, enabling fast inference in production environments. | ```python
scripted_model = torch.jit.script(model)
scripted_model.save("optimized_model.pt")
``` |
```

3.3.3 Data Preparation Techniques

Before training text classifiers, proper data preprocessing ensures optimal performance. The key steps include the following:

1. **Tokenization:** Converts text into numerical representations that models can understand.

   ```
   tokenizer = BertTokenizer.from_pretrained("bert-base-uncased")
   tokens = tokenizer("Example text for classification", padding=True, truncation=True)
   ```

2. **Padding and truncation:** Models require input sequences of the same length. Padding ensures that shorter sequences match the longest in the batch, while truncation shortens overly long sequences.

   ```
   tokens = tokenizer("Example text", padding="max_length", max_length=128, truncation=True)
   ```

3. **Handling imbalanced data:** In real-world datasets, classification categories may not be evenly distributed. Methods to address imbalance include:

   - **Class weighting:** Adjusts loss function to assign higher importance to underrepresented classes.

   - **Data augmentation:** Generates synthetic samples to balance datasets.

   - **Oversampling/undersampling:** Replicates minority class samples or reduces majority class samples.

   This example applies class weighting in PyTorch:

```
from torch.utils.data import WeightedRandomSampler
import torch

# Define class distribution
class_counts = [1000, 200]  # Example: 1000 samples in class 0, 200 in class 1
```

```
class_weights = [1/class_counts[i] for i in range(len(class_counts))]
sample_weights = [class_weights[label] for label in dataset["train"]
["label"]]
sampler = WeightedRandomSampler(weights=sample_weights,
num_samples=len(sample_weights), replacement=True)
```

By implementing these best practices in data preparation, models can achieve improved accuracy and generalization in classification tasks.

3.3.4 Summary

This section detailed open-source approaches to text classification, covering fine-tuning Hugging Face Transformers, optimizing SLMs for efficiency, and data preprocessing techniques for enhanced model performance.

3.4 Industry Use Cases

Text classification is widely used across industries to automate decision-making and enhance customer interactions. The following sections explore four major industry use cases: Spam Detection, Customer Feedback Categorization, Social Media Sentiment Analysis, and Automated Customer Support.

3.4.1 Spam Detection

Spam detection is a crucial application in filtering out malicious or unwanted messages from email, SMS, and social media platforms. LLMs like RoBERTa offer high-accuracy spam detection, while SLMs such as DistilBERT enable real-time filtering with lower latency.

Implementation steps:

1. **Dataset preparation:** Use datasets like SpamAssassin or SMS Spam Collection.

2. **Preprocessing:** Tokenize, clean, and balance the dataset.

3. **Model selection:** Use RoBERTa for high-accuracy filtering or DistilBERT for low-latency environments.

4. **Fine-tuning:** Train on spam/non-spam labels.

5. **Deployment:** Optimize the model for inference using ONNX or TensorRT for efficiency.

Refer to these GitHub links for LLM and SLM Spam detection:

https://github.com/anvcse562/ModernNLP-LLMSLM/blob/main/Ch03/ch3.4/SpamDetectionWithRoBERTa.ipynb

https://github.com/anvcse562/ModernNLP-LLMSLM/blob/main/Ch03/ch3.4/SLM_Spam_Detection.ipynb

Optimization tips:

- Use DistilBERT or MobileBERT for lightweight spam filtering on mobile devices.
- Deploy with TorchScript for optimized inference on edge devices.

3.4.2 Customer Feedback Categorization

Companies rely on NLP-based classification to analyze customer reviews, categorize complaints, and prioritize service improvements.

LLMs vs. SLMs:

- **LLMs (e.g., BERT, RoBERTa)** provide high accuracy and context-aware categorization.
- **SLMs (e.g., DistilBERT, TinyBERT)** enable faster, cost-effective classification at scale.

See this link for more information:

https://github.com/anvcse562/ModernNLP-LLMSLM/blob/main/Ch03/ch3.4/CustomerFeedbackCategorization.ipynb

Deployment considerations:

- Use SageMaker for scalable classification in production.
- Optimize using ONNX Runtime for faster inference in customer service applications.

3.4.3 Social Media Sentiment Analysis

Analyzing sentiment in tweets, comments, and posts helps brands monitor reputation and predict trends.

LLMs vs. SLMs:

- **LLMs** (e.g., RoBERTa, DeBERTa) for detailed sentiment analysis with subtle context.

- **SLMs** (e.g., TinyBERT, DistilBERT) for real-time, mobile-friendly sentiment tracking.

See this link for more information:

https://github.com/anvcse562/ModernNLP-LLMSLM/blob/main/Ch03/ch3.4/SentimentAnalysisonTwitterData.ipynb

Performance tips:

- Fine-tune using multilingual models for global social media monitoring.

- Deploy optimized models on AWS Lambda for real-time processing.

3.4.4 Automated Customer Support

Chatbots and virtual assistants use text classification for intent recognition and automated responses.

LLMs vs. SLMs:

- **LLMs (e.g., GPT-4, BERT)** for handling complex customer queries with detailed responses.

- **SLMs (e.g., TinyBERT, DistilBERT)** for fast, cost-efficient intent classification.

For this use case, we are using TinyBERT_General_4L_312D, which is excellent for understanding natural language tasks. Its distinguishing feature is its compact size, being 7.5 times smaller than BERT-base, which enhances deployment efficiency in limited-resource settings.

See this link for more information:

https://github.com/anvcse562/ModernNLP-LLMSLM/blob/main/Ch03/ch3.4/IntentClassificationinChatbots.ipynb

Deployment tips:

- Use ONNX quantization for efficient chatbot inference.
- Deploy models in serverless environments (AWS Lambda, Azure Functions) for scalability.

By leveraging LLMs for accuracy and SLMs for efficiency, industries can deploy NLP-powered text classification solutions that are scalable, fast, and cost-effective.

3.5 Summary

In this chapter, you explored text classification using LLMs like BERT and RoBERTa, and SLMs like DistilBERT and TinyBERT. You learned how to apply these models to tasks such as spam detection, sentiment analysis, and text categorization, understanding when to use LLMs for complex tasks and SLMs for real-time, efficient solutions.

You also gained practical knowledge on deploying these models using AWS and open-source tools like Hugging Face Transformers and explored industry use cases such as automated customer support and sentiment analysis.

Good job! You now have a strong understanding of text classification and model deployment. The next chapter dives into another key NLP task: named entity recognition (NER).

CHAPTER 4

Named Entity Recognition (NER) with LLMs and SLMs

This chapter explores Named Entity Recognition (NER) using LLMs like BERT and RoBERTa and SLMs like DistilBERT and TinyBERT. We compare their performance in various NER scenarios, including identifying diverse entity types and handling contextual variations.

This chapter focuses on real-world applications such as resume parsing, legal document analysis, and customer data extraction. It also covers implementation strategies, including AWS services (Amazon Comprehend and SageMaker) for scalable deployments, and Hugging Face Transformers for open-source solutions.

By the end of this chapter, you will understand when to use LLMs for deep semantic understanding in complex NER tasks, and when to choose SLMs for efficient, cost-effective performance in simpler applications.

This chapter covers:

- **Introduction to NER:** Importance of entity extraction, entity types, and model complexity vs. performance tradeoffs.

- **NER implementation on AWS:** Using Amazon Comprehend for pretrained NER models, training custom models on SageMaker, and deploying SLMs for efficient, real-time entity extraction.

- **Open-source implementation:** Fine-tuning LLMs and SLMs with Hugging Face Transformers, optimizing training data, and evaluating model performance.

- **Low-level LLM exploration: Small LLM cost-effective alternatives:** Cost-effectiveness of SLMs in NER, accuracy vs. efficiency tradeoffs, and optimal use cases.
- **Industry use cases:** NER in resume parsing, legal document analysis, and customer data extraction.

This chapter provides theoretical insights and hands-on guidance for implementing effective NER solutions in real-world applications.

4.1 Introduction to Named Entity Recognition (NER)

In Chapter 3, we explored *text classification*—assigning labels or categories to whole texts (e.g., classifying an email as "spam" or "not spam"). This chapter shifts from labeling an entire document to identifying and labeling specific pieces of information within the text. This task is known as *Named Entity Recognition (NER)*. While text classification gives a broad overview (one label for the whole text), NER dives deeper into the text to find and categorize particular words or phrases. In other words, instead of asking *"What is this document about?"* we ask, *"Which words in this document refer to important real-world entities, and what are their types?"*. This transition from whole-text labels to token-level labels will build on your understanding of NLP tasks and sets the stage for more fine-grained information extraction.

4.1.1 What Is Named Entity Recognition?

NER is a fundamental NLP task that automatically locates and classifies key information ("named entities") in text into predefined categories. These named entities are the real-world objects and concepts mentioned in the text—for example, people names, organization names, locations, dates, and product names, among others. By detecting these, NER transforms unstructured text into structured data. For instance, consider this sentence:

> *"John works at Google in New York."*

An NER system will detect "John" as a Person, "Google" as an Organization, and "New York" as a Location. In this way, NER breaks down a sentence to reveal the "who, what, and where" embedded in the words.

Behind the scenes, NER is often treated as a sequence labeling problem. This means the model looks at each word (or token) in a sentence and determines if it is part of an entity and what type it is. Models usually produce tags for each token, often using schemes like BIO (Begin, Inside, Outside) tags to mark the start and continuation of entities. For example, in "Larry Page," the model might tag "Larry" as B-PER (Beginning of a Person name) and "Page" as I-PER (Inside a Person name), whereas a non-entity word gets an O (Outside) tag. Modern NER systems handle these tagging details internally; the user usually just sees the final entities.

Common entity categories that NER models recognize include these:

- **PERSON:** Individual people's names (e.g. Alice Johnson).
- **ORG:** Organizations or companies (e.g. Google, United Nations).
- **LOC or GPE:** Geographic locations or geopolitical entities (e.g. New York, France).
- **DATE/TIME:** Dates and time expressions (e.g. January 23, 2026, 5:00 PM).
- **PRODUCT:** Product names (e.g. iPhone 13).
- **EVENT:** Named events (e.g. Olympics 2024).
- **MONEY:** Monetary values (e.g. $1,000).

These are just examples—the exact categories depend on the application and training data. Some domains use custom entity types (for instance, medical texts might label Disease or Medication entities). NER is a flexible technique: you define the entity types you care about, and the model learns to tag those in the text.

Why is this useful? By identifying entities, you can turn unstructured text into structured data that can be counted, linked to databases, or used in downstream processes. NER is like a linguistic highlighter that marks up raw text with valuable metadata, making it easier for other systems (or humans) to understand what the text is about at a glance. It's a foundation for many complex NLP applications, as you'll see next.

4.1.2 Why NER Matters: Real-World Applications

NER might sound abstract, but it has powerful real-world applications across industries. Here are a few examples where NER is used to automate and enhance data processing:

- **Resume parsing:**

    Organizations receive thousands of resumes and need to quickly identify key details. NER can extract names, contact info, education, skills, and job titles, streamlining HR workflows by structuring resume data for easy search and filtering.

    Example: `John Doe` (PERSON), `Master's in Computer Science` (EDUCATION), `Google` (ORG), `Java, Python` (SKILL).

- **Legal document analysis:**

    Legal texts contain critical entities like names of parties, dates, clauses, and locations. NER helps extract this information, making legal research faster and more efficient.

    Example: `Alice Smith` (PERSON), `ACME Corp` (ORG), `Section 5.2` (CLAUSE), `New York State` (LOCATION).

- **Customer data extraction:**

    In e-commerce and support, valuable data is hidden in unstructured text. NER can pull out customer names, order numbers, products, and addresses to automate responses and update records.

    Example: `12345` (ORDER_ID), `iPhone 13` (PRODUCT), `456 Elm St, Springfield` (ADDRESS).

- **Healthcare and medical records:**

    NER can extract diseases, symptoms, medications, and patient info from medical notes and reports, aiding in knowledge base creation and alert systems.

    Example: `diabetes` (DISEASE), `Metformin` (DRUG).

NER is a powerful tool for extracting specific facts from text across domains. By customizing entity types to fit the context—like resumes, legal, or healthcare—NER enables smarter search, analytics, and automation, making it a vital part of many NLP pipelines.

4.1.3 Techniques for NER: Traditional Models vs. LLMs

Like many NLP tasks, NER can be tackled with different types of models. Broadly, we can categorize approaches into "traditional" NER models and LLMs. Traditional NER models are usually smaller, task-specific models trained on labeled data (we call them SLMs for NER). The LLMs are massive general-purpose models that can perform NER via prompting or minimal tuning. Both approaches aim to do the same thing—identify entities—but they differ in how they're built and used. The following sections break down each approach and compare them side by side.

4.1.3.1 Traditional NER Models (SLMs)

Traditional NER models are typically specialized, small models trained *explicitly* for the NER task. Historically, NER began with rule-based systems and statistical models. For example, early systems used handcrafted rules (like *"Mr. X" is likely a person*) or dictionaries of names (gazetteers) to identify entities. These were followed by statistical machine learning models such as Hidden Markov Models (HMMs) and Conditional Random Fields (CRFs), which learned patterns from annotated text.

The state-of-the-art then shifted to deep learning approaches. One common architecture is the BiLSTM-CRF: A bidirectional LSTM network that learns context on the left and right of each word, combined with a CRF layer on top to ensure coherent tagging sequences. More recently, transformer-based models like BERT (and its variants) have excelled at NER. By fine-tuning a pretrained language model (like BERT, RoBERTa, etc.) on an NER dataset (e.g., the CoNLL-2003 dataset), we get a high-accuracy NER model that's still relatively compact (usually tens or hundreds of millions of parameters, not billions). These models are what we refer to here as SLMs in the context of NER—they are focused, efficient, and trained for the specific purpose of entity recognition.

Key characteristics of traditional NER models:

- **They are trained on labeled examples:** We provide texts with entities annotated (e.g., "Larry [PER]Page[/PER]" marked as a person). The model learns from these examples to predict tags on new text.

- **They have a fixed set of entity types:** The model outputs one of the known labels (person, org, etc.) or "O" for non-entity. They don't normally identify entities outside the categories they were trained on.

- **They are relatively small and efficient:** Compared to giant LLMs, these models are smaller. This means they can often run on commodity hardware or in real-time systems without too much latency.

- **They can be domain-tuned:** If you have niche data (say, biomedical text), you can train or fine-tune a model on that data. The model will then excel in that specific domain, picking up domain-specific entities that a general model might miss.

There are many off-the-shelf NER solutions, like spaCy and Hugging Face's pretrained models (e.g., BERT for newswire data). These SLMs are well-suited for NER, often showing strong accuracy, especially in specific domains. They're also easier to deploy and less resource-intensive than large models.

4.1.3.2 A Small Model Example: Using spaCy for NER

One of the easiest ways to use a traditional NER model is via spaCy's pretrained pipelines.

Refer to this GitHub link for the full code: https://github.com/anvcse562/ModernNLP-LLMSLM/blob/main/Ch04/ch4.1/spacy_for_new_4.1.3.1.py.

In this code, en_core_web_sm is a small English model that has NER capability. After processing the text, doc.ents contains the entities. If you run this, you would see the following rough output:

```
Google ORG
Larry Page PERSON
Sergey Brin PERSON
California GPE
```

SpaCy identified Google as an Organization (ORG), Larry Page and Sergey Brin as Persons (PERSON), and California as a geopolitical entity (GPE, a type of location). This shows how a pretrained small model can quickly tag entities with high accuracy. Under the hood, spaCy's model has learned these patterns from training data. We didn't have to train it ourselves—it works out-of-the-box for general text.

4.1.3.3 An Example Using Hugging Face Transformers for NER

Another common approach is to use a transformer-based NER model via Hugging Face's Transformers library. Hugging Face provides a simple pipeline for NER:

https://github.com/anvcse562/ModernNLP-LLMSLM/blob/main/Ch04/ch4.1/huggingface_transformers_for_ner_4.1.3.1.py

Here, pipeline("ner") downloads a default NER model (often a BERT or RoBERTa model fine-tuned on a standard NER dataset). You set grouped_entities=True to group tokens into whole entity names (so "Larry Page" comes as one entity rather than two tokens). The result might look like this (simplified for clarity):

```
[
  {'entity_group': 'ORG', 'word': 'Google', 'score': 0.999},
  {'entity_group': 'PER', 'word': 'Larry Page', 'score': 0.998},
  {'entity_group': 'PER', 'word': 'Sergey Brin', 'score': 0.998},
  {'entity_group': 'LOC', 'word': 'California', 'score': 0.996}
]
```

The output lists entities with their type, as well as confidence score (e.g., Google = ORG, Larry Page and Sergey Brin = PER, California = LOC). It matches spaCy's results, which confirms consistency between the models. Hugging Face's pipeline makes it easy to use powerful transformer models for NER without writing model code.

4.1.3.4 Advanced Example: Building a Custom NER Model with TensorFlow

For advanced users, it's possible to build custom NER models from scratch using frameworks like TensorFlow or PyTorch. An example is a simplified TensorFlow Keras model using a BiLSTM (bidirectional LSTM) architecture, assuming input data is prepared as sequences of word and tag indices.

See this GitHub link for the full code: https://github.com/anvcse562/ModernNLP-LLMSLM/blob/main/Ch04/ch4.1/custom_ner_model_tenserflow_4.1.3.1.py.

In this snippet, the model learns to predict an entity tag for each word in the input sequence. We use an Embedding layer to convert word IDs to vectors, a bidirectional LSTM to capture context, and a TimeDistributed dense layer to output a probability for each possible entity label at each position. In practice, you could add a CRF layer on top for better sequence consistency, but that requires a bit more complex setup. After training such a model on, say, a corpus of resumes annotated with entities, it would learn to identify entities in new resumes.

Traditional NER models train on labeled data to create a model that tags new text efficiently in production. Libraries like spaCy, Hugging Face, and TensorFlow offer tools with varying abstraction levels, from pipelines to custom models. For most use cases, pretrained models or fine-tuning with Hugging Face's Trainer API are sufficient—no need to build from scratch.

4.1.3.5 LLMs for NER

LLMs like GPT-3, GPT-4, PaLM, LLaMA, and FLAN-T5 are massive, general-purpose models trained on vast text corpora. Although not specifically built for NER, they can execute it effectively through prompting.

LLMs act as powerful text predictors. When prompted correctly (e.g., "Identify the named entities in the following text and list their types"), they can extract entities by leveraging their broad language understanding. This is an example of *zero-shot learning*, where the model handles tasks it wasn't explicitly trained for.

LLMs can also do *few-shot learning* for NER. This means you give a couple of examples in the prompt to show the model what format you want. For instance:

> **User prompt to LLM:** "Extract the entities from each sentence. Example 1: 'John works at Google in New York.' ➤ PERSON: John; ORG: Google; LOC: New York. Now your turn: 'Alice joined Microsoft in 2022.' ➤"

By providing this pattern, the LLM will likely continue and output something like: "PERSON: Alice; ORG: Microsoft; DATE: 2022.". Few-shot prompting essentially guides the LLM by demonstration, without any parameter updates—the model adapts on the fly.

Consider these key characteristics of using LLMs for NER:

- **Zero-shot capability:** No task-specific training or labeled data is needed—LLMs leverage their broad training to perform NER via prompting.

- **Flexibility:** They aren't limited to predefined entity types and can extract specific or unusual entities if clearly instructed.

- **Language and format versatility:** LLMs handle multiple languages and varied text styles, and they can output results in custom formats like JSON or tables.

- **Higher-level reasoning:** They use context and world knowledge to disambiguate tricky cases, making smarter guesses than smaller, fine-tuned models.

Using an LLM for NER typically involves an inference API call or integration rather than embedding it in your pipeline like a small model. For example, using OpenAI's API for GPT-4, you might do something like this example from GitHub:

https://github.com/anvcse562/ModernNLP-LLMSLM/blob/main/Ch04/ch4.1/llm_for_ner_4.1.3.2.py

If prompted correctly, the response might be:

```
Barack Obama - PERSON
Paris - LOCATION
July 2017 - DATE
```

This illustrates zero-shot NER via an LLM. We didn't need any specialized NER model—we just tapped into the general intelligence of the LLM.

However, there are some considerations when using LLMs for NER:

- **Consistency and accuracy:** LLMs can hallucinate—adding entities not in the text or incorrectly formatting outputs if prompts aren't clear. Precise instructions and constraints (e.g., JSON output using exact text) help improve reliability.

- **Cost and latency:** LLMs, especially via API, can be slower and more expensive per request than local small models. For large-scale processing, this makes them less cost-effective than purpose-built NER models.

- **Domain knowledge vs. real-time adaptability:** LLMs lack real-time knowledge updates unless fine-tuned or given context. Small models can be retrained quickly with new data, making them more adaptable to recent or domain-specific changes.

Despite some limitations, LLMs excel in zero-shot scenarios. When no training data is available or a quick prototype is needed, they can often perform NER immediately through prompt engineering—no model training is required.

To summarize, LLMs provide a fast, flexible approach to NER, ideal for diverse entity types, multiple languages, and tasks needing deeper contextual understanding. The tradeoff is less efficiency and control compared to smaller, purpose-built models.

4.1.3.6 Comparing NER with SLMs vs. LLMs

Now that we've described both approaches, Table 4-1 compares SLMs (traditional NER models) and LLMs side by side in the context of NER.

Table 4-1. Comparing Traditional NER Models and LLMs in the Context of NER

Aspect	Traditional NER Models (SLMs)	Large Language Models (LLMs)
Model size and resources	Relatively small (millions of parameters). Can often run on CPUs or modest GPUs. Efficient at inference for tagging text.	Huge (billions of parameters). Require powerful hardware or cloud APIs. Inference is slower and more memory-intensive.
Training data needs	Require task-specific labeled data for training/fine-tuning. For example, need examples of text with annotated entities to achieve high accuracy.	Pretrained on vast general text (no task-specific labels needed for usage). Can perform NER out-of-the-box (zero-shot) or with a few examples (few-shot) without additional training.
Domain adaptability	Very tunable—can be trained or fine-tuned on domain-specific corpora (finance, law, medicine) to recognize specialized entities. Excellent for targeted domains.	Broad knowledge by default (trained on Internet-scale data). Knowledge of many domains, but may not use domain-specific terminology consistently unless guided. Fine-tuning an LLM on a new domain is possible but extremely resource-intensive.
Accuracy and precision	High accuracy on the entity types it's trained for, especially if ample annotated data is available. Limited to those types—won't catch an entity of a novel type it wasn't trained on.	Competent accuracy even without explicit training for common entity types (person, place, etc.). Can sometimes identify uncommon entity types if asked. However, may occasionally make a mistake or hallucinate if the prompt is ambiguous.

(continued)

Table 4-1. (*continued*)

Aspect	Traditional NER Models (SLMs)	Large Language Models (LLMs)
Output format	Returns structured outputs (e.g., lists of tokens and tags). Integration into pipelines is straightforward since the output is machine-friendly (token-indexed).	Output is freeform text unless specifically instructed. Needs careful prompting to output in a structured format (JSON, CSV, etc.). More flexible in format (can explain its reasoning if asked, for instance).
Speed and scalability	Fast inference—can process large volumes of text quickly. Ideal for batch processing (e.g., processing thousands of documents on a single server).	Slower inference—processing is computationally heavy. Often done via API calls that have rate limits or cost, making it less ideal for very large-scale batch jobs without significant compute.
Resource cost	Cost-effective: After initial training/fine-tuning (which can be done on a smaller scale), running the model is cheap. Can be deployed on-premises, no ongoing API cost.	Expensive: Using an API (OpenAI, etc.) incurs cost per token. Running your own LLM requires investment in GPU infrastructure. Not cost-efficient for simple NER if done at massive scale.
Maintenance	You control the model and can update it with new training data, fix errors by retraining, and so on. Models can be versioned and improved iteratively for your use case.	You often rely on a third-party model (if using API) and cannot tweak the internals. Prompt engineering is the main way to improve performance. Model knowledge is fixed to its training data (e.g., it won't know about events after its training cutoff).
Examples	SpaCy NER, fine-tuned BERT (e.g., BERT-base-cased fine-tuned on CoNLL2003), Stanford NLP NER, and so on. These are specialized tools focused on NER.	GPT-3, GPT-4, Claude, LLaMA, and so on—general AI models that can do NER among many other tasks when prompted. No single "NER model," but the same model that writes essays can also extract entities.

As you can see in Table 4-1, LLMs offer broad capability and convenience, while smaller NER models provide greater efficiency, control, and easier deployment. With enough training data, fine-tuned small models can match or surpass LLM accuracy in specific domains—at much lower computational cost. But when data is scarce or a quick, multilingual solution is needed, an LLM is incredibly useful.

These approaches aren't mutually exclusive. We're seeing workflows where an LLM is used to generate training data or weak labels for NER, which are then used to train a smaller model. This hybrid approach leverages the LLM's knowledge to bootstrap a cheaper model—essentially getting the "best of both worlds." For example, an LLM could label a thousand unlabeled sentences with entities (perhaps with some noise), and then you fine-tune a BERT-based NER model on those labels to get a refined system.

4.1.4 LLM Agents and Automated NER Pipelines

A recent advancement in NLP is the emergence of LLM agents. These agents integrate the power of LLMs with additional tools and workflows, forming autonomous, multi-step systems capable of complex reasoning and dynamic task execution.

What are LLM agents:

- LLM agents are systems that use LLMs to understand and generate text while coordinating various subtasks. They can extract entities, follow up with queries, and refine outputs based on new context.

- For instance, an LLM agent might perform zero-shot NER on a legal document and then use a specialized NER model (SLM) to verify uncertain entities, balancing both breadth and precision.

LLM agents in action:

- **Dynamic pipelines:** Instead of a single static model, an LLM agent can decide which tools to invoke based on the input text. For example, if the text is highly technical, the agent might route the data to a specialized model, fine-tuned on legal or medical jargon.

- **Interactive debugging:** LLM agents can provide explanations for their outputs, making it easier for human operators to understand why certain entities were tagged. This is especially useful in sensitive domains such as legal or in the healthcare industry.

- **Adaptive learning:** By integrating feedback loops, LLM agents can update their strategies in real time, adjusting prompts or switching models to improve accuracy.

4.1.4.1 Example: Using LangChain to Build an LLM Agent for NER

This GitHub link leads to a simplified example using the LangChain framework, which helps build agents that can chain multiple operations:

https://github.com/anvcse562/ModernNLP-LLMSLM/blob/main/Ch04/ch4.1/llm_agent_for_new_using_langchain_4.1.4.py(LLM agents using langchain)

In this example:

- We define a tool (perform_ner) that leverages a transformer pipeline for NER.
- The LLM agent, built with LangChain, is then capable of calling this tool when prompted.
- This setup allows for an adaptive system that can integrate multiple models or operations—typical of LLM agents that are rapidly emerging in industry.

Benefits of LLM agents for NER:

- **Flexibility:** Agents can switch between different models (LLMs and SLMs) based on the complexity and domain of the input.
- **Interactivity:** They can explain their decisions and even ask clarifying questions if the input is ambiguous.
- **Scalability:** By integrating with cloud APIs and on-demand tools, LLM agents enable scalable NER pipelines that can handle variable loads and diverse document types.

NER is a vital NLP task that turns raw text into structured information. We covered its connection to text classification, core concepts, and examples from resumes to medical records. We also compared using small NER models versus LLMs, highlighting their strengths and tradeoffs.

The next section explores practical strategies to improve NER performance and dives into real-world implementation, helping you transition from understanding NER to building and integrating solutions.

4.2 Implementation on AWS with SLMs and LLMs

The previous section laid the groundwork for NER and explored basic approaches for entity extraction. This section shifts focus to cloud-based NER implementation using Amazon Web Services (AWS), which simplifies scaling and deploying NER solutions. AWS provides managed services like Amazon Comprehend for pretrained NLP models and Amazon SageMaker for custom ML model training, allowing easy integration with existing data pipelines.

AWS offers scalability, reliability, and cost-effective solutions for handling large datasets and high throughput. Comprehend automatically extracts entities, key phrases, and sentiments from millions of documents, while SageMaker enables fine-tuning for domain-specific models. This section dives into AWS-based NER implementation, from using Amazon Comprehend's out-of-the-box models to deploying custom solutions, addressing deployment considerations such as cost, performance, and security.

4.2.1 Using Amazon Comprehend for NER Tasks

The previous section established the foundations of NER and examined initial approaches to identifying entities in text. Building on that, this section focuses on cloud-based implementation using AWS. AWS simplifies and scales NER solutions, enabling a move from prototype to production with minimal infrastructure overhead. AWS provides scalability, reliability, and integration with data pipelines.

AWS offers scalable NER solutions through services like Amazon Comprehend, a fully-managed NLP API with pretrained models, and Amazon SageMaker for custom model training and deployment. These services make complex text analytics, like NER, available on-demand, enabling quick use of pre-built models and later customization. Comprehend efficiently handles large data volumes, extracting entities, key phrases, and sentiments from millions of documents. This section explores NER implementation on AWS, using Comprehend's models to train custom models on SageMaker, and covers deployment considerations like cost, performance, and security.

4.2.1.1 Pretrained NER Models and Their Advantages

Pretrained NER models identify entities like PERSON, LOCATION, DATE, and so on, providing confidence scores (e.g., "John" as PERSON). Advantages include:

- **No training required:** Immediate results without needing to label data; recognizes general entity patterns.
- **Broad coverage:** Trained on diverse text to identify many common and uncommon entities.
- **Scalable:** Managed service that automatically handles scaling and batch processing.
- **Multi-language support:** Supports several languages for NER.
- **AWS ecosystem integration:** Works with other AWS services like Lambda and supports PrivateLink.

4.2.1.2 Using the Comprehend API: Python Example

Using Amazon Comprehend for NER is straightforward with the AWS SDK (e.g., Boto3 for Python). The GitHub link shows a simple example of using Python to call Comprehend's detect_entities API and extract named entities from text:

https://github.com/anvcse562/ModernNLP-LLMSLM/blob/main/Ch04/ch4.2/comprehend_api_4.2.1.py

If you run this code, the output might look like this:

```
PERSON: Barack Obama (Confidence: 0.99)   ; LOCATION: Hawaii (Confidence: 0.99)
TITLE: President (Confidence: 0.98)   ; ORGANIZATION: United States (Confidence: 0.99)
```

In the example sentence, Amazon Comprehend accurately identified "Barack Obama" (PERSON), "Hawaii" (LOCATION), "President" (TITLE), and "United States" (ORGANIZATION). This shows the speed and ease of getting NER results with minimal code using AWS's pretrained model, where AWS manages the underlying infrastructure.

4.2.1.3 LLM Use: Handling Complex Documents

Amazon Comprehend's NER is effective for many tasks, but complex documents like legal contracts or technical manuals can pose challenges. In such cases, domain knowledge and context are crucial. While Comprehend's pretrained model can handle common entities like names, dates, and locations, it might struggle with highly specialized terms or novel contexts. For example, it could miss recognizing "Section 4.3 of the Data Privacy Act" as a legal statute if it's not part of its predefined entity types. The following are suggested for these tough cases:

- **Bigger brains for better understanding:** Using more specialized models can improve NER for complex documents. AWS offers Amazon Comprehend Medical for clinical texts, which recognizes medical terminology. For legal documents, while there's no dedicated "Legal Comprehend," a general LLM like GPT models via AWS services (Amazon Bedrock or OpenAI integration) can perform zero-shot entity extraction. LLMs can understand complex language and context better but come with higher computational costs and may require careful prompt engineering.

- **Chunking long documents:** Comprehend's real-time API has a text size limit (e.g., 5,000 bytes per call). For long documents, you may need to break the text into smaller chunks and process them separately. AWS also offers asynchronous batch processing for large documents like PDFs and Word files, using Textract to read them. While this helps process long documents, maintaining context across chunks may require post-processing to merge entities or resolve duplicates.

In summary, Comprehend's pretrained (LLM-powered) service is often good enough for complex text. For very specialized documents, customization might be needed, but it is suggested to try Comprehend's general model first due to its broad training. Its LLM foundation provides a strong starting point for entity extraction, even in technical or scientific documents.

4.2.1.4 SLM Use: Lightweight Models on AWS Lambda

While large models offer accuracy and context handling, they can be costly and overkill for simpler or high-volume NER tasks. In such cases, SLMs provide a more efficient alternative. Instead of relying on external APIs like Comprehend for each text, you can deploy a lightweight NER model, such as a distilled or compact version of BERT (e.g., DistilBERT), on AWS Lambda. Each Lambda invocation takes a text input, runs the NER model, and returns entities—offering a serverless, cost-effective solution.

AWS Lambda's recent improvements, including up to 10 GB of memory and container image support, make it feasible to host small NER models (50-100 MB). The advantage of this approach is lower cost, as you pay only for execution time (duration X memory cost) rather than idle server costs. This is particularly useful for spiky or low-volume usage. For instance, instead of paying per 100 characters as with Comprehend, you pay for milliseconds of Lambda execution. However, for very large or continuous traffic, an always-on server might be cheaper than Lambda's per-request pricing.

For example, by fine-tuning a DistilBERT model for NER (discussed in the "Training Custom NER Models on SageMaker" section), you could deploy it on Lambda. DistilBERT, being smaller and faster than BERT, allows rapid inference (tens of milliseconds per request). The containerized model, using the Hugging Face Transformers library, tokenizes the input text, runs the NER model, and returns entities. This approach offers a serverless NER microservice, balancing cost and efficiency.

The tradeoffs of the SLM-on-Lambda approach are important to note:

- **Model management:** You're responsible for packaging and updating the model, unlike AWS-managed Comprehend.

- **Entity types:** The recognized entities are determined by your model's training.

- **Accuracy:** May be slightly lower than Comprehend's large model on general text but can excel in specific domains.

- **Privacy/control:** Enhanced privacy as data stays in your environment (though Comprehend can be VPC-attached); full model control for compliance.

In summary, Comprehend offers quick, powerful NER. Custom small models on Lambda suit specialized, high-volume, cost-sensitive needs. Start with Comprehend and then consider custom models later for control and cost.

4.2.2 Training Custom NER Models on SageMaker

Although Amazon Comprehend's pretrained NER is effective for general entities, domain-specific entities often require custom NER models. For example, in medical, legal, or financial domains, specialized entities like "HbA1c" or "CUSIP" need accurate identification. AWS SageMaker is a powerful service for building and deploying these custom models. It allows you to fine-tune state-of-the-art models like BERT or RoBERTa using labeled datasets and provides scalable training and easy deployment.

SageMaker is a comprehensive ML platform that simplifies the process of training and deploying custom models. With built-in support for Hugging Face models, you can fine-tune pretrained models like BERT or DistilBERT on your NER dataset. The typical workflow involves preparing your dataset in S3, configuring and running the training job on SageMaker's infrastructure, and deploying the trained model as an endpoint or batch transform job. SageMaker handles the complexities of distributed training and serving, eliminating the need to manage EC2 instances or Docker containers manually.

4.2.2.1 Step-by-Step Guide: Fine-tuning a BERT-based NER Model

This section walks through the steps to train a custom NER model using SageMaker, focusing on a BERT example for a specialized domain—in this case, legal documents:

1. **Prepare labeled data:** Create a labeled dataset with entities like "CaseNumber" and "Statute," alongside standard entities. Use formats like CoNLL or JSON. SageMaker Ground Truth can help label the data or upload the existing data from S3. Split data into training and validation sets.

2. **Choose a pretrained model:** Decide on a base model to fine-tune. For instance, BERT-base-cased or RoBERTa-base are good starting points for English NER. If the documents are in a different language, you might choose a multilingual BERT (mBERT) or an XLM-R model. The Hugging Face Transformers library provides many such models. Using a pretrained model is crucial—it already understands general language, and you just teach it to recognize your specific entity classes. This approach (transfer learning) is far more efficient than training from scratch.

3. **Set up the SageMaker training job:** SageMaker can run training via the Estimator API in the Python SDK. This example uses the Hugging Face integration. You need a training script (e.g., `train.py`) that reads the data, loads the pretrained model, sets up the training loop, and writes out the model. Hugging Face's Transformers library can simplify a lot of this (you can use the `Trainer` class, for example). SageMaker provides pre-built Docker containers with Hugging Face Transformers and PyTorch. You would define a `HuggingFace` estimator with appropriate hyperparameters. Key parameters:

   - **Instance type:** Such as ml.p3.2xlarge (GPU) for large datasets, ml.g4dn.xlarge (GPU) for medium, or ml.m5.xlarge (CPU) for testing.

   - **Hyperparameters:** Epochs, batch size, learning rate (e.g., 2e-5), model name (e.g., `bert-base-cased`).

   - **IAM Role:** SageMaker's role for S3 access.

   - **Input channels:** S3 locations for data.

   - **Output path:** S3 location for the model.

   Specify `entry_point='train.py'` for a custom script. SageMaker also offers built-in Hugging Face training options. For clarity, use the following GitHub link to see how you might configure the estimator in the code:

   `https://github.com/anvcse562/ModernNLP-LLMSLM/blob/main/Ch04/ch4.2/sagemaker_training_4.2.2.py`

   In this setup, `train.py` includes the logic to load the model, process the data, and begin training. SageMaker automatically handles infrastructure provisioning and shuts down the instance after training. Logs (e.g., loss values) are viewable in CloudWatch. The trained model is saved to your specified S3 bucket. You're billed only for the training time (e.g., 30 minutes on a `p3.2xlarge` = 0.5 hours of compute cost, plus S3 storage).

4. **Fine-tuning and evaluation:** During training, the model learns to identify your custom entities. You would monitor metrics like F1 score or accuracy on the validation set after each epoch. If results aren't satisfactory, adjustments (e.g., learning rate or early stopping) can be made. SageMaker supports easy reruns and even hyperparameter tuning to improve performance.

5. **Deployment via SageMaker endpoint:** Once the model performs well, you deploy it to a SageMaker endpoint for real-time inference. This turns the model into an API. Instance type depends on usage—GPU for low-latency or high-load, CPU for smaller models or lower traffic. Deployment can be done directly from the training estimator.

   Refer to this GitHub link for the full code: `https://github.com/anvcse562/ModernNLP-LLMSLM/blob/main/Ch04/ch4.2/deploying_model_4.2.2.py`

   The output format will depend on how the `train.py`/inference code is written. If you used the Hugging Face `pipeline` for NER under the hood, the result could be a list of entities with their labels (e.g., `[{ "entity": "ORG", "score": 0.99, "word": "Apple Inc."}, {...}, ...]`). You can then parse this in your application. SageMaker endpoints by default autoscale down to 0 instances only if you delete them, but you can configure Auto Scaling to add more instances if traffic increases. For example, if the service gets 100 requests/second, you might scale out to multiple `ml.m5.xlarge` instances behind an endpoint. SageMaker will load-balance requests to them.

6. **Testing and iteration:** After deployment, you test the endpoint with sample texts to verify entity recognition. If results are off, you can adjust the training data, switch models (e.g., to RoBERTa or BERT-large), or tweak hyperparameters-hyperparameters[`'model_name'`]. Thanks to SageMaker's modular setup, the pipeline stays consistent—Data in S3 ➤ Training ➤ Endpoint—making iteration simple.

Through these steps, we have taken a general model and specialized it for our domain. A fine-tuned model can significantly outperform a generic model on niche entity types. For instance, a BERT model fine-tuned on biomedical text will pick up drug and gene names far better than the generic Comprehend model. As an anecdotal comparison: a generic NER might label BRCA1 as an OTHER entity with low confidence, whereas a custom model trained on biomedical papers would know BRCA1 is a Gene entity with high confidence.

4.2.2.2 Python Implementation for Training on SageMaker

The `train.py` script, which runs within the SageMaker training container, typically uses the Hugging Face Transformers library to perform the model training.

Refer to this GitHub link for the full code: https://github.com/anvcse562/ModernNLP-LLMSLM/blob/main/Ch04/ch4.2/complete_training_script_4.2.2.py

This code outline shows what might happen inside the training job. SageMaker's Hugging Face Estimator will execute this, and the final model is saved to `/opt/ml/model`, which is mapped to S3 output. The details of data loading may vary (we might read from an S3 path directly), but the principle is that we use a high-level Trainer API to simplify fine-tuning. This results in a custom NER model artifact.

4.2.2.3 Real-Time Inference via SageMaker Endpoint

After deployment, using the model is similar to using Comprehend, except you call your own endpoint. You can write a small client function or use Boto3's `invoke_endpoint` to send data. The benefit of a SageMaker endpoint is that it can be integrated within your AWS environment (you can put it in a VPC, attach an Application Load Balancer if needed, or restrict access via IAM). The latency of a SageMaker endpoint for NER is typically tens of milliseconds (depending on model size and instance) plus a bit of overhead for network. This is suitable for real-time applications like processing a chat message to highlight entities or enriching data on-the-fly in a pipeline.

4.2.2.4 SLM Implementation: Training and Deploying DistilBERT/TinyBERT

In many cases, the *efficiency* of the model at inference time is a major concern, especially if you're deploying at scale or on cost-constrained environments. This is where small, optimized models come into play. Two popular strategies are using DistilBERT and TinyBERT (or more generally, model distillation and compression techniques):

- **DistilBERT:** A streamlined version of BERT that runs ~60% faster with ~40% fewer parameters yet retains ~97% of its performance. It's easy to fine-tune in SageMaker by changing the model name to something like `distilbert-base-uncased`. Training is quicker, and the model is lighter (~250 MB vs. BERT's ~440 MB), making it suitable for CPU endpoints (e.g., `ml.c5.large`) or even serverless deployments like AWS Lambda.

- **TinyBERT:** Even smaller and faster than DistilBERT (e.g., a 4-layer TinyBERT can be 7.5x smaller and 9.4x faster than BERT-base with similar performance). It uses aggressive distillation, a two-stage training process. While complex to set up, it yields very small models (e.g., ~15 million parameters) for extreme efficiency. In SageMaker, you'd likely need a custom script for knowledge distillation; DistilBERT is a simpler alternative.

From an SLM deployment perspective, a model like DistilBERT fine-tuned for NER can be deployed in various low-cost ways:

- **On a single CPU instance** (e.g., an `ml.m5.large`, which costs significantly less per hour than a GPU instance). The throughput might be dozens of sentences per second per vCPU. If higher throughput is needed, you can scale out with more CPU instances, which is still often cheaper than a few GPU instances for the same load, depending on pricing.

- **On AWS Lambda**, as previously described. DistilBERT's smaller size means it will cold start faster and use less memory.

- **On edge devices or AWS IoT Greengrass** if needed, since model size is smaller.

It's important to note that the training of DistilBERT or TinyBERT for the specific task still happens on SageMaker or a similar training setup. You're not changing the SageMaker part drastically; you're changing the model architecture you train. The SageMaker training job might even complete faster with a smaller model, which is a bonus.

To give a concrete comparison: Suppose your fine-tuned BERT-base model achieved 90% F1 on your custom entities. A DistilBERT version might achieve ~88% F1 on the same test, but run ~2x faster and cost less to host. You must decide if that small loss in accuracy is acceptable for the business case. Often, for production, a slight drop in accuracy is acceptable if it dramatically lowers latency and cost, especially if that accuracy is still above a required threshold.

SageMaker makes it fairly easy to experiment with this—you can train both versions and compare the results. You might even deploy both models to endpoints and run A/B tests on a sample of traffic to see the real-world performance difference.

In summary, SageMaker enables training custom NER models tailored to your domain. Fine-tuning large models like BERT offers high accuracy, while smaller models like DistilBERT provide speed and cost efficiency. These models can be deployed flexibly—via endpoints or serverless functions—to power real-world applications. The next section looks at how to make a choice between Amazon Comprehend and custom models and explores deployment best practices.

4.2.3 Deployment Considerations on AWS

Deploying a NER model in production involves considering cost, performance, scalability, and security. AWS offers Amazon Comprehend (managed service) and hosting custom models (from SageMaker) on endpoints or Lambda. The following section compares these options and outlines deployment best practices. Table 4-2 provides a cost and performance comparison.

Table 4-2. Comparison of Using Amazon Comprehend vs a SageMaker-trained Custom NER Model

Aspect	Amazon Comprehend (Pretrained NER API)	Custom NER Model on SageMaker
Setup effort	Minimal: No ML training needed. Just call the API and get entities.	Moderate/high: Requires preparing data and training a model (e.g., fine-tuning BERT).
Entity types	Pre-defined entity types (PERSON, ORG, DATE, etc.). Cannot natively add new types without using Comprehend's custom entity training.	Fully customizable. You define the entity schema and train the model accordingly (e.g., add CaseNumber, ChemicalName, etc.).
Accuracy	High for common entities in general-domain text. Continually improved by AWS on the backend. May struggle with niche domain terms without explicit training.	High for the domain it's trained on (can surpass Comprehend for niche vocabulary). Accuracy depends on quality/quantity of training data and the chosen model.
Scalability	Automatically scales as a managed service. Can handle thousands of requests concurrently; for batch jobs, it will parallelize under the hood.	You control scalability. Can deploy multiple instances or auto-scaling endpoints. Needs planning to handle spikes (auto-scaling policies or using AWS Lambda as serverless scaling).
Cost model	Pay-per-use, based on text size. For example, about $0.0005 per 100 characters processed (with a 300-char minimum per request). No idle cost—if not used, you pay nothing.	Pay for infrastructure. Training cost (e.g., ~$3 per hour for an ml.p3 GPU instance during training) + inference cost (hourly cost for running endpoints, or per-ms cost if using Lambda). A single `ml.m5.large` (CPU) might be ~$0.11/hour, which is ~$80/month if running 24/7 regardless of queries.

(continued)

Table 4-2. (*continued*)

Aspect	Amazon Comprehend (Pretrained NER API)	Custom NER Model on SageMaker
Cost at scale	For very large volumes of text, pay-per-char can become substantial (e.g., processing 1 million characters ~ 10K units = $5 at $0.0005/unit). Volume discounts may apply at 100M+ units.	If utilization is high (constant traffic), a dedicated instance might be cheaper (e.g., an `ml.m5.xlarge` ($0.23/hour) running full tilt could handle many requests for a fixed cost, which over a month at $170 might handle more than the equivalent $170 worth of Comprehend units.) However, if usage is sparse or unpredictable, you might waste money on idle time.
Maintenance	AWS handles model maintenance, updates, and optimization. You just maintain the integration. (If using Comprehend's custom entity feature, you might retrain when your data changes, but the service still hosts it.)	You are responsible for updating the model. If data drifts, you should collect new examples and retrain in SageMaker. Also responsible for deploying new model versions to endpoints.
Security and privacy	Data is sent to a managed service. Can use TLS (HTTPS) by default. For additional security, use VPC endpoints to keep traffic within the AWS network. AWS is responsible for safeguarding the model and infrastructure.	You have full control. You can deploy the model in a private VPC, restrict access via IAM roles or security groups, and encrypt data at rest (S3, EBS) and in transit. No third-party (including AWS service teams) sees the data except for from the underlying EC2 hosting your endpoint. If using Lambda, the code runs in your account's isolated environment.

(*continued*)

Table 4-2. (*continued*)

Aspect	Amazon Comprehend (Pretrained NER API)	Custom NER Model on SageMaker
Inference latency	Low latency for single API calls (tens of milliseconds plus network). However, there's overhead in calling a public API. For batch jobs, you incur some startup/wind-down time.	Potentially very low latency if deployed close to your data/users (e.g., same VPC or region). A custom model on a powerful instance might answer in <10 ms for short text. Lambda cold starts can add latency occasionally. You have flexibility to optimize (e.g., model quantization for speed).
Integration	Easy—just an API call. No need to manage scaling and fits well in serverless workflows (Lambda can call Comprehend as part of a Step Function, etc.).	Requires an inference endpoint or Lambda function. A bit more complex to integrate (need the endpoint name, invoke it, handle scaling). But can be containerized and deployed in many ways (SageMaker, ECS, EKS, etc.).

The best choice depends on the use case: for quick setup and broad coverage, Comprehend is the clear winner; for domain-specific accuracy and control, a custom model is superior. Cost tradeoffs hinge on usage patterns and volume.

Best practices for deploying NER models on AWS:

- **Start simple, then customize:** Begin with Comprehend for prototyping and early production to understand requirements. Later, assess the need for a custom model to avoid premature optimization.

- **Monitor and optimize cost:** Track usage and cost using CloudWatch to right-size your solution (e.g., smaller instances, serverless, provisioned concurrency). Compare Comprehend's per-character cost with custom endpoint costs; high load favors fixed endpoints, sporadic usage favors pay-per-call.

- **Auto scaling and load testing:** For SageMaker, use auto-scaling policies. Load test endpoints to ensure they scale within latency limits. SageMaker Serverless Inference offers auto-scaling. For Comprehend, check and potentially request quota increases for high throughput.

- **Security considerations:** Encrypt data in transit (HTTPS). Use KMS for S3 encryption with Comprehend. Deploy SageMaker endpoints in a Private VPC. Use IAM roles to restrict access. Secure Lambda deployment packages and EFS access. Log and audit requests with CloudTrail.

- **Model storage and versioning:** Store custom model artifacts in versioned S3 buckets for rollback. Automate SageMaker endpoint deployments with CI/CD pipelines. Maintain model history and metrics. For Comprehend Custom Entities, version-train datasets and document the model.

- **Inference optimization:** For real-time inference, batch multiple records per call (use Comprehend's `BatchDetectEntities` or implement in custom endpoints). Consider model quantization for custom models. Explore AWS Neuron or NVIDIA TensorRT for high performance.

- **Fallback strategies:** Use a hybrid approach: primary extraction with a custom model, fall back to Comprehend or an LLM for uncertain cases. This enhances reliability but requires orchestration.

- **Using AWS Lambda for low-cost, real-time inference:** Develop best practices specifically for the Lambda deployment scenario:

  - **Model size management:** Keep the deployment package small. If using Transformers, consider stripping out unnecessary files or using the `transformers` pipeline in a minimal way. You might not need the entire model config if you hardcode some aspects. Using a tool like `onnxruntime` to run the model could also improve speed—you can convert your PyTorch model to ONNX and load it in Lambda for faster CPU inference.

  - **Provisioned concurrency:** If your Lambda will handle steady traffic (like a few requests per second consistently), enabling Provisioned Concurrency can keep containers warm and avoid cold-start latency. This will add a baseline cost (as if a certain amount of memory is always allocated), but still might be cheaper than an equivalent EC2 running 24/7.

- **Concurrency limits:** Lambda scales rapidly by default, which is good for bursts. However, ensure your account's concurrency limit is sufficient for large spikes. Implement throttling or queues (SQS/SNS) to smooth out traffic and avoid overwhelming systems or unexpected cost increases.

- **Logging and monitoring:** Use CloudWatch Logs to monitor Lambda execution times and errors; adjust memory or optimize code as needed.

- **Graceful scaling down:** Architect your solution modularly (e.g., with an "NER inference" interface) to easily switch among Comprehend, SageMaker, or Lambda as needed. This way you can swap the backend easily without changing the rest of the pipeline.

Following these practices allows for scalable, secure, and cost-effective NER deployment on AWS, leveraging its diverse services for tailored and adaptable solutions.

4.2.4 Example Use Case: Real-World Deployment

To cement these concepts, this section walks through a real-world deployment scenario: an enterprise implementing NER for automated document processing. Consider a large financial services company that receives thousands of documents every day—these could be loan applications, contracts, insurance claims, emails, and so on. The goal is to extract key information (entities) from these documents and feed it into downstream systems for automated decision-making, with minimal human intervention. This section outlines how AWS services can be combined to achieve this, and how LLM agents can be integrated on top for smarter automation.

Use case scenario: ACME Financial processes loan applications, which include unstructured texts like customer letters, credit reports, and legal agreements. They want to extract entities such as Customer Name, Account Number, Income Amount, Property Address, Loan Type, Date, and any mention of Legal Clauses (e.g., bankruptcy or litigation references). Some of these (Name, Date, etc.) are standard, but others like specific legal clauses are domain-specific. ACME decides to build an AWS-based pipeline:

1. **Document ingestion:** All documents (PDFs, Word files, etc.) are uploaded to an Amazon S3 bucket by an intake system. This triggers an AWS Lambda function (via S3 event notification).

2. **Text extraction:** The Lambda function uses Amazon Textract to extract raw text from the documents (especially for scanned PDFs or images). Textract returns machine-readable text and basic structure for digital text like an email in text form. This step might be skipped.

3. **NER pipeline:** The extracted text is sent through the NER system, where ACME has several options. For a quick start, they may use Amazon Comprehend, which can be called directly from Lambda to extract entities. However, if Comprehend's pretrained model fails to recognize domain-specific entities like tagging "Chapter 7" as OTHER instead of BankruptcyType, ACME decides to create a custom NER model. They use SageMaker to fine-tune a RoBERTa model on a labeled dataset from their documents, with input from their compliance department to label terms and clauses. This model is then deployed to a SageMaker endpoint.

   In the live pipeline, the Lambda now calls the SageMaker endpoint (via the AWS SDK) to extract entities instead of relying on Comprehend. This solution remains serverless in architecture, with the only persistent component being the managed SageMaker endpoint. For handling high throughput, ACME could introduce an asynchronous approach, where Lambda enqueues jobs to Amazon SQS or AWS Step Functions. These services then trigger the model, helping to manage spikes in S3 events and avoid overwhelming the endpoint.

4. **Post-processing and validation:** The raw entities extracted are then processed. For instance, if the model returned "John Doe" as a Person and "123-456-789" as an AccountNumber, the system might cross-verify the account number against a database for validity. If certain critical fields are missing (maybe the model didn't find an Income Amount), the pipeline flags the document for human review. This is where AWS Augmented AI (A2I) could

be used—it allows low-confidence predictions to be sent to a human reviewer in a UI, and the corrected result can be fed back to improve the model over time.

5. **LLM agent integration for decision-making:** ACME Financial aims to extend its automation beyond simple entity extraction to decision-making and summarization, leveraging LLMs. After extracting key data points via NER, such as customer names, income amounts, and legal history, ACME uses an LLM to construct a structured summary of the loan application. The LLM can be prompted with these extracted entities, enabling it to assess whether the application is complete and identify any potential risk factors, offering assessments or recommendations. For this, ACME might use Amazon Bedrock to access a foundation LLM, with the LLM agent being implemented through AWS Lambda to interact with the extracted data and original text, answering specific questions such as whether the loan application mentions any legal issues like bankruptcy.

In practice, ACME can integrate NER with prompt engineering for decision-making. For example, after extracting facts such as a bankruptcy history from the loan application, the system might query the LLM: "Should this application be flagged for manual review?" The LLM will evaluate the extracted data based on company policies, providing a recommendation like, "Yes, it should be flagged because the bankruptcy was within the last seven years." This showcases how structured data (entities) and unstructured reasoning (LLM) combine to automate decision-making processes, ensuring alignment with business rules and policies.

LLMs can also be used to summarize loan applications more efficiently. After extracting entities, ACME can use the LLM to generate a concise summary, helping managers quickly assess key details such as loan amount, income, and potential risk factors like bankruptcy. For example, the LLM might summarize the document with: "Applicant John Doe, requesting a home equity loan of $50k, has stable income but a past bankruptcy. Recommend higher interest rate or collateral." By combining precise NER with a generative LLM model, ACME can automate document analysis, ensuring both accuracy and efficiency in scaling its document processing workflows.

6. **End-to-end workflow and scaling:** All these components (Textract, Lambda, Comprehend/Endpoint, LLM, etc.) are orchestrated likely via AWS Step Functions, which can define a state machine: S3 event -> Lambda Extract -> Parallel invoke NER and maybe other checks -> LLM agent -> Outcomes. The entire pipeline is either serverless or managed. It can scale to handle bursts of documents during business hours and scale down when idle. ACME can also leverage AWS's robust monitoring: CloudWatch for metrics, X-Ray for tracing the pipeline latency, and CloudTrail for auditing who accessed what (important for financial data compliance).

7. **Results and continuous improvement:** In deployment for example, ACME's system may be able to automatically process about 90% of applications with no human in the loop, flagging 10% for review. Over time, they review those and realize some new entity or pattern needs to be extracted—they can then update their NER model by labeling a few more examples and retraining (SageMaker makes it easy to run periodic training jobs as new data labels come in). They could even use the LLM agent in a feedback loop. Occasionally, the LLM's output highlights something that wasn't captured as an entity, which might suggest a new entity type to add to the NER model's next version.

4.2.5 Key Takeaways and Practical Insights

This use case highlights how an AWS-based NER solution is deployed to streamline document understanding. The combination of OCR (Textract), NER (Comprehend or custom SageMaker models), orchestration (Lambda/Step Functions), and an optional LLM creates a robust pipeline for extracting and processing data. By leveraging AWS services, ACME avoids maintaining servers for each component, instead relying on scalable, managed services. The integration of the LLM agent adds cognitive automation, allowing the system to not only extract data but also to reason about it, mimicking human analysis at high speed.

A key advantage of this approach is that the LLM works with structured data from the NER model, reducing errors by grounding its outputs in factual information. This reduces the likelihood of hallucinated or incorrect details. Many enterprise AI architectures follow a similar pattern, using NER or information extraction to feed into an LLM or knowledge base, which then generates final answers. For companies that prefer not to use external LLMs due to cost or data sensitivity, fine-tuning large models on proprietary data via SageMaker is another option, although it is resource-intensive. A managed LLM via Bedrock or an API offers a simpler, more flexible alternative.

Outcome: The result of this setup is impressive efficiency gains for ACME Financial. Tasks that previously took analysts 30 minutes per document now only take about a minute, with human intervention reserved for exceptional cases. The scalable AWS architecture allows ACME to easily handle increased document volume without needing to redesign the system. This case study demonstrates the synergy between NER models and LLMs: the NER structures the data, while the LLM provides reasoning and narrative, creating an intelligent automation system that exceeds the capabilities of individual components.

4.2.6 Summary of NER Implementation with AWS Services

This section explored NER implementation with AWS services, starting with Amazon Comprehend for quick, scalable entity extraction using pretrained models. For specialized domains like legal or medical text, we turned to custom NER models on Amazon SageMaker, fine-tuning transformer models such as BERT and RoBERTa for higher accuracy. We also compared large pretrained models (LLMs) like BERT with smaller, cost-efficient models like DistilBERT/TinyBERT, suitable for low-latency deployments on AWS Lambda.

We discussed deployment considerations, comparing the performance and cost of using Comprehend versus custom endpoints and provided best practices for ensuring robustness, security, and scalability in production. A real-world case study demonstrated how integrating an LLM agent with NER outputs can enable intelligent decision-making automation, streamlining processes and increasing efficiency.

In conclusion, AWS offers a versatile toolkit for building NER systems, blending managed services for fast deployment with custom models for specialized needs. The next section explores alternative NER solutions outside AWS, including open-source tools and optimizations, helping you evaluate the best approach for your needs.

4.3 Open-Source Implementation

4.3.1 From AWS Solutions to Open-Source Alternatives

The previous section explored AWS-based solutions for NER (e.g., Comprehend, SageMaker), which offer a convenient, scalable infrastructure. However, many projects benefit from open-source frameworks for greater control, customization, and cost efficiency. This section transitions to open-source NER implementations using Python, runnable on local machines or cloud VMs (like AWS EC2). These tools enable fine-tuning models on domain-specific data (e.g., financial documents) for better accuracy and deploying NER systems without cloud vendor lock-in. We'll focus on Hugging Face Transformers, spaCy, and Flair, demonstrating how to fine-tune LLMs (like BERT and RoBERTa) and SLMs (like DistilBERT and TinyBERT) for NER. Finance examples are used.

4.3.2 Fine-Tuning with Hugging Face Transformers (BERT, RoBERTa, etc.)

Hugging Face's Transformers library is a popular tool for NER with pretrained LLMs like BERT and RoBERTa. Fine-tuning involves training a general language model on an NER-labeled dataset (e.g., financial news with annotated entities). This updates the model to predict entity labels, specializing it for the NER task. The Transformers library simplifies loading models, preparing data, and fine-tuning with the Trainer API. Listing 4-1 demonstrates fine-tuning a BERT model for NER, assuming training and validation sets have already been prepared.

CHAPTER 4   NAMED ENTITY RECOGNITION (NER) WITH LLMS AND SLMS

Listing 4-1. Fine-tuning a BERT-based NER Model Using Hugging Face Transformers

```
from transformers import AutoTokenizer, AutoModelForTokenClassification, Trainer, TrainingArguments

# 1. Load a pre-trained model and tokenizer for token classification (NER)
model_name = "bert-base-uncased"   # can be "roberta-base", etc. for
                                   different LLMs
tokenizer = AutoTokenizer.from_pretrained(model_name)
# Initialize model for NER (assuming num_labels is the number of entity tags in our dataset)
model = AutoModelForTokenClassification.from_pretrained(model_name, num_labels=num_labels)

# 2. Prepare training and validation data (already tokenized & formatted as needed)
# train_dataset = ...  (e.g., a list of tokenized inputs with labels)
# val_dataset = ...

#..

# 4. Initialize Trainer with model, data, and training params
trainer = Trainer( 3. Set up fine-tuning parameters
training_args = TrainingArguments(
    output_dir="ner_model_output",
    learning_rate=2e-5,
    num_train_epochs=3,
    per_device_train_batch_size=16,
    per_device_eval_batch_size=16,
    evaluation_strategy="epoch",      # evaluate at end of each epoch
    logging_steps=50,
    weight_decay=0.01,
    save_strategy="epoch",
    load_best_model_at_end=True
)
```

```
# (Optional) Define a metrics function for evaluation (e.g., computing
precision, recall, F1)
# def compute_metrics(eval_pred): .
    model=model,
    args=training_args,
    train_dataset=train_dataset,
    eval_dataset=val_dataset,
    tokenizer=tokenizer,
    compute_metrics=compute_metrics    # e.g., uses seqeval to compute
                                       entity-level F1
)

# 5. Fine-tune the model
trainer.train()

# 6. Evaluate the fine-tuned model on the validation set
metrics = trainer.evaluate()
print(metrics)   # e.g., outputs overall precision, recall, F1 scores
```

In this code, we start by loading a pretrained BERT model and tokenizer, then set training hyperparameters like learning rate and batch size—these can be tuned for optimal results. The Hugging Face `Trainer` API handles training, evaluation, and checkpointing. Although not shown, a typical `compute_metrics` function uses tools like seqeval to compute precision, recall, and F1 scores by comparing predicted and actual tags. After training for a few epochs (e.g., 3), we evaluate the model on a validation set. Models like BERT or RoBERTa often achieve strong results—BERT-base, for example, can reach 92–94% F1 on standard datasets like CoNLL-2003.

After fine-tuning, the model is ready for inference. Hugging Face provides a convenient `pipeline` for NER that wraps tokenization, model inference, and result formatting (see Listing 4-2). This allows easy deployment on new financial documents, enabling accurate recognition of key entities—such as names, amounts, or legal terms—tailored to domain-specific content.

CHAPTER 4 NAMED ENTITY RECOGNITION (NER) WITH LLMS AND SLMS

Listing 4-2. Using the Fine-Tuned Model for NER Inference on a Finance Sentence

```
from transformers import pipeline

# 7. Load the fine-tuned model into an NER pipeline for inference
ner_pipeline = pipeline("ner", model=trainer.model, tokenizer=tokenizer,
aggregation_strategy="simple")

# 8. Use the NER pipeline on new text (finance-related example)
text = "Apple Inc. reported Q1 revenue of $89.6 billion in 2023."
entities = ner_pipeline(text)
for ent in entities:
    print(f"{ent['word']} -> {ent['entity']}")
# Expected output (entity predictions):
# Apple Inc. -> ORG
# Q1 -> MISC
# $89.6 billion -> MONEY
# 2023 -> DATE
```

In this inference example, the Hugging Face `pipeline` breaks the sentence into tokens, runs them through the model and groups tokens into full entity mentions. For example, in the sentence *"Apple Inc. reported Q1 revenue of $89.6 billion in 2023,"* a well-trained model might tag "Apple Inc." as an ORG, "Q1" as DATE or MISC, "$89.6 billion" as MONEY, and "2023" as DATE. This shows how open-source models can extract structured financial data—similar to AWS Comprehend but with the added flexibility to define custom entities or train on proprietary datasets.

Switching to models like RoBERTa is easy—just change the checkpoint name (e.g., `"roberta-base"`). RoBERTa-base is similar in size to BERT-base and performs comparably when fine-tuned. You can also explore domain-specific models (e.g., FinBERT) from the Hugging Face Model Hub for better performance on financial texts. Experimenting with different models often yields the best NER results for your specific use case.

4.3.3 Embracing Smaller Models: DistilBERT and TinyBERT

While large transformers like BERT and RoBERTa offer state-of-the-art accuracy, they are computationally heavy. SLMs like DistilBERT and TinyBERT provide cost-effective alternatives. These models are distilled versions of BERT, where a smaller "student" model mimics the larger "teacher" model, retaining much of the knowledge with fewer parameters.

DistilBERT, a lighter model from Hugging Face, has ~40% fewer parameters than BERT-base and runs ~60% faster while preserving over 95% of BERT's performance. Fine-tuning DistilBERT for NER is similar to BERT, and it uses less memory, making it suitable for training on modest GPUs or for faster inference. The tradeoff is a slight drop in accuracy: DistilBERT might achieve around 90% F1, compared to BERT's 92%. This efficiency gain often justifies the slight drop in accuracy.

TinyBERT compresses further with a two-stage distillation process. A 4-layer TinyBERT model is ~7.5× smaller than BERT-base and up to 9× faster, while retaining about 96% of BERT's performance. Fine-tuning TinyBERT for NER follows the same procedure. It's ideal for real-time or edge-device applications, particularly for cost-sensitive deployments like processing financial news on a budget or running NER on-device in a finance app.

Table 4-3 provides a comparison between some transformer models in terms of size, speed, and NER performance. The F1 scores are illustrative for a generic NER task (such as CoNLL-2003 English) to give a sense of accuracy differences.

Table 4-3. Comparison of Large vs. Distilled Models for NER (Illustrative Performance on CoNLL-2003). Distilled Models Achieve Nearly the Accuracy of Their Larger Counterparts at a Fraction of the Size and Latency

Model	Parameters	Inference Speed	NER F1 Score (approx.)
BERT-base (uncased)	110M	1.0× (baseline)	~92% F1
RoBERTa-base	125M	1.0× (baseline)	~92% F1
DistilBERT	66M	~1.6× faster	~90% F1
TinyBERT-4L	~14M	~7–9× faster	~88–89% F1

In summary, Hugging Face Transformers allow you to select a model based on your accuracy and efficiency requirements. Large models like BERT and RoBERTa provide maximum accuracy, while smaller models like DistilBERT and TinyBERT offer cost-effective efficiency. The open-source nature allows experimentation to find the optimal balance (e.g., using BERT for a performance baseline and then trying DistilBERT for efficiency).

4.3.4 spaCy: Industrial-Strength NER Pipelines

spaCy is a popular open-source NLP library known for its ease of use and efficiency. It includes pretrained pipelines with a built-in NER (e.g., en_core_web_sm for English) that recognizes various entity types. While useful for tasks like financial document processing, fine-tuning is often needed for domain adaptation or custom entity types. spaCy allows updating existing NER models or training new ones by providing example texts and annotations. Listing 4-3 shows a simplified example of fine-tuning spaCy's small English NER model.

Listing 4-3. Fine-tuning spaCy's NER Model with a Custom Example, then Extracting Entities from a New Sentence

```
import spacy
from spacy.training import Example

# 1. Load a pre-trained small English NER model
nlp = spacy.load("en_core_web_sm")

# 2. Define training data (text with entity spans and labels)
TRAIN_DATA = [
    ("Goldman Sachs is based in New York City.", {"entities": [(0, 14,
    "ORG"), (28, 41, "GPE")]})
    # You can add more examples, e.g., ("Apple raised $5 million.",
    {"entities": [(0,5,"ORG"), (13,22,"MONEY")]})
]

# 3. Add new entity labels to the NER pipeline (if any new labels not in
pretrained model)
ner = nlp.get_pipe("ner")
for _, annotations in TRAIN_DATA:
```

```
    for ent in annotations.get("entities"):
        ner.add_label(ent[2])  # ensure the label is recognized by the
        pipeline

# 4. Fine-tune the model on the new data
optimizer = nlp.resume_training()
for epoch in range(30):
    losses = {}
    for text, annotations in TRAIN_DATA:
        example = Example.from_dict(nlp.make_doc(text), annotations)
        nlp.update([example], drop=0.5, losses=losses)
    # (In practice, use nlp.initialize before training and adjust epochs,
    dropout, etc., as needed)

# 5. After training, test the updated model on new text
doc = nlp("Apple is looking at buying a U.K. startup for $1 billion.")
for ent in doc.ents:
    print(ent.text, ent.label_)
# Expected entities (with the fine-tuned model):
# Apple ORG
# U.K. GPE
# $1 billion MONEY
```

In this code, we load spaCy's en_core_web_sm pipeline, which already recognizes entity types like "ORG" and "GPE." It provides a training example ("Goldman Sachs" as ORG, "New York City" as GPE), adds these labels if needed, and uses nlp.update to adjust the model's weights. After training (e.g., 30 epochs), the model can identify entities in new text (e.g., "Apple" as ORG, "U.K." as GPE, "$1 billion" as MONEY). Real-world fine-tuning requires much more data, especially for specific domains. SpaCy v3 allows configurable training and transformer-based encoders, like the en_core_web_trf pipeline using RoBERTa.

huggingface.co

SpaCy's transformer-based pipeline (en_core_web_trf) offers a significant accuracy increase but at the cost of speed and higher memory usage. SpaCy's models, particularly the smaller ones, are optimized for production, being fast and lightweight for processing large amounts of text. You might choose spaCy's small model for speed with acceptable accuracy tradeoffs or its transformer model for high accuracy with available GPU resources. SpaCy's user-friendly API makes it a practical choice for integrating NER into real-world finance applications, like analyzing financial reports or news.

4.3.5 Flair: Simple and Flexible NER Training

Flair is another open-source NLP library, developed by Zalando Research, that provides a simple interface for sequence labeling tasks like NER. Flair is built on PyTorch and is particularly known for its contextual string embeddings (called Flair embeddings) and an easy training API. It allows you to mix and match different word embeddings (Glove, BERT, Flair's own, etc.) and use a BiLSTM-CRF architecture by default for NER. You can often train a custom NER model or fine-tune a pretrained one in Flair with just a few lines of code, making it very user-friendly for experimentation.

Flair offers pretrained NER models (e.g., flair/ner-english or SequenceTagger. load("ner"), trained on CoNLL-03 data). These models can be used directly or finetuned on domain-specific data, like financial texts. Listing 4-4 provides an example of training a new NER model using Flair (using a built-in corpus for demonstration). The code then shows how to predict entities in a sample sentence after training.

Listing 4-4. Training a Custom NER Model with Flair and Using It for Inference

```
from flair.datasets import CONLL_03
from flair.embeddings import WordEmbeddings
from flair.models import SequenceTagger
from flair.trainers import ModelTrainer
from flair.data import Sentence

# 1. Load a corpus (CoNLL-03 NER dataset as an example; replace with your finance dataset)
corpus = CONLL_03()
label_dict = corpus.make_label_dictionary(label_type="ner")

# 2. Initialize embeddings (here we use GloVe word embeddings for simplicity)
embeddings = WordEmbeddings("glove")

# 3. Initialize a sequence tagger (BiLSTM-CRF by default) for NER
tagger = SequenceTagger(hidden_size=256,
```

```
                embeddings=embeddings,
                tag_dictionary=label_dict,
                tag_type="ner",
                use_crf=True)

# 4. Train the model (fine-tuning if starting from a pre-trained tagger is also possible)
trainer = ModelTrainer(tagger, corpus)
trainer.train("outputs/ner-model",
              learning_rate=0.1,
              mini_batch_size=32,
              max_epochs=5)

# 5. Load the trained model and run inference on a new sentence
trained_tagger = SequenceTagger.load("outputs/ner-model/final-model.pt")
sentence = Sentence("Barclays PLC reported a 10% increase in Q2 revenue.")
trained_tagger.predict(sentence)
for entity in sentence.get_spans("ner"):
    print(entity.text, entity.get_label("ner").value)
# Example output:
# Barclays PLC ORG
# 10% PERCENT
# Q2 MISC
```

In the training code, the CoNLL-2003 dataset is loaded using CONLL_03() and a label dictionary is created. A SequenceTagger model is defined using GloVe embeddings, a hidden layer, and a CRF layer for "ner" tags. The model is trained for five epochs, with the ModelTrainer handling optimization. After training, the model is used to predict entities in a sentence. A model trained on general data identifies "Barclays PLC" as ORG and "10%" as a percentage. Fine-tuning on financial data could teach it to recognize "Q2" as a specific entity type (e.g., QUARTER).

Flair's interface is highly adaptable. You could easily swap in TransformerWordEmbeddings (e.g., embeddings = TransformerWordEmbeddings('bert-base-uncased')) to use BERT embeddings instead of GloVe, which would likely boost accuracy. In fact, Flair's official NER English model uses a combination of Flair's own contextual embeddings and GloVe, yielding an F1 over 93% on CoNLL.

huggingface.co

You can fine-tune those embeddings by simply setting fine_tune=True on the embeddings (as shown in Flair's docs for transformer fine-tuning). The ease of trying different embeddings and models makes Flair a powerful open-source option. It's particularly friendly for researchers or practitioners who want to train custom NER models without writing a lot of boilerplate code.

For a financial project, you can start with Flair's pretrained model and fine-tune it on a financial dataset. The code is similar to this example, but you'd use ColumnCorpus or construct a Corpus from Sentence objects instead of CONLL_03(). After training, Sentence and tagger.predict simplify integrating the model into applications, allowing easy extraction of entities (e.g., company names, monetary figures) from text like news articles.

4.3.6 Model Evaluation and Optimization

After implementing NER models using the previous open-source tools, evaluating their performance is crucial to ensure they meet the project's needs. The primary evaluation metrics for NER are Precision, Recall, and F1 score (the harmonic mean of precision and recall). These are usually computed at the entity level. For example, Precision asks, "of all the entity mentions the model predicted, how many were correct?" and Recall asks, "of all the actual entity mentions in the text, how many did the model successfully find?" An F1 score combines these into a single number. Typically, you would evaluate on a held-out test dataset with ground-truth annotations. All three frameworks have utilities for evaluation:

- In Hugging Face, you might use the evaluate library with the seqeval metric or call trainer.evaluate() which returns Precision, Recall, F1, and so on.

- In spaCy, you can use nlp.evaluate() or the built-in Scorer to get per-entity-type scores and overall stats on an evaluation set.

- In Flair, the ModelTrainer prints evaluation metrics at each epoch if you provide a test set (and you can use trainer.test() on a model and corpus to get a detailed classification report).

When evaluating on finance-specific data, you may pay attention to particular entity types. For instance, you might find that your model has high precision on ORG (organizations) but lower recall on MONEY (monetary values), indicating that it misses some money amounts mentioned in text. Such insights can guide further improvement.

4.3.6.1 Optimization Techniques

You can improve an open-source NER model's performance or efficiency using several techniques:

- **Hyperparameter tuning:** Fine-tuning hyperparameters, such as learning rate, batch size, and dropout rate, can significantly enhance NER performance. For instance, a lower learning rate or more training epochs can boost the F1 score if the model was undertrained. A development or validation set should be used to test various settings, and early stopping can help avoid overfitting by halting training when no improvements are seen.

- **Data augmentation:** If your annotated dataset is small, you might augment it to improve the model's generalization. In the finance domain, this could involve generating variant sentences (e.g., replacing company names with other companies, or currencies with other currency symbols) to expose the model to more examples. This must be done carefully to maintain data realism.

- **Domain-specific pretraining:** Starting with a model that has been pretrained on domain text can give you a head start. For instance, a language model pretrained on financial news or SEC filings might better identify finance jargon. There are models like FinBERT (for financial sentiment) or others that you could fine-tune for NER. If such a model is available, using it as your base instead of generic BERT could improve entity recognition of, say, rare financial terms (e.g., stock ticker symbols or specific financial instruments).

- **Knowledge distillation and smaller models:** We already discussed DistilBERT and TinyBERT, which are outcomes of knowledge distillation. If you have a very high-performing large model, you can distill it into a smaller model yourself using open-source tools (Hugging Face provides recipes for distilling task-specific models). This way, you create a lightweight model tailored to your dataset. Distillation can be seen as an optimization: you maintain accuracy as much as possible while optimizing model size and speed.

- **Quantization:** Model quantization reduces the precision of model weights, such as converting 32-bit floating-point values to 8-bit integers. This can significantly reduce memory usage and increase inference speed with minimal loss in accuracy. Tools like PyTorch and Hugging Face's Optimum or Intel Neural Compressor help automate this process. Quantization is especially useful for high-throughput environments, like processing large-scale financial transaction data, where reduced computational costs are crucial.

- **Pruning and other compression:** Pruning involves removing less important parts of the model, such as neurons or attention heads, to make it more efficient. Research shows that large transformers often contain redundant parameters, which pruning can eliminate. Combined with fine-tuning, pruning can lead to models that maintain near-optimal performance with fewer parameters.

- **Pipeline optimization:** Outside the model itself, you can optimize the inference pipeline. For example, batching multiple texts together can improve GPU utilization. If using spaCy in production, you might use its built-in `nlp.pipe()` to process texts in batches efficiently. Caching the tokenizer results for frequently seen vocabulary (important in finance if the same terms appear often) is another practical trick.

4.3.6.2 Example: Optimizing for a Budget

Suppose you initially fine-tuned BERT for your NER task on financial reports and achieved an F1 of 92%. For deployment, you need to handle thousands of documents per minute, and BERT's inference speed is a bottleneck. You have a few options:

1. **Fine-tune DistilBERT on the same data** and see if you still get ~90% F1. If so, you might accept a 2% loss in accuracy for roughly 1.6× throughput improvement.

2. **Quantize the DistilBERT model to 8-bit.** This could further speed it up (perhaps another 2×) with minimal impact on accuracy (maybe dropping to ~89% F1). Now you have a model that is ~4× faster than original BERT, with only ~3 points lower F1—a good tradeoff for your needs.

CHAPTER 4   NAMED ENTITY RECOGNITION (NER) WITH LLMS AND SLMS

3. **If even more efficiency is needed, you could try TinyBERT or a pruned DistilBERT.** Perhaps you get to ~88% F1 but with a model that can run on CPU in real-time with ease. This might be acceptable depending on your application (for example, if you are doing preliminary entity extraction to be verified by humans, 88% F1 might be fine).

Throughout this optimization process, open-source tools give you full visibility and control. You can measure performance at each step with the test set and decide based on empirical results. You might summarize results of these experiments in a table for clarity (like Table 4-3 did for model stats) or track them in a spreadsheet to pick the best candidate model for production.

Before moving on, it's worth noting that optimization isn't only about speed; it can also be about improving accuracy. Using cross-validation to better utilize limited data or assembling multiple NER models (e.g., taking the union of entities predicted by two different models) could improve recall. These approaches, however, add complexity and are chosen based on specific project requirements.

This section demonstrated how open-source frameworks can implement NER for finance applications. We fine-tuned large transformer models like BERT and RoBERTa for high accuracy and showed how DistilBERT and TinyBERT can offer similar performance with lower computational costs. We explored spaCy's pipeline approach and Flair's flexible training interface, providing practical examples of each. The tradeoffs between model size and performance were discussed, alongside techniques for optimizing NER models for real-world use.

Open-source frameworks allow practitioners to experiment with models tailored to their data, whether it's customizing entity categories for financial regulations or optimizing a model for mobile device usage. The choice of tool and model depends on the specific context: Hugging Face offers state-of-the-art transformers, spaCy delivers speed and integration, and Flair simplifies training with mixed embeddings. Many teams combine tools (e.g., fine-tuning with Hugging Face, then exporting to spaCy or ONNX for deployment).

4.3.7 Quantizing the Model for Efficient Inference

To deploy SLMs like DistilGPT or TinyGPT efficiently on mobile devices or edge computing, it is common to use quantization techniques to reduce the model size and speed up inference.

Refer to this GitHub code snippet: https://github.com/anvcse562/ModernNLP-LLMSLM/blob/main/Ch04/ch4.3/4.3.7DistilBERT_Quantization_ONNX_Deployment.ipynb

This code demonstrates how to export a quantized model to ONNX, a cross-platform format ideal for running inference in mobile apps using ONNX Runtime on Android and iOS devices. By leveraging fine-tuning, quantization, and ONNX export, these models offer a balance between performance and efficiency, delivering fast, accurate results in NER while minimizing computational overhead.

Adopting SLMs for real-time NER on mobile or edge platforms helps companies lower costs and improve user experience without compromising performance.

Having covered high-level NER implementations with large and small models, this chapter now turns to more low-level explorations of cost-effective alternatives. The next section goes beyond fine-tuning pre-existing models to explore building or using even smaller models and techniques for extreme cost-efficiency. You'll examine the limits of compact models, their performance, and how these alternatives apply when budget/resource constraints are paramount. This shift provides a complete perspective on NER implementation, from powerful cloud solutions to bespoke lightweight models.

4.4 Low-level LLM Exploration: Small LLM Cost-Effective Alternatives

4.4.1 Cost-Effectiveness of SLMs in NER

In the realm of natural language processing (NLP), NER involves extracting key entities such as names of people, organizations, dates, and locations from raw text. While LLMs like GPT-3 and BERT have significantly advanced the field, their large size and computational demands often make them impractical for real-time or resource-constrained applications. This is where SLMs like TinyGPT, DistilGPT, and ELECTRA provide a more cost-effective and efficient solution for NER tasks.

SLMs are designed to be resource-efficient, making them suitable for environments with limited computing power, such as mobile devices and edge computing platforms. These models offer notable advantages in memory consumption, inference speed, and latency, making them ideal for real-time applications where traditional, large models would be too slow or require excessive computational resources. This section explores the technical advantages and challenges of using SLMs for NER and explains some cutting-edge optimization techniques for deploying these models on resource-constrained devices (see Table 4-4).

CHAPTER 4 NAMED ENTITY RECOGNITION (NER) WITH LLMS AND SLMS

Table 4-4. SLM Optimization Techniques for NER

Optimization Technique	Code Snippet	Short Description	Benefit

Quantization	`from torch.quantization import quantize_dynamic`
	`quantized_model = quantize_dynamic(model, {torch.nn.Linear}, dtype=torch.qint8)`
	`quantized_model.save_pretrained("./quantized_model")`
		Reduces the precision of the model weights to lower bit-width (e.g., 8-bit integers).	Reduces memory usage and improves inference speed without significant loss in accuracy.

Pruning	`from torch.nn.utils import prune` `prune.random_unstructured(model.transformer.layer.attn, name="weight", amount=0.3)`	Removes less important parameters or neurons from the model.	Decreases model size, accelerates inference, and saves computational resources.

(*continued*)

Table 4-4. (*continued*)

Optimization Technique	Code Snippet	Short Description	Benefit

Knowledge distillation	`teacher_model = GPT2LMHeadModel.from_pretrained("gpt2")`
	`student_model = DistilBertForTokenClassification.from_pretrained("distilbert-base-uncased")`
	`# Implement distillation loss and training loop manually`
		Transfers knowledge from a larger "teacher" model to a smaller "student" model.	Allows smaller models to perform nearly as well as large models, reduces computational load.

Layer reduction	`model.config.num_hidden_layers = 6`
	`# Reducing from 12 to 6 layers`
	`model = DistilBertForTokenClassification(config=model.config)`
		Reduces the number of layers in the model's architecture.	Reduces memory footprint and increases inference speed by simplifying the model.

Token clustering	`Use embeddings clustering to reduce token representations` `# Example code would use clustering libraries (e.g., KMeans from sklearn) to group similar tokens`	Groups token representations into clusters for better processing.	Optimizes token handling, reduces computation by grouping similar tokens, and reduces model complexity.

Dynamic computation	`# Implement Mixture of Experts (MoE) in the model architecture` `# Use expert gating mechanism to activate only part of the model`	Activates only a subset of the model based on the input complexity.	Improves efficiency by utilizing only necessary parts of the model, reduces computation overhead.

Low-rank factorization	`import torch.nn as nn`
	`U, S, V = torch.svd(model.transformer.embeddings.word_embeddings.weight)`
	`# Apply low-rank approximation to embeddings`
		Decomposes weight matrices (like token embeddings) into lower-rank approximations.	Reduces memory usage and speeds up computations while maintaining performance.

Attention head sharing	`model.config.num_attention_heads = 4  # Reducing the number of heads` `# Model fine-tuning will use this configuration`	Reduces the number of attention heads in the model's architecture.	Decreases model size and increases processing speed by reducing the number of attention heads.
Early stopping	`from transformers import TrainerCallback` `trainer.add_callback(TrainerCallback())`	Stops training early if the model's performance stops improving.	Saves computational resources and avoids overfitting by halting training once convergence is reached.

(*continued*)

Table 4-4. (*continued*)

Optimization Technique	Code Snippet	Short Description	Benefit
| Mixed precision training | ```
from torch.cuda.amp import
autocast, GradScaler
with autocast():
    loss = model(input_ids,
labels=labels)
scaler.scale(loss).backward()
``` | Uses lower precision (e.g., FP16) during training to reduce memory usage. | Speeds up training and reduces memory usage without sacrificing significant model accuracy. |
| Model quantization-aware training (QAT) | ```
from torch.quantization import
prepare_qat, convert
model.train()  # Enable QAT
model = prepare_qat(model)
model = convert(model)
``` | Prepares the model for quantization by incorporating quantization during training. | Maintains model accuracy while enabling efficient deployment on resource-constrained devices. |

4.4.1.1 Lower Memory Consumption and Faster Inference

SLMs are designed with an optimized architecture that reduces their memory footprint and improves inference speed compared to larger models. This is crucial when deploying NER models to resource-constrained environments, such as mobile devices and edge computing, where computational resources (e.g., CPU, memory, and battery life) are limited.

Optimized architectures:

- **TinyGPT and DistilGPT:** Reduce the number of layers, attention heads, and model parameters while maintaining a substantial portion of the model's language generation capabilities. This leads to faster inference and reduced memory consumption, making them ideal candidates for NER tasks on edge devices or low-latency environments.

- **ELECTRA:** With its unique replaced-token detection strategy, ELECTRA allows more efficient pretraining by using all tokens to make predictions, which not only reduces training time but also lowers memory consumption during inference. This makes it a great choice for resource-constrained settings.

Memory optimization techniques:

- **Quantization:** This involves converting model weights to lower precision (e.g., from 32-bit floating point to 8-bit integers). Quantization reduces memory usage and improves inference speed with minimal impact on accuracy.

- **Pruning:** This technique removes less important neurons or parameters from the model that have little effect on performance. It reduces the model size and enhances inference speed.

- **Knowledge distillation:** Involves transferring knowledge from a larger "teacher" model (e.g., GPT-2) to a smaller "student" model (e.g., DistilGPT). Distilled models use fewer resources while performing nearly as well as larger models on many NLP tasks.

4.4.1.2 Cost Savings and Efficiency

The key advantage of SLMs is their ability to reduce computational costs while maintaining reasonable performance on tasks such as NER. TinyGPT and DistilGPT are smaller versions of GPT models, and ELECTRA is designed with a more efficient training process, which leads to lower inference time and reduced memory requirements.

SLMs often operate in resource-constrained environments where larger models cannot be deployed due to their high memory and compute requirements. These models allow for:

- **Lower latency:** Faster inference speeds make them ideal for real-time applications like chatbots, virtual assistants, and mobile apps.

- **Lower memory usage:** They can be deployed on devices with limited memory (e.g., smartphones, embedded systems, edge devices).

- **Cost savings:** With smaller models, you save on cloud compute costs when running models for large-scale production workloads.

4.4.1.3 Tradeoffs and Challenges

While SLMs bring significant cost savings, they come with certain tradeoffs:

- **Reduced accuracy:** Compared to larger models like GPT-3 or BERT, SLMs may have reduced accuracy in complex tasks like NER, especially on challenging or ambiguous inputs. The reduced model size limits their ability to capture nuances and intricate relationships in text.

- **Model complexity:** Training or fine-tuning these models for specific tasks (e.g., NER) might not always yield results as strong as those obtained from larger models. However, with careful training strategies like distillation and quantization, their performance can be enhanced.

- **Generalization:** Smaller models may struggle with generalization on unseen data or out-of-distribution samples due to their limited capacity.

CHAPTER 4 NAMED ENTITY RECOGNITION (NER) WITH LLMS AND SLMS

Despite these challenges, the tradeoff between model size, accuracy, and performance is often acceptable when real-time execution and resource constraints are prioritized.

4.4.2 Use Case: Small LLMs for Real-Time NER

Small LLMs are particularly suited for real-time NER tasks on mobile devices or edge computing platforms, where processing power and memory are limited. For example:

- **Mobile applications:** Real-time text recognition for entity extraction in apps that require low latency, such as news aggregators, personal assistants, and healthcare apps.

- **Edge computing:** Deploying NER models on IoT devices or edge servers for quick, on-device text processing without needing to send data to a cloud server, thus reducing latency and increasing privacy.

- **Chatbots and virtual assistants:** TinyGPT and DistilGPT can perform NER efficiently in interactive environments like customer support or personal assistants.

SLMs like TinyGPT and DistilGPT offer great promise for such use cases, enabling fast processing while maintaining satisfactory performance for real-time operations.

4.4.3 Fine-Tuning SLMs for NER Tasks

To deploy SLMs for NER tasks, models need to be fine-tuned to perform well on specific datasets. As an example of how to fine-tune DistilGPT for NER using the Hugging Face Transformers library, refer to this GitHub link for the full code: https://github.com/anvcse562/ModernNLP-LLMSLM/blob/main/Ch04/ch4.4/4.4.1.3-TinyBERT_NER.ipynb.

This example demonstrates fine-tuning a DistilBERT model (similar to DistilGPT) on a NER task using the CoNLL-03 dataset. The tokenization process ensures that the labels are aligned with the tokenized words, and the fine-tuning proceeds with a `Trainer` object, where we specify the learning rate, batch size, and other relevant parameters.

4.5 Industry Use Cases

4.5.1 Resume Parsing

We run this use case to extract key information (entities, skills, experience, and alignment with job postings) from resumes and online profiles to automate tasks such as talent matching, resume tailoring, and interview preparation.

This notebook uses a `BERT-based NER pipeline` (`dbmdz/bert-large-cased-finetuned-conll03-english`) from Hugging Face to extract named entities from resumes. The focus remains on efficient use of pretrained models for downstream NLP tasks.

Code highlights:

- **NER pipeline:** The code uses a pretrained NER model to identify entities (e.g., names, companies) and custom regex to extract specific skills from the resume text.

- **Resume parsing:** PDFs are read using `PyPDF2`, with parsed text passed through the NER model for structured extraction.

- **LangChain and CrewAI agents:** Four distinct agents (`Tech Job Researcher`, `Personal Profiler`, `Resume Strategist`, `Interview Preparer`) work together using a task pipeline to:
    - Analyze job postings via web search.
    - Parse and analyze the candidate's resume, GitHub, and statement.
    - Generate a tailored resume.
    - Prepare interview materials aligned to the job.

- **Search and LLM integration:** The agents use DuckDuckGo search and OpenAI for summarization, profiling, and content generation, coordinated via LangChain's `Tool` abstraction.

- **Orchestration:** The entire workflow is driven by a `Crew` object executing tasks sequentially, enabling an automated pipeline from resume ingestion to interview prep.

Refer to the GitHub code here: https://github.com/anvcse562/ModernNLP-LLMSLM/blob/main/Ch04/ch4.5/resumeparsing.ipynb(Resume parsing).

4.5.2 Legal Document Analysis

We will run this use case to extract key information (entities, clauses, relationships) from legal documents for tasks like contract review, due diligence, and legal research.

The ELECTRA SLM model is explored for this task, as it offers a good balance of performance and efficiency compared to other models like BERT.

Benefits of tailored SLMs:

- **Efficiency:** SLMs require fewer computational resources than larger models, leading to faster inference and lower costs.

- **Domain specificity:** Fine-tuning SLMs on a legal document dataset allows them to learn the nuances of legal language and improve accuracy for specific tasks.

- **Reduced risk:** SLMs are generally less prone to generating incorrect or misleading information (hallucinations) than larger, more complex models.

Quantization benefits:

- **Further speed and efficiency:** Quantization reduces model size and speeds up inference, making the analysis even faster and more resource-efficient.

- **Deployment flexibility:** Quantized models can be deployed on a wider range of devices, including edge devices and mobile platforms.

In essence, the code fine-tunes an ELECTRA model for legal NER, quantifies it for efficiency, and then uses it to extract entities from legal documents.

Refer to GitHub 4.5 for the full code: https://github.com/anvcse562/ModernNLP-LLMSLM/blob/main/Ch04/ch4.5/Legal_Document_Analysis.ipynb.

4.5.3 Customer Data Extraction

We will run this use case to extract key customer information (names, IDs, transaction details, dates, totals) from scanned or digital PDFs. This supports tasks like customer onboarding, KYC compliance, invoice processing, and analytics.

This notebook does not use a transformer-based SLM directly. Instead, it combines rule-based text extraction using Python tools like `pdfplumber` with regex and custom logic, ideal for structured and semi-structured documents.

Code highlights:

- **PDF text extraction:** Uses `pdfplumber` to extract text from multiple pages in a PDF. This method retains the layout structure better than many OCR tools.

- **Batch processing:** Includes a utility to extract from multiple PDFs at once, storing each file's content in a structured dictionary.

- **Regex-based parsing:** Although not all parsing logic is shown in the early cells, this typically involves identifying entities like customer name, account or invoice number, date of transaction, and amount or totals.

- **Logging and robustness:** Includes logging for each step—useful for auditing and debugging in enterprise workflows.

- **Data export and visualization (expected):** Later cells likely include logic for JSON export or visual summaries using `matplotlib`, based on import statements.

Refer to GitHub 4.5 for the full customer data extraction code: https://github.com/anvcse562/ModernNLP-LLMSLM/blob/main/Ch04/ch4.5/Customer_Data_Extraction.ipynb.

4.6 Summary

In this chapter, you explored NER using LLMs and SLMs. You learned how to apply these models to extract key information from unstructured text, identifying entities such as names, locations, organizations, and dates. You also gained practical knowledge on deploying these models using AWS and open-source tools like Hugging Face Transformers and explored industry use cases such as resume parsing, legal document analysis, and customer data extraction.

You should now have a strong understanding of NER and model deployment. The next chapter dives into another key NLP task: sentiment analysis.

CHAPTER 5

Sentiment Analysis with LLMs and SLMs

This chapter explores sentiment analysis using LLMs like GPT-4, DeBERTa, and T5, and SLMs like DistilBERT, TinyBERT, and MobileBERT. It compares their performance in various sentiment analysis scenarios, including understanding nuanced sentiment, handling context, and classifying sentiment in real time.

We focus on real-world applications such as brand monitoring, social media sentiment analysis, and customer satisfaction insights. We also cover implementation strategies, including AWS services (Amazon Comprehend and SageMaker) for scalable deployments, and Hugging Face Transformers for open-source solutions.

By the end of this chapter, you will understand when to use LLMs for deep semantic understanding in complex sentiment analysis tasks, and when to choose SLMs for efficient, cost-effective performance in simpler applications.

This chapter covers the following topics:

- **Introduction to sentiment analysis:** Importance of sentiment extraction, sentiment types, and model complexity/performance tradeoffs.

- **Implementation on AWS:** Using Amazon Comprehend for pretrained sentiment analysis models, training custom models on SageMaker, and deploying SLMs for efficient, real-time sentiment extraction.

- **Open-source implementation:** Fine-tuning LLMs and SLMs with Hugging Face Transformers, optimizing training data, and evaluating model performance.

- **Low-level LLM exploration/small LLM cost-effective alternatives:** Cost-effectiveness of SLMs in sentiment analysis, accuracy/efficiency tradeoffs, and optimal use cases.

- **Industry use cases:** Sentiment analysis in brand monitoring, social media sentiment analysis, and customer satisfaction insights.

5.1 Understanding Sentiment Analysis

The previous chapter explored how even small-scale language models can efficiently perform tasks like Named Entity Recognition (NER) at low cost. Building on that foundation of leveraging SLMs for practical NLP, this chapter turns to another crucial text analysis task: sentiment analysis. *Sentiment analysis* (also called *opinion mining*) is the process of determining the emotional tone or subjective opinion expressed in natural language. This section delves into what sentiment analysis entails, why it's challenging, and how LLMs (such as GPT-4, DeBERTa, T5) and SLMs (like DistilBERT, TinyBERT, MobileBERT) can both be used to classify sentiment. We use customer reviews as a running example to illustrate concepts and code, bridging theory with practice in an academic yet practical tone. By the end of this section, you will understand the fundamentals of sentiment analysis and how to implement it with modern language models, setting the stage for the next section, which focuses on deploying these solutions using AWS tools.

5.1.1 Sentiment Analysis: Definition and Significance

Sentiment analysis involves classifying text based on the expressed sentiment (e.g., positive, negative, or neutral). This subsection defines sentiment analysis and highlights its importance in modern NLP applications, using customer reviews as a motivating example.

At its core, *sentiment analysis* is the task of identifying and categorizing opinions or emotions in a piece of text. Typically, this means determining whether a text's sentiment is positive, negative, or neutral. For example, a product review stating, "I absolutely love this phone's camera quality!" would be classified as positive, whereas "The battery dies too quickly, very disappointed." would be negative. Sentiment analysis can be binary (positive/negative), multi-class (e.g., positive/neutral/negative, or a five-star rating scale), or even charted using a regression score (e.g., sentiment intensity from −1 to +1). It can be applied at different levels of granularity: a whole document (overall sentiment of a review), a sentence, or even an aspect (sentiment toward specific features within the text, like sentiment about "camera quality" vs. "battery life" in a phone review).

In today's data-driven world, understanding sentiment is immensely valuable. Businesses and researchers use sentiment analysis to gauge public opinion and consumer satisfaction. Some real-world applications include:

- **Customer feedback analysis:** Companies analyze product reviews and support tickets to determine customer satisfaction. For instance, thousands of app store reviews can be automatically labeled to determine if users feel good or bad about a new feature.

- **Social media monitoring:** Brands monitor tweets and Facebook posts to measure audience sentiment toward marketing campaigns or breaking news. This helps in reputation management and marketing strategy adjustments.

- **Market research:** Financial analysts assess news articles or forum discussions for sentiment about companies to inform stock trading decisions (assuming positive news correlates with stock upticks, etc.).

- **Healthcare and surveys:** Sentiment analysis can also be applied to patient feedback or survey responses, to quickly summarize emotional tones (e.g., detecting frustration vs. satisfaction in survey comments).

In all these cases, the ability to quickly distill how people feel from text data provides a competitive and operational advantage. For example, an e-commerce company might automatically flag a sudden surge of negative reviews for a product release, prompting an immediate investigation. Without automated sentiment analysis, they would have to manually read through large volumes of text—an impractical solution at scale.

Earlier sentiment analysis used lexicons or classical machine learning, which struggled with nuance. Modern methods use pretrained language models, greatly improving accuracy by capturing context and subtle cues (e.g., negation, slang). Fine-tuning or prompting LLMs leads to more reliable results. The following sections explore the challenges of sentiment analysis and how different language model sizes address them.

5.1.2 Challenges in Analyzing Sentiment

Accurately detecting sentiment is not always straightforward. This subsection outlines key challenges—such as sarcasm, context dependency, and domain-specific language—that make sentiment analysis difficult, especially for automated models.

Despite its apparent simplicity, sentiment analysis faces several linguistic and practical challenges that require sophisticated modeling:

- **Sarcasm and irony:** Sarcasm can invert the literal sentiment of words. For example, "Yeah, great job on delivering my package a week late!" is clearly negative, despite the positive phrasing ("great job"). Models relying on literal word cues may misclassify such sarcastic remarks as positive. Detecting sarcasm requires understanding tone or context—something humans infer from voice or situation, but that is absent in plain text. It's like reading between the lines—the model must infer that the writer means the opposite of what's said. This remains a major challenge in sentiment analysis.

- **Negation handling:** Negation can flip sentiment, and it's tricky to get right. A phrase like "not bad" means good, while "not happy" means unhappy. A naive model might focus on "happy" and tag it as positive. More complex structures like double negation or softeners—e.g., "I don't dislike this"—imply mild positivity. Simple models often misinterpret phrases like "It was not unpleasant," mistaking them as negative due to the word "unpleasant." Robust sentiment models must learn these linguistic nuances.

- **Context and ambiguity:** The same phrase can express different sentiments in different contexts. Words with multiple meanings add complexity: "critical" could be negative (critical feedback) or positive (critical acclaim). Similarly, "the plot was terribly good" uses "terribly" as a positive intensifier. Longer texts introduce further challenges: "The first half was okay, but the rest was awful" might sound positive in isolation, but context reveals otherwise. Accurate sentiment analysis requires understanding both sentence-level and broader contextual cues.

- **Domain-specific language and slang:** Sentiment expressions vary across domains. A word like "volatile" is bad in a product review but neutral or even expected in a financial context. Slang adds another layer: "This burger slaps!" means it's excellent, but an out-of-domain model may miss the meaning. Domain adaptation or fine-tuning on relevant data is essential for models to learn vocabulary, tone, and idioms specific to contexts like movie reviews, medical feedback, or social media.

- **Intensity and nuance:** Sentiment isn't always clearly positive or negative—many texts express mixed or nuanced views. For instance, "The camera quality is fantastic, but the phone is overpriced." A simple classifier may struggle to assign a single label. Even human annotators may disagree on subtle or mixed sentiment. This subjectivity sets a natural ceiling on model performance, especially when the ground truth itself is ambiguous.

- **Data and multilingual challenges:** Most sentiment models are trained on English, limiting their effectiveness in other languages. Sentiment expression is often language-specific. Emojis and emoticons—common in social media—also carry sentiment: a ☺ emoji can shift the tone of a message. While advanced models are trained to interpret emojis, many still fall short. Applying a model trained on formal data (e.g., movie reviews) to informal text (e.g., tweets) requires careful data curation or additional fine-tuning.

These challenges mean that a sentiment analysis system must have a deep understanding of language—far beyond simple keyword spotting. LLMs, with their vast training on diverse data, have an edge in handling many of these issues (for example, an LLM may have seen countless sarcastic remarks and learned subtle cues). However, even smaller fine-tuned models have made great strides by learning from large, labeled datasets. The next section explores how the size of a model—large versus small—influences its capabilities in sentiment analysis, and how recent research has enabled smaller models to punch above their weight in understanding sentiment.

5.1.3 LLMs in Sentiment Analysis

LLMs like GPT-4, DeBERTa, and T5 are powerful tools for sentiment analysis. This section discusses how LLMs approach sentiment tasks, their advantages (such as understanding nuance and performing well with minimal task-specific data), and considerations when using them for classifying text like customer reviews.

LLMs refer to transformer-based models with a very large numbers of parameters (often hundreds of millions to billions) trained on massive corpora. Examples include OpenAI's GPT-4, the T5 (Text-to-Text Transfer Transformer) series from Google, BERT-derived large models like DeBERTa (by Microsoft), and others. These models have a high capacity to learn complex language patterns. For sentiment analysis, LLMs can be used in two main ways: (1) Zero-shot or few-shot classification via prompting, and (2) fine-tuning on a sentiment dataset.

LLMs shine in scenarios requiring understanding of subtle context and rare linguistic phenomena. Because they have been trained on extremely diverse data (from Internet forums to literature), they often already "know" about sarcasm, idioms, and other nuances of sentiment. For instance, GPT-4 has enough world knowledge and language understanding to recognize a sarcastic sentence or an unusual idiom and infer the correct sentiment without being explicitly trained on that specific example. In fact, state-of-the-art LLMs like GPT-4 or large BERT variants can achieve very high accuracy on sentiment tasks. OpenAI's GPT-series models demonstrated that with appropriate prompts, a large model can classify sentiment with near-human accuracy in many cases, even with no additional training. This is extremely useful when you have little labeled data: you can ask GPT-4 in plain English "Is the following review positive or negative?" and it will usually respond correctly, leveraging its extensive pretraining. Such zero-shot or few-shot learning was virtually impossible with smaller models in the past.

Moreover, fine-tuned large models currently top most sentiment analysis benchmarks. For example, Microsoft's DeBERTa (a large transformer with improved attention mechanisms) achieved state-of-the-art results on the Stanford Sentiment Treebank (SST-2) benchmark. Likewise, Google's T5-11B (a model with 11 billion parameters) has been reported as achieving top accuracy on SST-2 when fine-tuned. These models can correctly classify even very tricky cases that confuse simpler models. They handle contextual sentiment (e.g., understanding the whole paragraph's tone) and intra-sentence nuances, and can even detect shifts in sentiment within a text. An LLM can identify that a customer review starting with "I had high hopes, but…" signals a pivot from positive to negative. Thanks to their sheer size and training, they capture many linguistic patterns automatically.

It's also worth noting that LLMs like GPT-4 or T5 are often versatile. T5, for instance, treats every NLP task (including sentiment classification) as a text-to-text problem: you give it a review as input and it can generate a label like "positive" or "negative" as output (after fine-tuning). This unified text-to-text approach means the same model architecture can be used flexibly for classification by phrasing it as a generation task. In practice, researchers have fine-tuned T5 to output words like "positive" or "negative" for sentiment, and it achieves excellent results.

The power of LLMs comes with significant resource requirements. GPT-4, for example, is only accessible via cloud API and runs on specialized hardware due to its size (on the order of ~170 billion parameters, although exact number is not public). Fine-tuning such a model on new data is often not feasible for most users; instead, you would leverage its zero-shot abilities or use prompt-based few-shot learning. Other large models like DeBERTa-large or T5-XXL (with billions of parameters) can be fine-tuned if you have a high-end GPU or TPU and enough memory. They also require careful handling to avoid overfitting when data is limited, although techniques like prompt tuning or parameter-efficient fine-tuning (e.g., adapters) are emerging to address this.

Another aspect is inference speed. Classifying the sentiment of millions of tweets with an LLM might be slow and costly. For example, running a 1.5 billion-parameter model on each text can introduce latency that is unacceptable in real-time systems (like a live social media dashboard). Therefore, in practice, LLMs are used when top-notch accuracy is required and resources permit, or when doing analysis in batches offline. They're also used when the language is particularly complex. For instance, analyzing sentiment in a literary text or a very nuanced customer feedback might benefit from an LLM's deeper understanding.

To summarize, LLMs bring unparalleled accuracy and robustness to sentiment analysis, handling tricky language cases that simpler models might miss. They are like the "expert linguists" of the NLP world—given enough computational budget, they will generally produce the most accurate sentiment classifications, even with minimal task-specific data. However, not every application can afford to use an LLM, which leads to consider smaller models and how they too have been optimized for sentiment tasks.

5.1.4 SLMs in Sentiment Analysis

SLMs are lightweight counterparts to LLMs, designed to run efficiently while still providing strong performance. This section examines examples of SLMs (DistilBERT, TinyBERT, MobileBERT) and how they perform sentiment analysis. It discusses their advantages in terms of speed and deployment, and explains how they manage to retain high accuracy on tasks like customer review classification despite their size.

While large models push the boundaries of accuracy, SLMs offer a more practical solution for many real-world sentiment analysis systems. SLMs are models with significantly fewer parameters (often in the tens of millions, rather than billions) that have been compressed or distilled from larger models. Notable examples include DistilBERT, TinyBERT, and MobileBERT, which are all derived from the original BERT architecture through various compression techniques.

The primary driving force for SLMs is the need for speed and efficiency. Techniques like knowledge distillation create a smaller "student" model that learns to emulate a larger "teacher" model's behavior. For instance, DistilBERT was created by distilling BERT base (110 million parameters) down to a model with about 40% fewer parameters (roughly 66 million) while preserving most of its language understanding capability. Similarly, TinyBERT distilled BERT into an even smaller form (TinyBERT with four layers has on the order of ~30 million params) and managed to retain over 96% of the teacher model's performance on GLUE benchmarks. MobileBERT took a different approach by redesigning BERT's architecture for efficiency, resulting in a ~25 million parameter model optimized for smartphones, with minimal loss in accuracy compared to BERT-base.

These models are trained on large corpora like their bigger counterparts and often further fine-tuned on task-specific data (e.g., a dataset of labeled positive/negative reviews). The surprising finding from research is that a well-distilled small model can achieve accuracy very close to the original large model. For example, DistilBERT retains about 95% of BERT's performance on tasks like sentiment classification, meaning if BERT got 92% accuracy on a sentiment test, DistilBERT might get around 90% with half the size. TinyBERT's six-layer version was reported to match BERT-base performance on GLUE sentiment tasks. These results indicate that much of the essential knowledge for sentiment analysis (like understanding polarity of words, context handling for negations, etc.) can be compressed into a smaller network.

The key benefit of SLMs is efficiency. They require far less memory and compute, making them ideal for production settings where resources or costs are constrained.

For example, running GPT-4 on a mobile phone is unrealistic—but models like MobileBERT were designed for this exact use case. MobileBERT can run inference in ~60 milliseconds on a modern smartphone, enabling real-time sentiment analysis on-device.

Similarly, DistilBERT can operate efficiently on standard CPU servers, making it well-suited for high-throughput tasks like labeling thousands of tweets per second—without relying on GPUs.

SLMs also offer faster training times. Fine-tuning a smaller model on a custom dataset (e.g., healthcare reviews) typically converges more quickly and costs less. They are easier to deploy across edge devices or in lightweight microservice architectures, where using a large model for every request would be impractical.

Naturally, there's a tradeoff. SLMs may be slightly less accurate than LLMs on nuanced tasks. A small four-layer model might miss sarcasm or subtle shifts in tone that a larger model could detect. However, thanks to knowledge distillation, the gap is often smaller than expected.

On standard sentiment tasks—such as product or movie reviews—models like DistilBERT and MobileBERT achieve accuracy within a few points of larger models like BERT-large, but with much faster runtime. In one benchmark, DistilBERT reached ~99% of BERT's accuracy on IMDb sentiment classification while running significantly faster.

These models also inherit strengths from their larger counterparts. DistilBERT, distilled from BERT, handles negation and context well. TinyBERT, with its two-stage distillation (general and task-specific), captures both broad and targeted sentiment cues. When fine-tuned on datasets like Amazon reviews, these models perform well—missing only the most ambiguous or sarcastic examples.

5.1.5 LLMs vs. SLMs: A Comparative View

The decision of which to use often comes down to the application requirements. If you need to analyze sentiment on a device or at low cost with slightly less nuance, an SLM is the go-to. If maximum accuracy and the ability to understand subtle context (and perhaps handle more diverse language inputs) is crucial, and you have the resources, an LLM might be justified. To make this comparison clearer, let's take a side-by-side look at LLMs versus SLMs for sentiment analysis.

LLMs and SLMs can both be used for sentiment analysis, but they differ in capability, resource requirements, and typical use cases. This section compares LLMs and SLMs head-to-head, summarizing their pros and cons in Table 5-1 and discussing scenarios where one may be preferred over the other.

CHAPTER 5   SENTIMENT ANALYSIS WITH LLMS AND SLMS

Table 5-1 summarizes some key differences and examples, focusing on models already discussed. Choosing the right model depends on the task requirements: for instance, a cloud service analyzing a complex news article might use an LLM, while a real-time mobile app uses an SLM.

Table 5-1. *Comparison of LLMs and SLMS for Sentiment Analysis*

Model (Example)	Type	Size (Parameters)	Notable Characteristics for Sentiment
GPT-4 (OpenAI)	LLM (Transformer)	~170B (estimate)	– Extremely high accuracy, understands subtle context and sarcasm. – Can perform sentiment classification with zero-shot prompts (no training needed). – Cons: Requires cloud API or specialized hardware; high inference cost/latency.
DeBERTa (Large)	LLM (Transformer)	400M – 1.5B (various)	– State-of-the-art fine-tuned sentiment performance on benchmarks. – Handles complex language; can surpass human performance in some benchmarks. – Cons: Large memory footprint; needs GPU for real-time use.
T5-11B (Google)	LLM (Seq2Seq)	11B	– Can be fine-tuned to *generate* sentiment labels (text-to-text format). – Achieved top results on SST-2 sentiment task. – Cons: Very heavy model, typically used on TPU/GPU; not deployable on edge devices.

(*continued*)

Table 5-1. (*continued*)

Model (Example)	Type	Size (Parameters)	Notable Characteristics for Sentiment
DistilBERT (Hugging Face)	SLM (Distilled)	66M	— Distilled from BERT; retains ~95% of BERT's accuracy on GLUE tasks. — Fast inference (about 2× speed of BERT); well-suited for real-time classification. — Cons: May miss some nuance that larger models catch (e.g., very subtle sarcasm).
TinyBERT (four-layer)	SLM (Distilled)	~30M	— Aggressively compressed BERT; ~96.8% of BERT's performance on GLUE. — 7.5× smaller and 9× faster than BERT base, excellent for mobile/edge deployment. — Cons: Slight accuracy drop in complex cases; requires careful fine-tuning for best results.
MobileBERT (Google)	SLM (Specialized)	25M	— Architected for mobile devices; 4× faster than BERT base, ~99% of its accuracy. — Can run on a smartphone (~60 ms per inference for 128 tokens). — Cons: Still needs quantization/pruning for older phones; limited by BERT-base baseline performance.

In practice, many applications employ a hybrid strategy. One common approach is to use SLMs for most routine sentiment analysis tasks (due to their low cost and sufficient accuracy), and reserve LLMs for the particularly hard cases or for offline analysis that demands extra nuance. For example, a customer feedback pipeline might automatically tag the obvious positive/negative comments using a DistilBERT-based classifier running cheaply on servers. However, if the system encounters an ambiguous or potentially sarcastic piece of text (or perhaps a very important piece of text, like a CEO's email that needs analysis), it could send it to a larger model or an ensemble for a second opinion.

To put the difference in perspective: using an LLM versus an SLM is like consulting a professor versus a well-trained undergraduate. The professor (the LLM) has deep knowledge and can catch subtle implications—perfect for tricky sentiment passages—but access is limited and may be excessive for simple tasks. The undergraduate (the SLM) is fast, efficient, and gets most answers right, but might miss rare nuances or edge cases. Both are valuable—the key is aligning the model to the task.

Having explored theory and tradeoffs, the next section grounds these ideas with a hands-on example. It walks through using a modern transformer model to analyze customer review sentiment—demonstrating both inference (using a pretrained model) and fine-tuning (adapting a model to new data) with practical code.

5.1.6 Example: Classifying Customer Reviews with Transformers

To illustrate sentiment analysis in practice, this section uses a real example of classifying customer reviews. We demonstrate how to load a pretrained sentiment model and run inference on sample reviews, then show you how you can fine-tune a model (using Hugging Face Transformers) on custom data. The example ties together concepts from previous sections, showing an implementation step-by-step.

Imagine that you have a collection of customer product reviews and you want to automatically determine which reviews are positive and which are negative. This example uses the Hugging Face Transformers library to quickly get a sentiment analysis model working. Hugging Face provides convenient tools and pretrained models that make this task accessible.

First, you load a pretrained model for inference. One of the simplest ways to start is to use a pretrained sentiment analysis pipeline. Hugging Face has a pipeline for sentiment analysis that by default loads a model fine-tuned on the Stanford Sentiment Treebank (SST-2), which classifies text as positive or negative.

Refer to this GitHub link for the full code to see how you can use it: https://github.com/anvcse562/ModernNLP-LLMSLM/blob/main/Ch05/Ch5.1/classify-customer-reviewes-5.1.6.py

Running this code will output something like the following.

Review: I bought this laptop two weeks ago, and I am absolutely delighted with its performance!

Predicted Sentiment: POSITIVE (score=0.999)

Review: The phone case broke after two days... really poor quality.
Predicted Sentiment: NEGATIVE (score=0.998)
Review: Not bad at all – the product was actually better than I expected.
Predicted Sentiment: POSITIVE (score=0.879)

You can see the model correctly identified the first review as positive (with a very high confidence score close to 1.0), the second as negative, and the third ("Not bad at all...better than expected") as positive. Note that in the third review, the phrase "Not bad at all" was understood in context to be positive. A simple keyword approach might have been tripped up by the word "bad," but this transformer-based model handles the negation appropriately. This highlights how these models leverage context to determine sentiment.

Behind the scenes, the pipeline we used is likely employing a model like DistilBERT fine-tuned on SST-2. It tokenizes each review, feeds it through the transformer, and the model's output layer (a classification head) produces a probability for each class (positive/negative). The pipeline abstracts these steps, but it's essentially doing what we discussed earlier: using a pretrained language model that has been taught to classify sentiment.

You can also fine-tune a model on custom data. Pretrained models work well out-of-the-box, but sometimes you have a specific domain of reviews and you want to fine-tune a model to get even better performance on that domain. For example, imagine your reviews are all about a very niche product, or in a mix of English and Spanish. Fine-tuning allows the model to adapt to your specific data distribution.

To demonstrate the steps you would take to fine-tune an SML (say, DistilBERT) on a custom sentiment dataset using Hugging Face's `Trainer` API, refer to this GitHub link for the full code: https://github.com/anvcse562/ModernNLP-LLMSLM/blob/main/Ch05/Ch5.1/fine-tune-distilBERT.py.

In this code:

- We loaded `distilbert-base-uncased` as our starting point. This is a pretrained DistilBERT model not yet fine-tuned for any specific task. We specify `num_labels=2` to adapt the classification head to binary sentiment classification (the model will initialize a fresh classification layer).

- We created a small dataset of `train_texts` and corresponding `train_labels`. In practice, you would have hundreds or thousands of labeled examples rather than four. We then tokenized the texts using the same tokenizer that was used to train DistilBERT (ensuring consistency in how text is represented as input IDs).

- We wrapped the tokenized data in a custom `ReviewsDataset`, which can be used by the `Trainer`. This class simply provides an interface to get tokens and labels by index.

- We set up `TrainingArguments`. This includes things like how many epochs to train, batch size, learning rate, and so on. We also specify an `output_dir` where the model checkpoints will be saved.

- Finally, we create a `Trainer` object with our model, data, and arguments and call `trainer.train()` to start fine-tuning.

During training, the model will adjust its weights to better fit the specific data. Essentially, it will learn that "absolutely fantastic" corresponds to the positive class (label 1), "terrible experience" to the negative class (0), and so on. After a few epochs, if the dataset were larger, we'd expect the model to achieve high accuracy on classifying similar reviews. We could then use `trainer.evaluate()` on a validation set to check performance and use `trainer.save_model()` to save the fine-tuned model for later use.

After fine-tuning, using the model is similar to the first inference example, but you would load your fine-tuned model instead of the generic one.

Refer to this GitHub link for the full code: https://github.com/anvcse562/ModernNLP-LLMSLM/blob/main/Ch05/Ch5.1/run-fine0tune-distil-bert.py.

This code outputs a sentiment prediction using the newly trained model. A fine-tuned model often captures domain-specific nuances—especially if the training data includes unique slang or product types specific to your users.

In this example, we saw both off-the-shelf sentiment analysis and customization via fine-tuning. Many developers start with a pretrained model (as we did using a pipeline) because it delivers instant, often sufficient results. Fine-tuning is done when domain-specific data is available or when extra accuracy is needed. Thanks to libraries like Transformers, fine-tuning a top-tier model can be done in just a few lines of code.

It's remarkable that what once required training from scratch on massive datasets can now be done in minutes. This is the power of *transfer learning*—the model's general language knowledge from pretraining can be adapted quickly to new tasks.

Before wrapping up, it's worth emphasizing *evaluation*. Metrics like Accuracy, F1 score, and ROC-AUC (for balanced binary classification) help verify that the model isn't just defaulting to "positive" or missing key patterns. Bias checks are also crucial—does the model consistently misread certain phrasings or user groups?

Training a model is just the start. The next step is deployment—getting the model into production, often via a cloud service or integrated API. We cover that next.

5.1.7 Next Steps: Implementing Sentiment Analysis on AWS

Having explored sentiment analysis and how to build models for it, the next logical step is *deployment*. This section sets the stage for the "Implementing Financial Sentiment Analysis with AWS Services" section, which covers how to implement sentiment analysis using AWS tools—bridging your understanding of models with real-world cloud deployment.

You've now seen how sentiment analysis works using LLMs and SLMs. The next question is: how do you deploy this in a production setting? In most cases, this means leveraging cloud platforms to host and serve models—or using pre-built sentiment APIs.

Amazon Web Services (AWS) offers several tools for this. For a no-code option, Amazon Comprehend is a managed NLP service that performs sentiment analysis (among other tasks) right out of the box. For more control, Amazon SageMaker allows developers to deploy custom models—like those we fine-tuned earlier—on scalable infrastructure. AWS also supports data pipelines and streaming integrations to apply sentiment models to incoming text data (e.g., live product reviews) in real time.

The next section dives into the hands-on side of implementing sentiment analysis with AWS: setting up endpoints, using Lambda or SageMaker for inference, and comparing custom deployments with AWS's managed services like Comprehend.

By connecting model development with deployment strategies, you'll gain the practical tools to build end-to-end sentiment analysis systems—from choosing the right model to making its insights actionable. Let's now move into the cloud and explore how to bring sentiment analysis to life with AWS.

5.2 Implementing Financial Sentiment Analysis with AWS Services

This section focuses on a code-first practical implementation of sentiment analysis using AWS, tailored to financial text (e.g., stock-related tweets or news headlines). We build a solution that leverages LLMs and SLMs through AWS services. The architecture will combine managed services and custom models—including Amazon Comprehend, AWS Lambda, Amazon SageMaker, and Amazon API Gateway—to perform real-time sentiment analysis. We use a consistent financial example throughout (e.g., "Tesla stock outlook is very strong this quarter") and emphasize how an LLM-based agent could enhance the workflow by routing complex cases to the appropriate backend. Before diving into code, we present an overview of the system architecture.

5.2.1 Overview and Architecture

To handle sentiment analysis at scale, AWS provides building blocks that you can assemble into a robust architecture. The key components are as follows:

- **Amazon API Gateway:** Exposes a RESTful API endpoint for clients to submit text for sentiment analysis.

- **AWS Lambda:** A serverless function that acts as the orchestrator (and can also host an SLM model for sentiment). It receives requests from API Gateway, processes the input, and returns the sentiment result. This Lambda function can route the request to different analysis backends based on logic or on an LLM agent's decision.

- **Amazon Comprehend:** A fully managed NLP service that can be invoked by Lambda for sentiment analysis. Comprehend's pretrained model quickly classifies the text sentiment as positive, negative, neutral, or mixed (see docs.aws.amazon.com/docs.aws.amazon.com).

- **Amazon SageMaker:** A managed machine learning service used here to train and host a custom sentiment model (for example, a fine-tuned financial domain BERT model). Lambda can forward requests to a SageMaker endpoint for analysis when higher accuracy or domain-specific handling is needed.

- *(Optional)* **LLM agent routing:** An advanced component (not an AWS service by itself, but a concept) where an LLM is used to decide which backend to use. For instance, an LLM could examine the text and determine that a simple case can be handled by Amazon Comprehend or a small model, whereas a complex or ambiguous sentence should be handled by the more powerful SageMaker model. This agent logic could be part of the Lambda function.

The workflow is as follows: A client (such as a web app or data pipeline) sends a sentiment request—for example, a tweet about a stock—to an API Gateway endpoint. This triggers an AWS Lambda function, which processes the input and routes it to the appropriate backend.

In a basic setup, the Lambda can directly call Amazon Comprehend to retrieve a sentiment label. In a more advanced, hybrid design, the Lambda can include logic to dynamically choose between backends. For instance, if the input contains financial jargon or if confidence from a quick analysis is low, it can route the request to a SageMaker endpoint hosting a fine-tuned model. Otherwise, it might default to a smaller on-the-fly model or continue with Comprehend.

Once a result is obtained, the Lambda formats it into a JSON response and sends it back to the client via API Gateway. This serverless and scalable architecture ensures flexibility and resilience. API Gateway and Lambda scale automatically to handle traffic bursts (e.g., a spike in tweets), while Comprehend and SageMaker handle the NLP workloads. Crucially, this setup decouples the client interface from the underlying sentiment logic—allowing you to swap models or adjust routing without impacting the frontend.

You can implement this with a single intelligent Lambda or multiple dedicated microservices, each handling a specific path (Comprehend, embedded model, SageMaker). The following sections walk through each of these approaches with code examples.

5.2.2 Using Amazon Comprehend for Financial Sentiment Analysis

Amazon Comprehend offers pretrained sentiment analysis as a simple API call, which makes it a great starting point. Without needing to train any model, you can plug Comprehend into your pipeline to analyze financial text. Under the hood, Comprehend's models will classify sentiment as positive, negative, neutral, or mixed, and provide confidence scores for each category (see docs.aws.amazon.com). This approach is especially useful if you want quick results with minimal setup.

Before writing the code, ensure you have AWS credentials configured (e.g., via aws configure or environment variables) and have the Boto3 library installed in your Python environment. In the Lambda function (or any Python script), you can use Boto3 to call Comprehend's detect_sentiment API. Here is an example of using Comprehend to analyze a piece of financial news text:

CHAPTER 5   SENTIMENT ANALYSIS WITH LLMS AND SLMS

```
import boto3
import json

# Initialize the Comprehend client
comprehend = boto3.client('comprehend', region_name='us-east-1')
# specify region

text = "Tesla stock outlook is very strong this quarter."
response = comprehend.detect_sentiment(Text=text, LanguageCode='en')

# The response is a dictionary with sentiment and scores
sentiment = response['Sentiment']            # e.g., 'POSITIVE'
scores = response['SentimentScore']          # e.g., {'Positive': 0.95,
                                             #   'Negative': 0.01, ...}

print(f"Sentiment: {sentiment}")
print("Confidence Scores:", json.dumps(scores, indent=2))
```

When running this code, Amazon Comprehend will determine the sentiment of the input text. For instance, given "Tesla stock outlook is very strong this quarter," Comprehend is likely to return POSITIVE with a high confidence in the positive score. The output dictionary (response) would look like this example:

json
CopyEdit

```
{
  "Sentiment": "POSITIVE",
  "SentimentScore": {
    "Positive": 0.982,
    "Negative": 0.003,
    "Neutral": 0.014,
    "Mixed": 0.001
  }
}
```

In this example, the "Positive" score is ~0.982, indicating strong confidence in a positive sentiment. Comprehend also shows low scores for the other categories, which makes sense given the clearly positive language.

Amazon Comprehend automatically handles preprocessing (like tokenization or language detection). We specified `LanguageCode='en'` since our text is English. Comprehend supports many languages for sentiment analysis (see docs.aws.amazon.com), so it's useful if your financial data includes multilingual content. Note that Comprehend has a size limit of 5 KB per input text, which is usually sufficient for tweets or short news headlines, but you need to chunk longer documents.

Using Comprehend in this architecture is straightforward: the Lambda function can call `detect_sentiment` and simply relay the result. This gives a quick solution but with some limitations:

- **Domain specificity:** The model is general-purpose. It may not capture nuances of financial language (e.g., sarcasm or domain-specific terms like "overvalued", "bearish divergence").

- **Limited customization:** You cannot tweak the model's behavior except by possibly using Amazon Comprehend Custom Classification (which is a separate feature to train on your data, not covered here).

- **Costs:** Amazon Comprehend charges per character of text analyzed. For large volumes of data, cost can accumulate, although it saves ML development time.

Despite these factors, Comprehend is a great baseline and often surprisingly effective even on domain text. It could serve as a first tier in a tiered sentiment system: fast and cheap for most inputs, while more complex cases get handed off to a custom model. You will see later how to integrate such decision logic. The next section explores deploying your own model on AWS Lambda for more control.

5.2.3 Deploying SLMs on AWS Lambda for Real-Time Sentiment

While Comprehend is convenient, there are cases where you want to bring your own model. For instance, you might have a lightweight Transformer model that slightly outperforms Comprehend on financial text, or you need to run entirely within your AWS environment without calling an external API. Here, deploying an SLM like DistilBERT on AWS Lambda is an attractive option for real-time inference.

DistilBERT is a distilled version of BERT that is 40% smaller and 60% faster while retaining about 97% of BERT's language understanding capabilities (see arxiv.org). A sentiment model based on DistilBERT (for example, the DistilBERT fine-tuned on SST-2 for positive/negative classification) can achieve high accuracy with low latency. You can fine-tune such a model on financial data or even use it out-of-the-box if general sentiment is acceptable.

However, deploying Transformer models on Lambda has challenges:

- The model files and libraries can be large (hundreds of MB).

- Lambda has a deployment package size limit of 50 MB for zipped packages. The best solution is to use Lambda's container image support, which allows images up to 10 GB. This is plenty for a model and the necessary ML libraries.

- Lambda has limited memory and no GPU. But SLMs like DistilBERT can run on CPU within a few hundred milliseconds or faster if the code is optimized and memory is sufficient (512 MB to 2048 MB typically).

5.2.3.1 Packaging the Model for Lambda

We create a Docker container that includes our model and inference code. AWS provides base images for Lambda—for example, `amazon/aws-lambda-python:3.9` as a starting point for Python 3.9. This Dockerfile would roughly:

1. Start from the AWS Lambda Python base image.

2. Install required Python packages (for instance, `transformers`, `torch` for PyTorch, or `tensorflow` if using TF backend).

3. Copy the sentiment analysis code (the Lambda handler script).

4. Download or include the pretrained model weights. We can use the Hugging Face Transformers API to download the model during the image build (so that at runtime the model is already present and we avoid downloading it each invocation).

5. Set the Lambda handler appropriately (using the AWS Lambda runtime interface).

CHAPTER 5 SENTIMENT ANALYSIS WITH LLMS AND SLMS

For example, a simple Lambda handler script (let's call it app.py) might look like this:

```python
# app.py - Lambda function code for sentiment inference
from transformers import AutoTokenizer, AutoModelForSequenceClassification
import torch
import json

# Load model and tokenizer globally (to reuse across invocations)
model_name = "distilbert-base-uncased-finetuned-sst-2-english"
# or a custom model path
tokenizer = AutoTokenizer.from_pretrained(model_name)
model = AutoModelForSequenceClassification.from_pretrained(model_name)

def lambda_handler(event, context):
    # Expect input text in event (from API Gateway, it might be JSON)
    if 'body' in event:
        # If triggered by API Gateway HTTP API, text might be in the body
        body = json.loads(event['body'])
    else:
        body = event
    text = body.get('text') or body.get('message') or body
    # handle various possible keys

    # Tokenize and run inference
    inputs = tokenizer(text, return_tensors='pt')
    outputs = model(**inputs)
    # The model outputs logits for each class (positive/negative)
    scores = torch.softmax(outputs.logits, dim=1).tolist()[0]
    # convert to probabilities
    sentiment_label = 'POSITIVE' if scores[1] > scores[0] else 'NEGATIVE'
    confidence = round(max(scores), 4)

    result = {
        "text": text,
        "sentiment": sentiment_label,
        "confidence": confidence
    }
```

```
return {
    "statusCode": 200,
    "body": json.dumps(result)
}
```

In this code:

- We load the DistilBERT SST-2 model and tokenizer at global scope, so they load once when the container is cold-started. This model can classify text as positive/negative (two classes). For the financial scenario, this might be acceptable, or you could use a tri-class model (positive/negative/neutral) by fine-tuning accordingly.

- The `lambda_handler` expects the input text. If API Gateway passes the payload as JSON, we parse it (in many cases, API Gateway might pass a string in `event['body']`). We handle a couple of possible key names for flexibility.

- We run the model and then interpret the logits. DistilBERT SST-2 has two output logits (index 0 for negative, 1 for positive in the SST-2 configuration). We apply softmax to get probabilities and decide the label.

- We return a JSON with the sentiment and a confidence score (here, just the probability of the predicted class for simplicity).

5.2.3.2 Building and Deploying the Lambda

After writing `Dockerfile` and `app.py`, we would build the Docker image and push it to Amazon Elastic Container Registry (ECR). Then we create a Lambda function using that image. The Lambda will need an IAM role that allows it to be executed and (optionally) to access other services like SageMaker or logs.

5.2.3.3 Provisioning with Terraform

We can codify the deployment with Terraform (an infrastructure-as-code tool). Here is an example Terraform snippet that creates the Lambda function from an existing ECR image, an API Gateway endpoint to invoke it, and the necessary IAM role and permissions:

```
# IAM role for Lambda with basic execution permissions
resource "aws_iam_role" "lambda_role" {
  name = "lambda_financial_sentiment_role"
  assume_role_policy = jsonencode({
    "Version": "2012-10-17",
    "Statement": [{
      "Effect": "Allow",
      "Principal": { "Service": "lambda.amazonaws.com" },
      "Action": "sts:AssumeRole"
    }]
  })
}

resource "aws_iam_role_policy_attachment" "lambda_logs" {
  role       = aws_iam_role.lambda_role.name
  policy_arn = "arn:aws:iam::aws:policy/service-role/
  AWSLambdaBasicExecutionRole"
}

# Lambda function using a container image
resource "aws_lambda_function" "sentiment_lambda" {
  function_name = "financial-sentiment-lambda"
  role          = aws_iam_role.lambda_role.arn
  package_type  = "Image"
  image_uri     = "123456789012.dkr.ecr.us-east-1.amazonaws.com/financial-
                  sentiment:latest"
  timeout       = 10    # seconds, adjust based on model performance
  memory_size   = 1024  # MB, adjust as needed for the model
}

# API Gateway (HTTP API) to trigger the Lambda
resource "aws_apigatewayv2_api" "sentiment_api" {
  name          = "FinancialSentimentAPI"
  protocol_type = "HTTP"
}
```

```
# Integrate the Lambda with API Gateway route
resource "aws_apigatewayv2_integration" "sentiment_integration" {
  api_id             = aws_apigatewayv2_api.sentiment_api.id
  integration_type   = "AWS_PROXY"
  integration_uri    = aws_lambda_function.sentiment_lambda.arn
  integration_method = "POST"
}

resource "aws_apigatewayv2_route" "sentiment_route" {
  api_id    = aws_apigatewayv2_api.sentiment_api.id
  route_key = "POST /analyze"
  target    = "integrations/${aws_apigatewayv2_integration.sentiment_
              integration.id}"
}

# Permission for API Gateway to invoke the Lambda
resource "aws_lambda_permission" "api_invoke" {
  function_name = aws_lambda_function.sentiment_lambda.function_name
  action        = "lambda:InvokeFunction"
  principal     = "apigateway.amazonaws.com"
  source_arn    = "${aws_apigatewayv2_api.sentiment_api.execution_
                  arn}/*/*"   # allow any stage/route
}
```

In this Terraform configuration:

- We create an IAM role and attach `AWSLambdaBasicExecutionRole` (which allows writing logs to CloudWatch, etc.).

- The `aws_lambda_function` uses `package_type = "Image"` and references an ECR image URI (you would replace the example account ID and repository with your own). We allocate 1024 MB memory, which should be sufficient for DistilBERT. More memory can also improve CPU allocation speed.

- We set up an API Gateway v2 (HTTP API) with a POST route `"/analyze"` that triggers our Lambda. The integration is of type AWS_PROXY, meaning the entire HTTP request is passed to Lambda.

- We add a permission so API Gateway can invoke the Lambda.

With this in place, after applying the Terraform, we'd have a publicly accessible endpoint (the API Gateway URL) where we can POST a JSON like {"text": "Tesla stock outlook is very strong this quarter."} and get back a sentiment response from our Lambda-powered DistilBERT model.

5.2.3.4 Performance Considerations

A containerized Lambda with DistilBERT should handle a single inference in tens of milliseconds on warm invocations. Cold starts (when a new container spins up) might take a couple of seconds due to loading the model. You can mitigate this by provisioning concurrency or keeping the function warm. The model size (around 250MB with PyTorch and libraries) is fine within the 10 GB image limit, but do note that large images can increase cold start time. Still, this approach is cost-efficient: you only pay for invocation time and resources used, and you don't maintain any servers. For moderate traffic and real-time needs, this is often a sweet spot.

You have now seen two approaches: a fully managed API (Comprehend) and a custom model on serverless infrastructure (Lambda). For even more advanced needs, such as training a model specifically on financial data for better accuracy, we turn to Amazon SageMaker.

5.2.4 Custom Model Training and Deployment Using Amazon SageMaker

Amazon SageMaker is a managed service for the entire machine learning workflow, enabling training and deploying custom sentiment analysis models at scale. It allows fine-tuning models on domain-specific datasets (e.g., financial text) and hosting them behind real-time APIs, providing maximum flexibility and accuracy, but with more development effort and cost.

This scenario involves training a model (e.g., BERT or DistilBERT) on stock-related text labeled by analysts to capture financial language nuances (e.g., "expected to beat earnings"). FinBERT is mentioned as a good starting point.

5.2.4.1 Training on SageMaker

You can use your own code or SageMaker's built-in containers with Hugging Face Transformers integration for easy fine-tuning. A training script (train.py) handles loading data from S3, fine-tuning (e.g., with Hugging Face's TrainerAPI), and saving the model. Alternatively, SageMaker's Hugging Face estimator can directly use the Transformers library by specifying hyperparameters.

Here's a simplified example of launching a training job via Python code:

```python
import sagemaker
from sagemaker.huggingface import HuggingFace

# Define hyperparameters for fine-tuning
hyperparameters = {
    'model_name': 'ProsusAI/finbert',      # starting from FinBERT
                                           model weights

    'epochs': 3,
    'train_batch_size': 16,
    'learning_rate': 2e-5,
    'fp16': True                           # use mixed precision if
                                           supported
}

# Set up the HuggingFace estimator
huggingface_estimator = HuggingFace(
    entry_point='train.py',
    source_dir='src',                      # directory with train.py and any
                                           other code
    instance_type='ml.p3.2xlarge',         # GPU instance for training (V100 GPU)
    instance_count=1,
    role=sagemaker.get_execution_role(),
    transformers_version='4.17',
    pytorch_version='1.10',
    py_version='py38',
    hyperparameters=hyperparameters
)
```

```
# Start the training job (with training and validation dataset
locations in S3)
huggingface_estimator.fit({
    'train': 's3://my-bucket/datasets/fin-sentiment/train/',
    'validation': 's3://my-bucket/datasets/fin-sentiment/val/'
})
```

In this code:

- We use a pretrained FinBERT model (identified by "ProsusAI/finbert" on the Hugging Face hub) as model_name in the hyperparameters. Our training script train.py can load this model via AutoModelForSequenceClassification.from_pretrained(model_name) and then fine-tune on the data.

- We choose a GPU instance (p3.2xlarge) for faster training, since fine-tuning BERT on a decent dataset typically requires a GPU. We only need one instance for fine-tuning (distributed training is possible if the dataset is huge, but often not needed).

- The HuggingFace estimator automatically uses a Hugging Face Deep Learning Container appropriate for Transformers v4.17 and PyTorch 1.10, so we don't have to build a container from scratch. Our train.py will be executed inside that environment.

- The dataset is provided via S3 URIs for training and validation. SageMaker will stream data from S3.

After the job completes, the fine-tuned model artifacts (model weights, etc.) will be saved to an S3 path (usually s3://<bucket>/<prefix>/output/model.tar.gz). We can then deploy the model.

5.2.4.2 Deploying a SageMaker Endpoint

There are two ways to deploy:

1. Use the SageMaker SDK to call huggingface_estimator.deploy(...), which handles creating the endpoint.

2. Use infrastructure as code (Terraform or CloudFormation) to create the endpoint, which is useful for production deployments.

Using the Python SDK, for quick testing, we could do this:

```python
# Deploy the model to a real-time inference endpoint
predictor = huggingface_estimator.deploy(initial_instance_count=1,
instance_type='ml.m5.large')
# Now the model is live behind a REST endpoint. We can use predictor to
invoke it:
result = predictor.predict({"inputs": "Tesla stock outlook is very strong
this quarter."})
print(result)
```

The predictor.predict method sends the input to the endpoint and returns the model's inference result. Underneath, SageMaker has provisioned an instance (here ml.m5.large CPU, which is sufficient for DistilBERT/FinBERT for inference) and has set up a HTTPS endpoint for it. The result might be something like {'label': 'Positive', 'score': 0.99} depending on how train.py formats the output (if we use the default Transformers pipeline, it returns a label and score).

For a production setup, we can use Terraform to create the SageMaker model and endpoint. This gives more control over naming, scaling, and connecting to other infrastructure. Here's an example Terraform configuration that assumes the model artifact from training is in S3; we want to deploy it using the Hugging Face inference container:

```
# IAM role that SageMaker will use for inference (to access model artifacts
from S3, etc.)
resource "aws_iam_role" "sagemaker_role" {
  name = "sagemaker_financial_model_role"
  assume_role_policy = jsonencode({
    "Version": "2012-10-17",
    "Statement": [{
      "Effect": "Allow",
      "Principal": { "Service": "sagemaker.amazonaws.com" },
      "Action": "sts:AssumeRole"
    }]
  })
}
```

```
# Attach AmazonS3ReadOnlyAccess or a custom policy to allow access to
model data
resource "aws_iam_role_policy_attachment" "s3_access" {
  role       = aws_iam_role.sagemaker_role.name
  policy_arn = "arn:aws:iam::aws:policy/AmazonS3ReadOnlyAccess"
}

# SageMaker Model creation, referencing the model artifact in S3 and HF
inference image
resource "aws_sagemaker_model" "sentiment_model" {
  name                 = "financial-sentiment-model"
  execution_role_arn   = aws_iam_role.sagemaker_role.arn
  primary_container {
    image = "763104351884.dkr.ecr.us-east-1.amazonaws.com/huggingface-
      pytorch-inference:1.10.2-transformers4.17.0-cpu-py38-ubuntu20.04"
    # The above image is an AWS Deep Learning Container for HuggingFace
    inference (region-specific URI)
    model_data_url = "s3://my-bucket/finbert-output/model.tar.gz"
    # path to model artifact from training
  }
}

# Endpoint configuration (choosing instance type and initial count)
resource "aws_sagemaker_endpoint_configuration" "sentiment_ep_config" {
  name = "financial-sentiment-EndpointConfig"
  production_variants {
    variant_name           = "AllTraffic"
    model_name             = aws_sagemaker_model.sentiment_model.name
    initial_instance_count = 1
    instance_type          = "ml.m5.large"
  }
}

# Endpoint creation
resource "aws_sagemaker_endpoint" "sentiment_endpoint" {
```

```
    name                   = "financial-sentiment-endpoint"
    endpoint_config_name = aws_sagemaker_endpoint_configuration.sentiment_ep_
    config.name
}
```

In this Terraform:

- We create an IAM role for SageMaker that has access to S3 (to fetch the model artifact).

- We define the SageMaker model with the URI of the container image for the Hugging Face inference (the URI given is for us-east-1; it would differ by region and versions). We supply the `model_data_url`, which is the S3 path to our fine-tuned model `.tar.gz`.

- We set up an endpoint configuration by specifying the instance type (`ml.m5.large` for example) and model to use. We choose one instance to start.

- Finally, we create the endpoint itself. Once this is applied, SageMaker will spin up the instance and load our model for inference.

After deployment, the SageMaker endpoint can be invoked via the AWS SDK (Boto3 SageMaker-runtime `invoke_endpoint` API) or through any HTTP client if integrated with API Gateway or other routing. In our architecture, the Lambda function could call this SageMaker endpoint. For example, in the Lambda code, after detecting that the text should use the custom model, we can do this:

```
sm_runtime = boto3.client('sagemaker-runtime')
response = sm_runtime.invoke_endpoint(
    EndpointName='financial-sentiment-endpoint',
    ContentType='application/json',
    Body=json.dumps({ "inputs": text })
)
result = json.loads(response['Body'].read().decode())
# result could be {'label': 'Positive', 'score': 0.9876} depending on the
model's output format
```

We would then parse `result` and combine it into our Lambda's response to the client.

5.2.4.3 Benefits of the SageMaker Approach

Using SageMaker, we get a model tuned to our data, which often means higher accuracy on financial texts. We can also train on more classes (e.g., maybe we want to classify not just sentiment polarity but also an intensity or detect if something is an "uncertain" sentiment). With SageMaker, we control the whole lifecycle: retraining the model as new data arrives, A/B testing new models by deploying multiple variants, and scaling the endpoint (multiple instances or even deploying on GPU if we had a very large model that needs it).

5.2.4.4 Considerations

The tradeoff is cost and complexity. Running an `ml.m5.large` instance 24/7 has an ongoing cost, whereas Lambda only incurs cost per use. If traffic is sporadic, SageMaker endpoints could be turned off or switched to serverless endpoints (SageMaker has a Serverless Inference option as well) to save cost. Also, development is more involved: we need to prepare data and training scripts. However, for long-term projects where accuracy is paramount, this is often worth it.

5.2.5 Comparing Approaches: Comprehend vs. Custom Models vs. Lambda-deployed SLMs

We have discussed three approaches to sentiment analysis with AWS: using Amazon Comprehend, deploying a custom small model on Lambda, and fine-tuning a model on SageMaker. Each approach has its strengths and ideal use cases. Table 5-2 provides a comparative overview.

Table 5-2. Comparative Overview of the Three Approaches to Sentiment Analysis

Approach	Latency(typical)	Cost Model	Flexibility and Customization	Accuracy (Financial Domain)	Use Case Fit
Amazon Comprehend (managed API)	~100-200ms per call (includes network overhead)	Pay-per-request (priced per character of text)	Low: Pretrained model only (cannot tune weights; limited to provided features)	Good for general sentiment; may miss domain-specific nuances or sarcasm.	Rapid prototyping; multi-language support out-of-the-box; no ML expertise needed.
SLM on AWS Lambda (e.g., DistilBERT)	~50ms inference (after warm start); cold start 1-2s	Pay-per-invocation (milliseconds of execution time X memory); no idle cost	Medium: Can choose/preload any model that fits in memory; code can be customized (e.g., pre/post-processing)	Good if model is pretrained on relevant data (e.g., a generic Twitter sentiment model); moderate for highly specialized finance text unless fine-tuned.	Real-time applications with intermittent traffic; need low latency and full control over model and code; limited budget for constantly running instances.

(continued)

Table 5-2. (*continued*)

Approach	Latency(typical)	Cost Model	Flexibility and Customization	Accuracy (Financial Domain)	Use Case Fit
Custom model on SageMaker (e.g., fine-tuned FinBERT)	~30-100ms inference (on a dedicated instance, no cold start)	Pay per instance hour (or use serverless inference for scale-to-zero); training jobs incur hourly cost as well	High: Full control. Can train on custom data, choose model architecture, implement custom logic; and update model freely	Potentially highest if trained on sufficient domain data; can capture subtle finance-specific sentiment signals (e.g., "downside risk" context).	High-volume or mission-critical systems where accuracy is priority; willing to invest in training and infrastructure; need integration with ML ops (retraining, monitoring).

5.2.5.1 Latency

Comprehend calls go to a managed endpoint in AWS region, which is typically quite fast, but there is overhead of a service call. A Lambda with a loaded model, after initial load, can be very fast for small models. SageMaker endpoint runs continuously and can give low latency responses consistently, especially if using a GPU for large models, but for our scale (BERT on CPU) it's in the same ballpark as Lambda. For infrequent requests, Lambda might have occasional cold starts, whereas SageMaker endpoint is always warm (unless using a serverless inference, which can scale down).

5.2.5.2 Cost

Comprehend's pricing is straightforward per text size; it's cost-efficient for low volumes, but at very high volumes a dedicated model might be cheaper. Lambda cost depends on usage—it could be extremely cheap for low traffic, but if you have continuous high TPS (transactions per second), the cumulative compute time might approach or exceed the cost of an equivalent always-on server. SageMaker is best when you have steady load or require the advanced capabilities; its cost is the most predictable (you pay for the provisioned instance whether it's used or not, unless using the newer serverless option).

5.2.5.3 Flexibility

Comprehend is limited to what AWS provides (you can't change how it analyzes text, though you could preprocess text or use additional features like custom classification if you go beyond sentiment). Lambda gives you freedom to embed any model or logic, but within the constraints of memory, runtime (max 15 minutes per invocation, but that's more than enough for inference), and no GPU. SageMaker is like having your own server or cluster—you can train any model (even an LLM if you have the resources), use GPUs, distribute training, and so on, and you can implement custom inference logic or pipelines (e.g., use SageMaker Inference Toolkit to customize the endpoint handling).

5.2.5.4 Accuracy

In general, a model fine-tuned on in-domain data (SageMaker approach) will outperform a general model (Comprehend or a generic DistilBERT) on nuanced financial texts. If the financial data contains very domain-specific phrases, a custom model will likely catch sentiment signals better. That said, if the domain isn't drastically different from general language, Comprehend or a pretrained DistilBERT might do a decent job. It's always good to evaluate—you could test a sample of financial sentences on Comprehend versus a custom model to quantify the difference. In some cases, fine-tuning (SageMaker) yields only marginal gains, which might not justify the extra complexity.

5.2.6 Choosing the Best Approach

When to choose which:

- **Comprehend** is ideal for plug-and-play sentiment analysis. It supports multiple languages, requires no model training, and integrates easily via a simple API call. It's best suited for teams without ML expertise or when quick deployment is needed (e.g., dashboards analyzing multilingual tweets). Since AWS maintains the model, improvements roll out automatically.

- **Lambda + SLM** is great when you have a moderately sized model that you trust (perhaps open-source) and want the lowest latency and cost at low scale, or you need to keep data in your environment. It's also a good option if you want to avoid external API calls for privacy—the data stays in your AWS account. With this, you also have the agility to swap models (just deploy a new container) or add custom logic (e.g., do sentiment analysis plus some keyword flagging in one function).

- **A SageMaker custom model** should be chosen when you have special requirements, such as very high accuracy needs, a unique model architecture, or a large volume of data to process constantly. It shines in production ML workflows where you retrain models regularly, monitor performance, and perhaps use advanced features (A/B testing endpoints, scaling to multiple instances across regions, etc.). In the financial sentiment case, if an investment firm relies on these sentiment scores for decisions, they might invest in a SageMaker solution to squeeze out the best performance and reliability, possibly using an ensemble of models or an LLM for edge cases. An LLM-based agent could even be deployed via SageMaker (for example, a GPT-J or GPT-2 model fine-tuned for sentiment or routing) to analyze each input and decide the outcome or route it. However, running a true large-scale LLM in real-time can be expensive, so often a two-tier system (a fast small model versus a slower but more accurate model) is a pragmatic design.

To sum up, AWS provides a spectrum of options: from no-code managed services to full custom modeling. It's common to start with Comprehend for a quick win, then move to a custom approach if needed. You could also combine them—for instance, use Comprehend for the majority of straightforward cases and only trigger the SageMaker model for ambiguous cases (as identified by an LLM agent or a confidence threshold). This kind of smart orchestration can optimize cost and accuracy.

5.2.7 Conclusion and Next Steps

The section built a practical sentiment analysis system for financial text using AWS's NLP services and custom model hosting capabilities. We covered using Amazon Comprehend out-of-the-box, deploying a Hugging Face DistilBERT model on Lambda for serverless inference, and fine-tuning a domain-specific model on SageMaker for maximum accuracy. We also discussed how an LLM-based agent could play a role in directing traffic to the appropriate method. The next section explores implementing sentiment analysis using open-source tools, building on the foundation of deploying LLMs and SLMs in real-world applications established here.

5.3 Open-Source Implementation

This section explore how to implement sentiment analysis using open-source tools, specifically Hugging Face Transformers, which provides state-of-the-art pretrained models for LLMs and SLMs. It covers fine-tuning techniques for complex sentiment analysis with LLMs like GPT-4 and T5, as well as the use of resource-efficient SLMs such as DistilBERT and TinyBERT for real-time sentiment classification tasks. Additionally, it delves into data labeling techniques and challenges associated with sentiment classification.

5.3.1 Implementing Sentiment Analysis with Hugging Face Transformers

The Hugging Face Transformers library is a powerful, open-source platform that provides a collection of state-of-the-art models for a variety of NLP tasks, including sentiment analysis. It is widely used for both research and enterprise applications due to its ease of use, scalability, and support for a wide range of transformer models.

LLMs like GPT-4 and T5 can be fine-tuned for complex sentiment analysis, understanding nuances, irony, and context. T5, a text-to-text model, can be fine-tuned to generate sentiment labels, which is helpful for highly complex or ambiguous text.

SLMs like DistilBERT and TinyBERT are efficient alternatives for real-time, low-latency sentiment analysis in resource-constrained environments. TinyBERT is even smaller than DistilBERT and is designed for extreme resource constraints and efficient processing of smaller datasets. It's suitable for mobile or edge devices.

5.3.1.1 Fine-Tuning T5 for Sentiment Analysis

T5 is a text-to-text model, so we can frame the sentiment analysis task as a text generation problem.

```python
from transformers import T5ForConditionalGeneration, T5Tokenizer
from datasets import load_dataset
from transformers import Trainer, TrainingArguments

# Load dataset (for example IMDb for sentiment classification)
dataset = load_dataset("imdb")

# Load T5 tokenizer and model
model_name = "t5-small"
model = T5ForConditionalGeneration.from_pretrained(model_name)
tokenizer = T5Tokenizer.from_pretrained(model_name)

# Preprocessing function
def preprocess_function(examples):
    return tokenizer(examples['text'], truncation=True, padding=
    "max_length")

tokenized_datasets = dataset.map(preprocess_function, batched=True)

# Training arguments
training_args = TrainingArguments(
    output_dir="./t5-sentiment",
    evaluation_strategy="epoch",
    per_device_train_batch_size=8,
    per_device_eval_batch_size=8,
    num_train_epochs=3,
    logging_dir='./logs',
)

trainer = Trainer(
    model=model,
    args=training_args,
```

```
    train_dataset=tokenized_datasets['train'],
    eval_dataset=tokenized_datasets['test'],
)

trainer.train()
```

5.3.1.2 Fine-Tuning TinyBERT for Sentiment Analysis

TinyBERT is a smaller version of BERT and is highly optimized for mobile and edge devices. You can fine-tune TinyBERT as follows:

```
from transformers import BertForSequenceClassification, BertTokenizer
from datasets import load_dataset
from transformers import Trainer, TrainingArguments

# Load dataset
dataset = load_dataset("imdb")

# Load TinyBERT model and tokenizer
model_name = "google/tinybert-l4"
model = BertForSequenceClassification.from_pretrained(model_name, num_labels=2)
tokenizer = BertTokenizer.from_pretrained(model_name)

# Tokenizing the dataset
def tokenize_function(examples):
    return tokenizer(examples['text'], padding="max_length",
    truncation=True)

tokenized_datasets = dataset.map(tokenize_function, batched=True)

# Training arguments
training_args = TrainingArguments(
    output_dir="./tinybert-sentiment",
    evaluation_strategy="epoch",
    per_device_train_batch_size=8,
    per_device_eval_batch_size=8,
    num_train_epochs=3,
    logging_dir='./logs',
)
```

```
trainer = Trainer(
    model=model,
    args=training_args,
    train_dataset=tokenized_datasets["train"],
    eval_dataset=tokenized_datasets["test"],
)

trainer.train()
```

5.3.1.3 Optimizing Inference Using Quantization and Pruning (for SLMs)

You can use quantization and pruning to reduce the size of your model and speed up inference. Hugging Face offers integration with ONNX and TensorFlow Lite for deployment.

Here's an example of quantizing the model to reduce its size and increase inference speed:

```
from transformers import DistilBertForSequenceClassification
import torch
from torch.quantization import quantize_dynamic

# Load the fine-tuned model
model = DistilBertForSequenceClassification.from_pretrained("distilbert-sentiment")

# Quantize the model dynamically for inference optimization
quantized_model = quantize_dynamic(model, {torch.nn.Linear}, dtype=torch.qint8)

# Save the quantized model
quantized_model.save_pretrained("./quantized-distilbert")
```

5.3.2 MLOps Considerations for LLMs and SLMs in Sentiment Analysis

1. Model versioning and experiment tracking:

    - **Version control:** For LLMs and SLMs, it's essential to maintain proper versioning and experiment tracking. Tools like MLFlow, DVC (Data Version Control), and Weights & Biases help ensure that every model trained or fine-tuned is tracked, and previous versions can be easily rolled back or compared.

    - **Model registry:** Implementing a model registry allows you to store multiple versions of models and facilitates easy management, rollback, and deployment of the best-performing model versions.

2. Continuous integration and continuous deployment (CI/CD):

    - **CI/CD pipelines:** Automating the deployment process through CI/CD pipelines ensures faster and safer updates to the production environment. For LLMs and SLMs, CI/CD pipelines can automate tasks like model training, evaluation, testing, and deployment. Tools like Jenkins, GitLab CI, GitHub Actions, and CircleCI are commonly used to automate these workflows.

    - **Model monitoring:** Once deployed, it's crucial to continuously monitor the performance of sentiment analysis models, particularly in production. This involves tracking metrics like accuracy, latency, throughput, and model drift (i.e., how the model's performance changes over time due to changes in input data). Tools such as Prometheus, Grafana, and CloudWatch (AWS) can be used for this purpose.

3. Model retraining and auto-scaling:

    - **Auto-scaling:** Depending on the traffic, auto-scaling strategies are crucial for handling fluctuating workloads. For example, AWS offers Lambda functions, ECS, and SageMaker to auto-scale your sentiment analysis inference endpoints based on demand, ensuring lower latency and cost efficiency.

- **Model retraining:** As sentiment analysis systems evolve, it's important to monitor the model's performance and retrain periodically to adapt to new trends or patterns in data. This is particularly true when deploying LLMs, which may require larger training cycles. An automated retraining pipeline can be set up to trigger model updates when performance falls below a certain threshold.

5.3.3 Techniques for Deployment of LLMs and SLMs

1. Deployment of LLMs for sentiment analysis (e.g., GPT-4, T5, DeBERTa):

    - **Inference as a service:** Deploying LLMs like GPT-4 or T5 typically requires hosting the model via cloud services. These models can be deployed as APIs, where users send text inputs and the model returns sentiment predictions. Services such as AWS SageMaker, Google AI Platform, and Azure Machine Learning are commonly used for deploying large models in scalable, production-ready environments.

    - **Optimization for latency:** LLMs are typically large models with higher inference costs and latency. Techniques like model quantization, batch inference, and model distillation can help optimize the deployment. Additionally, deploying models on powerful GPU instances can further reduce latency.

2. Model deployment with AWS SageMaker:

    After fine-tuning the model, you can deploy it as an API endpoint for inference.

    ```
    import sagemaker
    from sagemaker import get_execution_role
    from sagemaker.huggingface import HuggingFaceModel

    # Set up the role and session
    role = get_execution_role()
    sagemaker_session = sagemaker.Session()
    ```

```python
# Deploy the fine-tuned model
huggingface_model = HuggingFaceModel(
    model_data='s3://your-bucket/path/to/model.tar.gz',
    role=role,
    entry_point='inference.py',   # Custom inference script
    framework_version='4.5.0',
    py_version='py36',
    instance_type='ml.m5.large',
    sagemaker_session=sagemaker_session
)

predictor = huggingface_model.deploy(instance_type="ml.m5.large",
initial_instance_count=1)
```

3. Deployment of SLMs for sentiment analysis (e.g., DistilBERT, TinyBERT, MobileBERT):

- **Lightweight deployment for real-time analysis:** SLMs like DistilBERT, TinyBERT, and MobileBERT are well-suited for deployment on edge devices or serverless platforms. These models are small, fast, and can be deployed on AWS Lambda, Google Cloud Functions, or even on-device in mobile applications.

    ```python
    import boto3

    client = boto3.client('lambda')

    response = client.invoke(
        FunctionName='sentiment-analysis-lambda',
        InvocationType='RequestResponse',
        Payload='{"input_text": "I love this product!"}'
    )

    print(response['Payload'].read().decode('utf-8'))
    ```

- **On-device inference:** SLMs are also deployed directly on mobile devices using frameworks like TensorFlow Lite or ONNX for on-device inference. This is crucial for real-time applications like sentiment analysis in mobile apps, where low latency is a key requirement.

Table 5-3 lists the various strategies for optimizing inferences.

Table 5-3. Inference Optimization Strategies for LLMs and SLMs

Strategy	Description	Example

Quantization	Converts a model's weights from floating-point to lower precision (e.g., int8), reducing size and improving inference speed. Especially effective for SLMs.	SLM example (DistilBERT):
		`from transformers import DistilBertForSequenceClassification`
		`import torch`
		`from torch.quantization import quantize_dynamic`
		`model = DistilBertForSequenceClassification.from_pretrained("distilbert-sentiment")`
		`quantized_model = quantize_dynamic(model, {torch.nn.Linear}, dtype=torch.qint8)`
		`quantized_model.save_pretrained("./quantized-distilbert")`

Model pruning	Removes weights that have little impact on model performance, leading to smaller models and faster inference.	General example: Use pruning techniques to remove redundant weights that don't contribute significantly to predictions.
Batch inference	Processes multiple sentiment classification requests in a single batch, reducing overhead and improving throughput.	Example: Group multiple inputs into a batch and process them together to optimize inference, especially in high-volume use cases.

(continued)

Table 5-3. (*continued*)

Strategy	Description	Example
Serverless architectures	Leverages platforms like AWS Lambda or Google Cloud Functions to automatically scale resources based on demand, making deployment cost-effective and efficient for fluctuating workloads.	Example: Deploy models as serverless functions on AWS Lambda to scale automatically based on traffic.
Model distillation	Creates smaller, faster models that approximate the behavior of larger models, useful for deployment on resource-constrained devices.	LLM example: Distill a T5 model to a smaller version with comparable performance for use in production environments with limited resources.
Caching	Stores results of common or repetitive sentiment analysis requests, reducing redundant processing and speeding up inference for frequently seen inputs.	Example: Cache sentiment results for common phrases or previously processed inputs, especially in high-frequency environments (e.g., customer feedback systems).

5.3.4 Data Labeling Techniques and Challenges for Sentiment Classification

Accurate sentiment classification heavily depends on well-labeled data. However, labeling data for sentiment analysis presents several challenges, especially when dealing with large-scale datasets or nuanced sentiments. The process requires not just basic sentiment labels (positive, negative, neutral), but also a clear understanding of contextual and emotional tones.

- **Manual labeling:** For high-quality sentiment labels, manual annotation by domain experts is often required, especially for complex datasets where sentiment is not immediately obvious. This is particularly useful when you need to account for subtle emotions like sarcasm, irony, or mixed emotions.

- **Automated labeling and pretrained models:** In some cases, pretrained sentiment models (like those available in Hugging Face) can be used to automatically label data. While this method is faster and more scalable, it can introduce noise in the labeled data, especially when the models have limitations in handling complex sentiments.

- **Crowdsourcing:** Platforms like Amazon Mechanical Turk can be used to outsource data labeling tasks to a large pool of workers. However, this method also requires quality control measures to ensure the reliability and accuracy of the sentiment labels. Aggregating multiple annotations for each example and using consensus models can help mitigate errors in crowd-labeled data.

- **Synthetic data generation:** In some cases, synthetic data can be generated to train sentiment models, especially for low-resource languages or underrepresented sentiment categories. However, this approach has its challenges in ensuring that the synthetic data adequately reflects real-world sentiments.

The challenges are as follows:

- **Nuanced sentiments:** Sentiment labels often need to capture a wide range of emotions beyond just "positive" or "negative." For example, sentiments like "angry" or "disappointed" may need to be tagged separately to provide useful insights.

- **Ambiguity in text:** Sentiment can often be ambiguous, particularly in texts that contain mixed sentiments (e.g., "I love the idea, but the implementation is terrible"). Handling this ambiguity requires careful labeling and sometimes a multi-label approach to capture the complexity.

- **Cultural differences:** Sentiment interpretation can vary across cultures and regions. What may be considered a positive sentiment in one culture might not be viewed the same way in another. Data labeling must take cultural contexts into account, especially with global applications.

This section covered deployment and optimization strategies for sentiment analysis models using LLMs (like GPT-4, T5, DeBERTa) and SLMs (such as DistilBERT, TinyBERT, MobileBERT). Implementing MLOps best practices—including model versioning, CI/CD pipelines, and model monitoring—is the key to ensuring models are performant and scalable. Optimization techniques like quantization, model distillation, and batch inference help boost inference speed and reduce resource consumption, enabling real-time, high-accuracy sentiment analysis at scale.

5.4 Industry Use Cases

5.4.1 Brand Monitoring

Brand monitoring leverages sentiment analysis to track public perception of a company or product across digital platforms. It helps organizations understand how consumers feel about their brand, enabling real-time decision-making, marketing optimization, and crisis prevention.

Companies can use this data to:

- Detect emerging issues.
- Understand customer feedback at scale.
- Compare sentiment around their brand versus competitor brands.
- Inform branding strategies and product development.

Refer to this GitHub link for the full code: https://github.com/anvcse562/ModernNLP-LLMSLM/blob/main/Ch05/Ch5.4/brand-monitoring.py.

5.4.2 Social Media Analysis

Social media sentiment analysis is widely used by businesses to gain insights into public opinion about their brand, products, and services. By analyzing the sentiment expressed in social media posts, companies can understand customer perceptions, track brand reputation, identify potential crises, and make data-driven decisions to improve customer satisfaction. This analysis helps them gauge customer reactions to marketing campaigns, product launches, and competitor activities. Benefits include enhanced brand monitoring, improved customer service, proactive issue management, and data-driven product development.

The example code in GitHub focuses on building and training a sentiment analysis model using the IMDb movie review dataset. It covers key steps like data loading, preprocessing, model initialization, training, and prediction. However, it doesn't cover aspects specific to social media data, such as handling different data formats (tweets, Facebook posts, etc.), dealing with noise and slang, and extracting relevant information like hashtags and mentions. To apply this code to social media sentiment analysis, you would need to adapt it to handle social media data sources and tailor the preprocessing and feature engineering steps accordingly.

The code in GitHub briefly touches on deployment options using AWS services (SageMaker, Elastic Beanstalk, Lambda) and open-source frameworks (Flask, Docker). These options enable you to make the trained model accessible for real-time sentiment analysis of new social media posts. AWS services provide managed infrastructure and scalability, while open-source options offer flexibility and control. The choice of deployment depends on factors like cost, infrastructure requirements, and desired level of customization.

Refer to this GitHub link for the full code: https://github.com/anvcse562/ModernNLP-LLMSLM/blob/main/Ch05/Ch5.4/Social_Media_Sentiment_Analysis_Deployment_Options.ipynb.

5.4.3 Customer Satisfaction Insights

Customer satisfaction insights leverage sentiment analysis and natural language processing techniques to gain deep, actionable understanding of customer opinions, emotions, and feedback from various digital channels, such as product reviews, social media, customer support interactions, and survey responses. Utilizing SLMs like DistilBERT and DistilGPT makes it practical and cost-effective, particularly for businesses with resource constraints or deployment scenarios requiring edge computing or mobile integration.

Businesses utilize this approach primarily to:

- Measure customer sentiment and satisfaction levels at scale.
- Identify common customer pain points and frequently raised issues.
- Enhance customer experience by proactively addressing concerns.
- Evaluate the effectiveness of product launches or updates.
- Facilitate rapid feedback loops between customers and development teams.

5.4.3.1 Implementation Overview

This implementation includes:

- **Data acquisition and preprocessing:** Customer feedback data (such as reviews or social media posts) are loaded and preprocessed to handle typical text challenges like noise, formatting inconsistencies, and linguistic variations.

- **Model selection and training:** A cost-effective SLM (DistilBERT) is fine-tuned specifically for sentiment classification, distinguishing positive and negative customer feedback. DistilBERT is chosen for its balance between performance and efficiency, ideal for resource-constrained environments.

- **Evaluation and validation:** The fine-tuned model undergoes rigorous evaluation using standard performance metrics (Precision, Recall, F1 score) to ensure reliability and accuracy in real-world scenarios.

- **Deployment and accessibility:** The model is deployed via a Flask-based RESTful API, making sentiment analysis services readily accessible to various client applications.

Dockerization instructions are provided for easy container-based deployment, ensuring consistent and scalable application hosting across different platforms and cloud environments.

- **Testing and validation of deployed API:** API endpoints are tested through example calls to ensure that real-time sentiment analysis is functional and performant.

5.4.3.2 Deployment Options and Scalability

The solution is designed for flexible deployment:

- **Local or edge computing:** Ideal for IoT devices, retail kiosks, and local servers requiring instant, local inference capabilities without heavy cloud dependencies.
- **Containerization with Docker:** Enables simple, portable, and consistent deployment across cloud environments (AWS, Azure, GCP), facilitating scalability and management simplicity.

5.4.3.3 Benefits of Customer Satisfaction Insights with Small LLMs

- **Cost efficiency:** Reduced operational costs due to lightweight models that require minimal compute resources.
- **Real-time insights:** Rapid inference capability, ideal for applications needing instant analysis of customer feedback.
- **Enhanced customer engagement:** Quickly identifying and addressing customer dissatisfaction can significantly improve customer loyalty and retention.
- **Data-driven decision making:** Empowers businesses to prioritize features and updates and support initiatives based on robust data insights.

5.4.3.4 Example Applications

- **Real-time customer sentiment monitoring** on websites or mobile applications.

- **Automated customer support systems** that immediately categorize and escalate negative feedback.

- **Market research and product management**, analyzing customer reactions to product changes or new feature introductions.

By integrating SLMs into customer satisfaction monitoring workflows, businesses can continuously track and improve their customer experiences with minimal overhead, achieving competitive advantage through agility, responsiveness, and efficiency.

Refer to this GitHub link for the full code: https://github.com/anvcse562/ModernNLP-LLMSLM/blob/main/Ch05/Ch5.4/Customer_Satisfaction_Insights.ipynb.

5.5 Summary

In this chapter, you explored sentiment analysis using LLMs and SLMs. You learned how to apply these models to determine the emotional tone or subjective opinion expressed in text, classifying it as positive, negative, or neutral. You also gained practical knowledge on deploying these models using AWS services like Amazon Comprehend and SageMaker, and open-source tools like Hugging Face Transformers, and you explored industry use cases such as brand monitoring, social media sentiment analysis, and customer satisfaction insights. You should now have a strong understanding of sentiment analysis and model deployment. The next chapter dives into another key NLP task: question answering.

CHAPTER 6

Question Answering (QA)

In this chapter, we delve into the construction of sophisticated question-answering systems, leveraging both LLMs and SLMs. We explore practical methodologies for both extractive and abstractive QA, providing a comprehensive guide to building intelligent systems capable of understanding and answering questions based on provided contexts or general knowledge.

This chapter covers:

- **Introduction to question answering (QA):** Defining QA systems, differentiating between extractive and abstractive approaches, and highlighting real-world applications.

- **Implementation on AWS:** Building and fine-tuning QA models using Amazon SageMaker, and deploying efficient, low-latency QA solutions with AWS Lambda.

- **Open-source implementation:** Utilizing the Hugging Face Transformers library to build and evaluate QA models, and exploring fine-tuning techniques.

- **Industry use cases:** Examining the application of QA systems in customer support, knowledge base querying, and the development of automated assistants.

By the end of this chapter, you will have a strong understanding of how to choose and implement the appropriate QA techniques and models for various applications, whether requiring precise extraction from a document or the generation of novel answers.

CHAPTER 6   QUESTION ANSWERING (QA)

6.1 Introduction to Question Answering

The previous chapter explored how *sentiment analysis* gauges the subjective tone and emotions in text. It concluded by noting the importance of understanding user sentiment—essentially reading *how* the user feels. Now, we transition from understanding feelings to fulfilling information needs. This section focuses on *question answering (QA)*, where the goal is to directly address users' queries with accurate answers. This shift moves us from analyzing *what* a user expresses to delivering *what information* a user is seeking. By harnessing LLMs, you can build QA systems that interpret a question and produce a helpful answer, making them a core component of applications like customer support chatbots, virtual assistants, and search engines.

QA is a broad field, but methods generally fall into two major paradigms: *extractive* and *abstractive* question answering.

- In **extractive QA,** the answer to a question is literally a substring (span) of some reference text (context).

- In **abstractive QA** (also called generative QA), the answer is generated in natural language, potentially combining or rephrasing information from the context (or from the model's own knowledge) rather than copying text verbatim.

We delve into each of these approaches, examining their model architectures (e.g. BERT/RoBERTa vs. T5/GPT-4) and use cases. We also introduce how LLM agents can perform multi-step reasoning to answer complex queries that go beyond a single context. Throughout this section, we use a running example of a fictional customer support knowledge base to illustrate how different QA methods would answer user questions. Each section begins with a brief summary, followed by technical details, examples, and code snippets using Hugging Face Transformers to demonstrate practical implementations. Finally, we summarize the key differences between extractive vs. abstractive and small vs. large models and pave the way to Section 6.2, which covers real-world QA deployment using AWS tools.

6.1.1 Extractive Question Answering

Extractive question answering finds an answer by selecting a span of text from a given document or context that directly answers the question. The model does not generate new phrasing; it literally pinpoints where the answer appears in the text and outputs that excerpt. This approach is akin to highlighting the relevant text in a passage. Extractive QA is well-suited for fact-based questions where the answer is explicitly stated in a reference text (for example, "What is the warranty period of the Gizmo Pro 3000?" answered from a product manual). Because the answer comes straight from the source, this method tends to preserve factual accuracy and context. However, its capability is limited to answering questions that have an answer explicitly present in the provided text and in the exact wording of that text (Rajpurkar et al., 2016; Li et al., 2023; Mallick et al., 2023).

6.1.1.1 Approach and Model Architecture

Most modern extractive QA systems are built on encoder-based Transformer models like BERT or RoBERTa. These models are pretrained on large corpora and then fine-tuned on QA datasets (such as SQuAD) to learn how to locate answer spans. Conceptually, the question and the context document are concatenated (often with a separator token) and fed into the encoder. The model produces a contextualized representation for each token in the question+context. A QA-specific output layer (typically a pair of linear layers in the case of BERT-like models) then computes a probability distribution over all context tokens for the start and end of the answer.

Training the model involves teaching it to assign high probability to the start and end positions that correspond to the ground-truth answer span in the training examples. At inference time, the model selects the span with the highest start and end scores as the predicted answer. For example, if the context is a support article that states "The Gizmo Pro 3000 comes with an 18-month warranty from the date of purchase," and the question is "What is the warranty period of the Gizmo Pro 3000?", an extractive model will identify the substring "18-month" (or "18-month warranty") as the answer.

For example, suppose you have the following context from a fictional customer support knowledge base:

"Product: Gizmo Pro 3000. Details: The Gizmo Pro 3000 needs to be charged for 2 hours before first use. If the device does not turn on, check the battery and hold the power button for 5 seconds to reset. The device comes with an 18-month warranty from the date of purchase."

Now consider a user question: *"How long is the warranty for the Gizmo Pro 3000?"* The answer ("18-month") is clearly stated in the context. An extractive QA model should retrieve exactly that span. We can demonstrate this using a pretrained extractive QA model from Hugging Face, found on GitHub at https://github.com/anvcse562/ModernNLP-LLMSLM/blob/main/Ch06/Ch6.1/QaPipelineHunggingFace.py.

Running this code will tokenize the question and context, pass them through the model, and print the extracted answer. In this case, the output would be:

18-month

The model has correctly identified "18-month" as the answer span within the context. Under the hood, the model likely produced high start and end scores around the token "18-month" in the context sequence. This highlights the strength of extractive QA: as long as the answer appears explicitly, the model can find it with high precision.

6.1.1.2 Use Cases

Extractive QA is ideal when you have a trusted body of text (documents, knowledge base articles, web pages, etc.) and you need to provide exact answers with evidence. Examples include FAQ retrieval (finding the exact answer snippet in an FAQ document), academic or legal QA (where quoting the precise text is important for credibility), and closed-domain QA (where the answer is guaranteed to lie in a specific document or passage). Because the answer comes directly from the provided text, it is easily verifiable—the user can be shown the source of the answer. In fact, many real-world systems combine extractive QA with information retrieval: first retrieve candidate documents relevant to the query (using a search engine or index), then use an extractive model to highlight the answer within those documents.

6.1.1.3 Technical Considerations

Models like BERT and RoBERTa fine-tuned for QA have achieved high accuracy on benchmarks like SQuAD, often surpassing human performance on extracting correct spans. These models typically have on the order of 110 million parameters (BERT-base) or 340 million (BERT-large), making them relatively "small" by today's LLM standards but efficient enough for real-time use. They require the question and context to fit within the model's input size (e.g., 512 tokens for BERT), so lengthy documents may need to be segmented or truncated. Additionally, if a question cannot be answered by the provided text, extractive models struggle—they might still extract some span that looks relevant (even if it's incorrect) unless trained to sometimes output a special "no answer" prediction (as done in SQuAD v2.0). Despite these limitations, the reliability of extractive QA is high when its assumptions are met: the answer is present and clearly stated. In many settings, extractive QA is considered more explainable and trustworthy than generative approaches, since the output is grounded in a known text.

6.1.1.4 Pros and Cons

To summarize the characteristics of extractive QA, Table 6-1 provides a quick comparison to abstractive QA (a detailed discussion of abstractive QA follows).

Table 6-1. *Comparison of Extractive vs. Abstractive QA*

Aspect	Extractive QA	Abstractive QA
Answer content	Selects a span from the context (no new words added)	Generates a novel answer phrasing (may not appear verbatim in context)
Model type	Typically encoder-only (e.g. BERT, RoBERTa) with a span prediction head.	Typically encoder–decoder or decoder-only (e.g. T5, BART, GPT-4) producing text output.
Strengths	– Answers are directly supported by the source (high fidelity, easy to verify). – Simpler models and training; often requires less data and compute. – Preserves context and wording, reducing risk of factual errors.	– Can provide natural, well-formed answers (more human-like). – Can synthesize information from multiple parts of text. – Handles questions that require rephrasing or slight inference (not just copy-paste).
Limitations	– Requires answer to be present in text (cannot answer implicit or absent information). – Limited to exact or near-exact wording in context (no rephrasing or adding explanation). – Struggles with questions needing reasoning across documents or missing info.	– Risk of generating incorrect or unsupported info (hallucination). – Harder to trace answer origin (less transparent). – Typically larger models with more computation, and more complex training (may need vast data).

The next section dives deeper into abstractive QA to further elaborate.

6.1.2 Abstractive Question Answering

Abstractive question answering generates an answer in natural language, rather than extracting it directly from a text. In this approach, the model may paraphrase, synthesize, or elaborate on information from one or more sources to produce a standalone answer. For example, given the same context about the Gizmo Pro 3000, an abstractive QA system might answer *the question "How do I reset the Gizmo Pro 3000 if it does not turn on?" with a sentence like: "If your Gizmo Pro 3000 won't turn on, you should check the battery and hold the power button for five seconds to reset it."* Even if the context contains that information verbatim, the answer here is formulated as a helpful instruction in the model's own words (notice the slight rephrasing from the context sentence). Abstractive QA is closely related to text generation and summarization, since the model must understand the context and then *write out* an answer, potentially combining facts or explaining them. This makes it powerful for open-ended questions (e.g., "why" or "how" questions) where a direct quote might not suffice, but it also introduces challenges in ensuring the answer's correctness.

6.1.2.1 Approach and Model Architecture

Abstractive QA systems rely on generative language models. Two common architectures are encoder–decoder transformers (such as T5 or BART), which take in the question (and usually a supporting context) as input and output a sequence of answer tokens, and large decoder-only transformers (such as GPT-3/GPT-4 in ChatGPT), which generate an answer given a prompt that includes the question (and possibly context or few-shot examples). Unlike extractive models, which output start/end positions, generative models treat QA as a *sequence-to-sequence* task: the input is some representation of the question and related text, and the output is a new text sequence (the answer). For instance, T5 ("Text-to-Text Transfer Transformer") is explicitly designed to accept a text prompt and produce a text output; it can be fine-tuned for QA by feeding it prompts like "**question:** <*Q*> **context:** <*C*>" and training it to emit the answer.

The T5 model uses an encoder to read the combined question+context and a decoder to generate the answer word by word. GPT-3/GPT-4, on the other hand, are usually not fine-tuned for specific QA datasets but are so large (with knowledge from massive pretraining) that they can often answer questions in a zero-shot or few-shot fashion by being prompted appropriately. These models generate answers autoregressively (one token at a time) using their decoder, conditioned on the prompt.

Encoder–decoder models for QA can also be fine-tuned on data like SQuAD, MS MARCO, or ELI5 (a dataset for long-form answers), learning to produce answers that are both correct and well-phrased (Raffel et al., 2020, Lewis et al.).

As an example, let's revisit the customer support scenario. Suppose a user asks: "My Gizmo Pro 3000 won't turn on. What should I do?" The context contains the line: "If the device does not turn on, check the battery and hold the power button for 5 seconds to reset." An extractive QA system would likely just extract the phrase "check the battery and hold the power button for 5 seconds" as the answer. An abstractive system, however, can frame it as a complete sentence.

We can try this using a generative model like T5. For a code snippet illustrating how you might use Hugging Face Transformers to generate an answer, see the following on GitHub: `https://github.com/anvcse562/ModernNLP-LLMSLM/blob/main/Ch06/Ch6.1/QAWithContext.py`.

In this code, we prompt the model with a formatted string containing the question and context. The model's output, which is stored in `generated_answer`, generates the following output:

```
Check the battery and hold the power button for 5 seconds to reset
the device.
```

This is an abstractive answer: it conveys the same solution as the context, but phrased as a direct instruction to the user. Depending on the model and prompt, the phrasing could vary (for example, "You should try checking the battery and holding the power button for five seconds."). The key is that the model generated a helpful sentence that *did not exist verbatim* in the input text.

6.1.2.2 When to Use Abstractive QA

Abstractive QA shines in scenarios where natural-sounding answers are desired or where the question requires slight inference or synthesis. Some use cases include:

- **Customer support chatbots:** Users expect answers in complete sentences, often combining information. A generative model can take pieces of information from various knowledge base articles and compose a single coherent answer.

- **Why/how questions:** These often require summarizing explanations. For example, "Why is my laptop overheating?" might not have a word-for-word answer in documentation, but a generative model can compile an answer from related facts that may include but not be limited to cooling and ventilation.

- **Summarizing multiple sources:** An abstractive QA system can pull together facts from multiple documents. For instance, answering "What are the features of product X and Y?" by reading two product descriptions and writing a comparative answer.

- **Conversational assistants:** Systems like Alexa, Siri, and ChatGPT use generative QA to have fluid dialogues. They don't just quote Wikipedia; they formulate answers, sometimes with additional context or politeness.

As noted earlier in Table 6-1, abstractive models offer flexibility and a human-like touch. They can rephrase content to be more concise or clearer, and they are not strictly bound to the exact wording of a source. Indeed, encoder–decoder models such as BART have been shown to paraphrase and summarize as part of answering questions, much like a miniature summary specific to the question's context. For example, if the context text is verbose, a generative model can distill just the relevant pieces into the answer. This is analogous to how a person might read a paragraph and then answer a question about it in their own words.

6.1.2.3 Challenges

The power of generative QA comes with challenges:

- **Hallucination and accuracy:** Because the model can generate new text, it might include information that is not in the context or even false. For instance, if a detail is not provided, the model might fill in a plausible-sounding answer from its prior knowledge, which is sometimes outdated or even incorrect. Ensuring that the answer is grounded in verified information is an active area of research. Solutions include grounding the model via evidence retrieval or adding checks for factual consistency.

- **Transparency:** It's harder to explain why a generative model answered a certain way. With extractive QA, you can point to the exact source text that was returned. With abstractive QA, you might need to highlight which parts of the context contributed to different parts of the answer—this is non-trivial and sometimes opaque.

- **Resource intensity:** Generative models, especially large ones, are computationally heavier. For example, GPT-4 requires far more compute time to run than a smaller BERT QA model. Even moderately sized encoder-decoder models like T5-base, which has ~220M parameters, or T5-large with ~770M are slower and need more memory than a BERT-base with ~ 110M. In practice, this means abstractive QA can be more costly to deploy and may need infrastructure that supports model parallelism or acceleration.

Despite these challenges, the trend in industry has been toward using large, generative models for QA because of the superior experience they can provide for complex queries. As an example, OpenAI's ChatGPT (powered by GPT-3.5/GPT-4) became famous for its ability to answer a wide range of questions in detail. Such LLMs can be thought of as knowledge interfaces: rather than the user searching and reading multiple documents, the LLM has effectively "read" a huge portion of the Internet during training and can distill relevant information into a concise answer. This works remarkably well for many queries, although caution is required for questions that need up-to-date or authoritative answers (where retrieval from a current source and extractive quoting might be safer).

To concretely illustrate differences, consider if the user asked a question that wasn't directly answered in one place in the context, such as: "Can I extend the warranty of the Gizmo Pro 3000 beyond 18 months, and how?" Suppose our knowledge base has one article stating the warranty is 18 months but no mention of extensions, and another article that says "Extended warranties for all products can be purchased through our website within 30 days of purchase" an extractive QA system might retrieve the 18-month fact from the first article, but it would miss the second part unless explicitly directed to that second article with a multi-turn process. An abstractive system that can search or that has seen such data could attempt an answer with the following: "The Gizmo Pro 3000 comes with an 18-month warranty by default, but you can buy an extended warranty within 30 days of purchase if you need additional coverage." This kind of synthesis across sources is where abstractive QA (especially when augmented with retrieval of multiple documents) demonstrates its value, especially when augmented with retrieval of multiple documents.

In summary, extractive and abstractive QA are complementary: the former excels at precision and trustworthiness, while the latter excels at fluency and flexibility. Many real-world QA systems actually use a hybrid approach, first retrieving evidence and then using a generative model to formulate the answer, thereby combining the strengths of both.

6.1.3 Large vs. Small Language Models in QA

Thus far, we have mentioned small transformer models like BERT or even T5, as well as large models like GPT-4. It is important to clarify the distinction and discuss how model size impacts QA systems. SLMs typically refer to models with a relatively modest number of parameters (on the order of 10^7 to 10^8 parameters, i.e., up to a few hundred million). They are often pretrained on sizeable text corpora but then fine-tuned on specific tasks or domains with narrower data. BERT-base and RoBERTa-base are classic examples of SLMs in this sense. LLMs have vastly more parameters (on the order of 10^{10} to 10^{12}, i.e., tens of billions or more) and are trained on very large-scale datasets (essentially reading the Internet). They are often further refined with techniques like reinforcement learning from human feedback. GPT-3, GPT-4, PaLM, and LLaMA are examples of LLMs. LLMs usually require distributed computing and specialized hardware due to their size, and they demonstrate emergent capabilities in understanding and generating language that smaller models do not have.

In the context of question answering, the differences between using an SLM and an LLM are significant:

- **Knowledge and context:** A small model like BERT has a limited "knowledge base" in its parameters. It mostly relies on the provided context to answer a question. If the context doesn't contain the answer, a BERT-based extractive model will simply fail to find a correct answer. By contrast, an LLM like GPT-4 has been trained on a vast amount of world knowledge. It can sometimes answer questions *without any external context*, drawing on its internal knowledge. This is why ChatGPT can answer trivia questions or common-sense queries even if you don't give it a paragraph to read. Effectively it has already read about it during training. However, LLMs may have outdated or incomplete knowledge (e.g., a model trained in 2021 won't know events from 2023 unless updated). For up-to-date or niche queries, retrieval is still needed, but an LLM can integrate retrieved information with its own knowledge. Small models, in contrast, usually need a curated context or a database lookup to get the necessary facts.

- **Adaptability:** SLMs are typically fine-tuned for a specific QA task or domain. For example, you might fine-tune a RoBERTa model on an internal company Q&A dataset (product manuals, policy documents, etc.). It will then be very adept at extracting answers from those documents. But if you suddenly ask a question outside its domain, it won't generalize well. LLMs, however, are often used in a zero-shot or few-shot manner for QA—you can prompt them with a question and they'll attempt to answer, even if the topic is something they were never explicitly fine-tuned on. They have this capability because broad pretraining gives them a prior ability. This means LLMs offer more out-of-the-box versatility. On the other hand, if you have a very specialized domain (say, medical questions on specific clinical protocols), a fine-tuned small model might outperform a general LLM on that niche, especially if the LLM wasn't trained on those specifics.

- **Reasoning abilities:** One striking difference is the ability to perform complex reasoning or multi-step inference. Large models have demonstrated emergent reasoning capabilities that small models lack, unless those small models are given special training. For instance, simply prompting an LLM to "think step by step" (chain-of-thought prompting) can lead it to solve multi-hop questions or math problems that stump smaller models. Studies have found that small models have limitations in both knowledge memorization and reasoning compared to LLMs. Without extensive support, a small model cannot easily emulate the reasoning chains an LLM might generate. However, researchers have developed techniques like knowledge distillation and specialized prompting to improve small models, sometimes having LLMs teach SLMs how to reason.

In practice, if a user asks: "If the Gizmo's warranty started in January 2022 for 18 months, is it still valid in August 2023?", an LLM can parse this, do the date arithmetic or logic in a chain-of-thought internally, and answer "No, the warranty would have expired by July 2023."
A small extractive model cannot do this, because the answer isn't explicitly stated in any text—it requires reasoning. You would have to explicitly program a reasoning component or break the question into parts for a small model to handle it.

- **Resource and deployment considerations:** SLMs are lightweight enough to deploy on-premises or even on edge devices in some cases. They respond quickly and have lower cost per query. LLMs usually run on cloud servers with GPU/TPU acceleration and can be expensive to use (both in terms of infrastructure and inference time). For a real-time QA system with high volume, using an LLM for every query might be impractical, whereas a smaller model or a cascade (using a small model first, and only falling back to a big model for difficult questions) could be more efficient. There is active engineering work in distilling LLM knowledge into smaller models to get the "best of both worlds," but as of 2025, fully replacing an LLM with an SLM without loss of quality remains challenging for open-ended QA. For example, a distilled model like DistilBERT may answer "What is photosynthesis?" accurately, but struggle with more open-ended questions like "Why is photosynthesis important for life on Earth?," where larger models provide more complete and coherent answers.

Table 6-2 summarizes some key differences between using a typical small model (like BERT) versus a large model (like GPT-4) for QA.

Table 6-2. *Differences Between Small and Large Language Models for QA Tasks*

Feature/Capability	SLM (e.g., BERT/RoBERTa)	LLM (e.g., GPT-4/PaLM)
Parameter scale	~10^8 (hundreds of millions). Fits on single GPU/CPU memory.	~10^{11} (tens of billions or more). Requires multi-GPU or TPU, often accessed via cloud API.
Knowledge	Limited to training data domain and provided context. Needs relevant text input to answer correctly.	Broad knowledge from web-scale training. Can often answer common questions without extra context, but may be outdated on specifics.
Fine-tuning	Often fine-tuned on specific QA datasets or company data for best performance. Each new domain may require a new fine-tuning pass.	Generally used in zero-shot or with prompt engineering. Fine-tuning (or instruct-tuning) done by provider on broad instructions. Can adapt to many domains via prompts, sometimes with few-shot examples.
Answer style	Typically extractive or short answers. Limited ability to generate long explanations (unless explicitly trained to do so).	Can generate long, well-formed answers and even engage in multi-turn dialogue. Good for descriptive or explanatory answers.
Reasoning	Struggles with multi-hop reasoning or complex logic unless a separate reasoning mechanism is provided.	Exhibits emergent reasoning; can perform chain-of-thought reasoning and tool use when prompted. Better at answering complex, composite questions.
Speed and cost	Fast inference, can be run on commodity hardware. Cost-effective for large volumes.	Slower and more expensive per query. Often leveraged for quality on hardest queries or via managed services (with usage costs).

(*continued*)

Table 6-2. (*continued*)

Feature/Capability	SLM (e.g., BERT/RoBERTa)	LLM (e.g., GPT-4/PaLM)
Transparency	More interpretable if extractive (you know where the answer came from). Model decisions are somewhat easier to debug due to smaller size.	Can be opaque in how it derived an answer (huge parameter space). Harder to debug or trace. Needs techniques to verify answers (like asking it to cite sources, which itself can be hit or miss).
Example scenario	A QA bot that answers from a specific document collection (e.g., internal Wiki). The bot finds the relevant doc and extracts the answer. Best for targeted, domain-specific Q&A.	A general assistant that can answer *any* question (trivia, advice, multi-domain queries). It might not require an external document for common knowledge and can articulate a thorough answer (e.g., "Explain how the Gizmo Pro 3000's warranty works and how to extend it" combining known info and provided hints).

In practice, many QA systems use a hybrid of small and large models. For example, an application might use a lightweight extractive model to handle straightforward factoid questions using a known context, but escalate to a powerful LLM for more complex or open-ended questions (possibly with a higher latency or cost). Another strategy is using LLMs to generate training data or reasoning steps to assist small models—for example, using GPT-4 to generate question-answer pairs or explanations and then fine-tuning a smaller model on that augmented data. This way, the small model learns some capabilities from the large model. The choice between SLM and LLM ultimately depends on the requirements: accuracy vs. cost, open-domain vs. closed-domain, and the importance of answer verifiability.

6.1.4 LLM Agents and Multi-step Reasoning in QA

So far, we have discussed QA in a single-step fashion: a question comes in, and a model (small or large) produces an answer, possibly referencing a given text. However, complex information needs often require multiple steps of reasoning, research, or interaction—something beyond the scope of a single forward pass of a QA model. This is where the concept of LLM-based agents for question answering comes into play. An *LLM agent* is an autonomous system that uses an LLM as its core, but augments it with the ability to plan, take actions (like calling tools or queries), and iterate in order to answer a question or fulfill a task. In the context of QA, an LLM agent can break down a complicated question into sub-questions, search for information, perform calculations, and then synthesize the final answer.

To illustrate, imagine a user asks, "I bought a Gizmo Pro 3000 in January 2022. It stopped working now in August 2023—is it still under warranty, and if not, what can I do?" Answering this thoroughly requires several steps:

1. Determine the warranty period for Gizmo Pro 3000 (from documentation, we know it's 18 months).

2. Calculate if 18 months from Jan 2022 extends beyond Aug 2023 (18 months from Jan 2022 is July 2023, so by Aug 2023, the original warranty is expired).

3. Determine if there's an extended warranty or service program, which may exist in another document or knowledge entry.

4. Formulate an answer combining these insights: that the standard warranty has lapsed, but perhaps suggesting contacting support for out-of-warranty service or checking if an extended warranty was purchased.

A single-pass QA model, even a powerful one, might not reliably perform Step 2 (date arithmetic) or know to look up Step 3. An LLM agent, however, could handle this by planning the steps. For instance, it might internally generate a chain-of-thought like so: "(Step 1) The question asks if it's under warranty. I know the warranty is 18 months. Jan 2022 + 18 months = July 2023. Now it's Aug 2023, so the warranty has expired. (Step 2) The question also asks what can be done. Perhaps check for extended warranty info." The agent could then perform an action to retrieve "extended warranty Gizmo Pro 3000" from the knowledge base (a search query), find that information, and then compose the final answer by incorporating the expiration and the suggestion.

Recent research has introduced prompting techniques and frameworks to enable such behavior. Chain-of-thought (CoT) prompting is one approach, where you prompt the LLM to generate a step-by-step reasoning path. For example, you might prefix the model prompt with "Let's think step by step" to encourage it to break the problem into sub-tasks. This has been shown to significantly improve accuracy on multi-step problems by making the reasoning explicit. Another approach, known as *ReAct* (*reasoning and acting*), interweaves reasoning steps with action steps. In ReAct, the model's output is structured to alternate between *thought* (a reasoning step in natural language) and *action* (a command to use a tool or look up something), then *observation* (the result of that action), and so forth. This turns the QA process into an interactive loop. For instance, a ReAct-enabled agent might output: "Thought: I should check the extended warranty policy. Action: Search('Gizmo Pro 3000 extended warranty policy')". The environment (tool) executes that search and returns a snippet, which the agent then incorporates: "Observation: Found a document stating extended warranty must be purchased within 30 days. Thought: Now I can answer the user by saying it's out of warranty and no extension available." Finally it outputs the composed answer. Figure 6-1 shows this process. The central agent (an LLM, highlighted in red) is complemented by modules for planning (breaking the query into sub-tasks, e.g., via chain-of-thought reasoning), memory (short-term for context in the current session, and long-term via external storage to remember information across queries), and tools for taking actions. This architecture enables handling more complex queries than a single-pass QA system does.

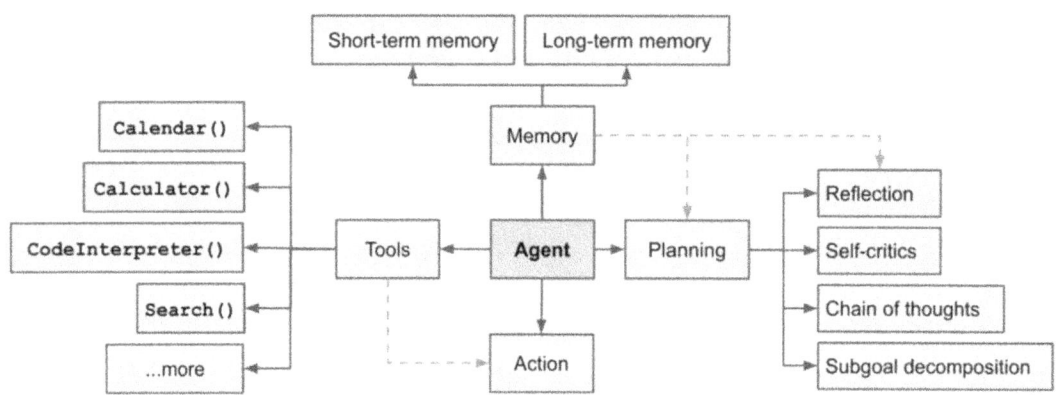

Figure 6-1. *An overview of an LLM-powered QA agent*

In implementing such agents, LLMs are typically used because they have the capacity to reason in natural language and can be instructed to follow the formats required for chain-of-thought or ReAct style prompts. They effectively serve as the "brain" of the agent, deciding what to do next at each step. The planning component may be implicit if is it emerging from the CoT prompt or explicit where the system enforces a planning step. The memory component often uses vector databases to store and retrieve relevant info from past interactions or large documents, which the LLM can query as needed. This is sometimes called *retrieval-augmented generation* in QA. Tool use is facilitated by integrating the QA agent with external APIs. For example, an agent might have a tool for document retrieval that can be used to get relevant text from a knowledge base, a tool for web search for open-domain questions requiring up-to-date information, or even a calculator for math. By calling these tools, the agent can fetch information beyond its internal knowledge and then reason about it.

It's worth noting that the idea of an LLM agent for QA blurs the line between "question answering" and "dialogue" or "problem solving." In effect, the agent is performing a mini-dialogue with itself or with external resources to arrive at an answer. For instance, multi-hop QA datasets like HotpotQA involve finding two or more pieces of evidence from different articles. They can be tackled by an agent that first finds one piece of evidence, then uses that to query for the next piece. Without an agent approach, a model might try to pack all evidence into one giant context and answer in one go, which is less efficient and may not even be feasible if the context is very large. Agents provide a way to decompose and conquer complex queries.

From an engineering perspective, LLM-based QA agents are more complex to implement than single-turn QA systems. They require managing the interaction loop, keeping track of the conversation or reasoning trace, and possibly safeguarding against the agent going off track. This is to prevent large models given free rein from generating irrelevant actions or getting confused. However, frameworks and libraries like LangChain in Python, which emerged to simplify development of such LLM-driven agent loops, have made it easier to build these systems. They handle the prompt formatting and tool API integration, allowing developers to focus on the knowledge sources and the specific tasks.

6.1.4.1 Practical Example

Going back to the support example, an LLM agent might handle a user query in the following way: "I've tried resetting my Gizmo Pro 3000 and it still won't work. It's out of warranty — what are my options?" The agent might perform the following actions in response:

- Recognize this is about a device not working and a warranty (perhaps plan to retrieve troubleshooting steps and warranty policy).
- Use a Search tool on the support knowledge base for "Gizmo Pro 3000 not working after reset".

- Find a document about hardware repair services.

- Use another tool or step to check the warranty policy (which it might recall is expired in this case).

- Finally, compose an answer suggesting the user contact technical support for out-of-warranty repair or consider purchasing a new device if repair is costly.

Such an answer is nuanced and might involve sentences such as, "Since you've already attempted a reset and the device is out of its 18-month warranty, your best option is to contact our support team for a potential repair or replacement. Although the standard warranty has expired, our team can guide you on out-of-warranty service options." This goes beyond a straightforward QA pair; it's more of an advice or resolution answer that requires integrating facts (warranty duration) with user context (already tried reset, device still not working) and company policy (support offers out-of-warranty repairs).

6.1.4.2 The Importance of LLM Agents

As user queries become more complex, especially in professional applications like customer support, business intelligence, and medical Q&A, the ability to do multi-step reasoning and tool usage becomes critical. LLM agents can also ask clarifying questions back to the user if needed, although that borders on interactive QA or dialogue, which might be beyond the scope. However, it is a logical next step. For example, if a question is ambiguous, an agent could respond with a question for the user: "Are you referring to the standard warranty or an extended warranty?" This sort of clarification can dramatically improve the quality of the QA system.

To conclude this subsection, LLM-based QA agents represent the frontier of QA systems—moving from static question-answer pairs to a more dynamic, reasoning-driven process. They leverage the full power of LLMs not just to answer questions, but to figure out *how* to answer them by breaking them down and using available tools and information. This approach aligns with how human experts might approach a hard question: by researching, analyzing, and then answering. It demonstrates the versatility of LLMs and sets the stage for building real-world QA applications that are robust and intelligent.

Now that we have covered the spectrum of QA techniques—from straightforward extractive methods to advanced LLM agents—you are ready to learn about how these can be implemented in practice. The next section explores the *real-world implementation* of QA systems, particularly focusing on how to deploy these models and agents using AWS tools and cloud services. You will see how services like Amazon's machine learning platform can be used to host QA models, how AWS can facilitate document retrieval (e.g., with Amazon Kendra for enterprise search), and how to integrate the components of a QA system into a scalable application. This will bridge the gap between the concepts you've learned and a production-ready QA solution.

6.2 Implementation of QA Systems on AWS

6.2.1 Extractive vs. Abstractive QA on AWS

To refresh your memory, *extractive QA* involves identifying and returning a span of text from a provided document or context that directly answers the user's question. The model does not generate new phrasing; it literally "extracts" the answer from the source text. For example, given a passage and a question, an extractive model (often based on BERT or RoBERTa architectures) pinpoints the exact segment in the text that answers the question. In AWS implementations, extractive QA typically uses smaller pretrained models fine-tuned on QA tasks (e.g. RoBERTa on SQuAD dataset) and can be deployed in a resource-efficient manner

Abstractive QA (also called *generative QA*) involves generating a novel answer phrased in natural language, which may not appear verbatim in the source text. These models can synthesize information from the context and produce a concise answer in their own words. Examples include architectures like T5 or LLMs like GPT-4. Abstractive QA on AWS typically leverages LLMs and requires more heavy-duty infrastructure.

6.2.1.1 Pros and Cons of Extractive vs. Abstractive QA

Both approaches have distinct advantages and trade-offs. Table 6-3 shows a comparison and decision guide to help you determine when to use extractive QA and when to use abstractive QA in an AWS deployment.

Table 6-3. When to Use Extractive QA vs. Abstractive QA in an AWS Deployment

Criteria	Extractive QA (Span Extraction)	Abstractive QA (Generative)
Model examples	BERT, RoBERTa, DistilBERT fine-tuned on QA (100M or fewer parameters).	T5, FLAN-T5, GPT-3/4, Claude (billions of parameters, LLMs).
AWS deployment	Can run on Lambda (for smaller models) or modest SageMaker instances. Uses CPU or small GPU. Example: RoBERTa on a CPU Lambda with EFS for model storage, AWS Lambda Quotas.[1]	Requires powerful hardware (GPU/Inferentia) or a managed service. Typically deployed on SageMaker GPU instances or via Bedrock API for fully managed services. Large memory footprint.
Answer source	Draws answers directly from input text—guarantees the answer is a substring of the provided context.[2] No new information is introduced (higher factual accuracy relative to content).	Generates answers in its own words—can rephrase or summarize information. May include relevant info not verbatim in text, but there is a risk of hallucination (model might fabricate info not in source) if not properly constrained.
Pros	– High fidelity to source data (answers are literally from the context, so accuracy is easier to verify). – Efficiency: Smaller model sizes yield fast inference (tens of milliseconds) and lower cost per query. – Deterministic output: Doesn't create new phrasing, so less variability across runs.	– Fluency and clarity: Can produce well-formed natural language answers, even combining info from multiple sentences or summarizing long text. – Flexibility: Can answer questions that require synthesis or an answer not explicitly present in text (e.g. summarizing a paragraph). – Context integration: Able to incorporate general world knowledge or reasoning beyond the exact text (useful for open-ended questions).

(continued)

---

[1] https://docs.aws.amazon.com/lambda/latest/dg/gettingstarted-limits.html#:~:text=50%20MB%20,larger%20files%20with%20Amazon%20S3
[2] https://medium.com/data-science/extractive-vs-generative-q-a-which-is-better-for-your-business-5a8a1faab59a

Table 6-3. (*continued*)

Criteria	Extractive QA (Span Extraction)	Abstractive QA (Generative)
Cons	– Limited expression: Cannot go beyond phrases in the source; may return awkwardly worded or partial-sentence answers if that's what the source contains. – Requires context retrieval: If answer isn't in provided text, the model can't generate it. Often paired with a document retrieval system for broader knowledge, which adds complexity. – Short answers only: Typically returns a brief span; not suitable for explanatory answers or summarization.	– Accuracy concerns: May introduce errors or facts not in the source. Requires careful prompt design or grounding techniques (e.g. retrieval-augmented generation) to ensure factuality. – Resource intensive: Large models have higher latency (hundreds of ms or seconds per query) and higher compute cost. Tuning and deploying these models is more complex. – Determinism: Generation is probabilistic; the same question might yield slightly different phrasings, which can be an issue for strict consistency requirements.

6.2.2 Model Selection Strategies for QA

Choosing the right model for a QA system involves balancing model size, capability, latency, cost, and customization needs. This section outlines criteria for selecting between LLMs and SLMs and maps common models to AWS deployment options.

When deciding between using a large model (like GPT-4 or other billion-parameter models) versus a smaller model (like DistilBERT or MiniLM), consider the following criteria:

- **Query complexity and knowledge scope:** Simple fact-based questions or those answerable from a specific document can often be handled by smaller extractive models. Complex queries that require reasoning, summarizing multiple sources, or general world knowledge might necessitate a powerful LLM. If your QA needs rarely go beyond the content of a single document or straightforward fact lookup, an SLM is usually sufficient. Conversely, if users may ask open-ended or analytical questions for example, "Compare these two policies and highlight differences," an LLM's broader reasoning ability is beneficial.

- **Latency requirements:** Smaller models typically have much lower latency. An SLM with tens of millions of parameters can return answers in tens of milliseconds on a CPU,[3] whereas an LLM with tens of billions of parameters might take hundreds of milliseconds to seconds and likely requires GPU acceleration. In real-time applications, such as an interactive website or voice assistant, a fast response is crucial. If you need sub-100ms responses or have strict SLA requirements, lean toward smaller models or highly optimized medium-sized models. LLMs might be reserved for asynchronous tasks or cases where an ~1-2 second response is acceptable.

- **Throughput and scalability:** Related to latency is how many queries per second (QPS) the system must handle. Smaller models with many parallel instances or Lambda functions can be scaled out easily at relatively low cost. LLMs are heavier—you might need multiple GPUs or instances just to handle a single request at a time for very large models. For high QPS scenarios, using a fleet of SLM-powered endpoints (or Lambda with provisioned concurrency) can be more cost-effective and easier to scale than an LLM deployment.

- **Inference cost:** Larger models incur higher compute costs per inference. They require more CPU/GPU time per query and often run on more expensive hardware. For example, GPT-4 via API has an associated cost of approximately $0.06 per 1,000 output tokens that is orders of magnitude higher per query than running a smaller open-source model on your own instance.[4] If your application must serve

---

[3] https://aws.amazon.com/blogs/machine-learning/achieve-four-times-higher-ml-inference-throughput-at-three-times-lower-cost-per-inference-with-amazon-ec2-g5-instances-for-nlp-and-cv-pytorch-models/#:~:text=16%20Full%20651%20158%2025,5X

[4] https://aws.amazon.com/blogs/machine-learning/achieve-four-times-higher-ml-inference-throughput-at-three-times-lower-cost-per-inference-with-amazon-ec2-g5-instances-for-nlp-and-cv-pytorch-models/#:~:text=16%20Full%20651%20158%2025,5X

thousands of queries per day (or more) on a tight budget, smaller models or optimized medium-sized models are preferable. On AWS, compare hosting an SLM on an EC2 or SageMaker instance (fixed hourly cost, but can serve many queries) versus paying per request for an LLM API. There is a breakeven point where beyond a certain traffic volume, hosting a model yourself on SageMaker is cheaper than paying per call for an external LLM.

- **Customization and domain adaptation:** If your QA needs are domain-specific, such as with medical documents or legal papers, you may require fine-tuning a model on your domain data for best results. Fine-tuning a 100M parameter model is far easier and cheaper (both in data needed and compute) than fine-tuning a 30B parameter model. SLMs are often sufficient when you have a well-defined domain and can fine-tune them to achieve high accuracy on in-domain questions. LLMs, on the other hand, often come pretrained with broad knowledge and sometimes can be used as zero-shot or with prompt engineering on domain context. If using an LLM via Bedrock or API, fine-tuning might not even be available or comes with additional cost. Some providers allow custom fine-tunes, but at a premium. In scenarios where data privacy or customization is critical, deploying your own mode, possibly an open-source LLM that you can fine-tune and run in-house on SageMaker might be necessary, versus sending data to a third-party API.

- **Memory and compute constraints:** This is a practical consideration on what your environment can support. AWS Lambda, for instance, maxes out at ~10 GB memory for a function and has no GPU, so it cannot host a model bigger than what fits in RAM or requires GPU acceleration. If your use case is constrained to Lambda in cases where you want a completely serverless architecture, you're inherently limited to smaller models or techniques like model quantization. In contrast, SageMaker can handle very large models using multi-GPU instances or even multi-node model parallel deployments (SageMaker supports model parallelism libraries for extreme scales). Thus, your infrastructure choice might dictate the model size threshold.

Table 6-4 shows how to map common QA-related models to AWS services, highlighting their sizes, typical performance, and where they can be deployed.

Table 6-4. Mapping Models to AWS Deployment Options

Model	Type	Size (Parameters)	Typical Latency	AWS Deployment Options	Notes
MiniLM/ DistilBERT Small BERT-derived models	Extractive SLM	~22M (MiniLM) ~66M (DistilBERT)	Very low (≈20–50 ms per query on CPU)	— Lambda (fits within memory; can use container image for ~200MB model) — SageMaker Serverless or small instance (e.g., `ml.t2.medium`)	Ideal for simple QA snippets and high QPS needs. Fine-tunable on domain data cheaply. Low cost per inference (fractions of a cent) due to small size.
RoBERTa-base and BERT-base variants	Extractive SLM	~125M	Low (≈50–100 ms CPU, <50 ms on GPU)	— Lambda (with EFS if package >250MB) for smaller BERT variants — SageMaker real-time on CPU (`ml.m5.xlarge`) or GPU (`ml.g4dn.xlarge`).	Strong general-purpose extractive QA model. Often pretrained on SQuAD. Might need ~1-2 GB RAM. On AWS, can use Hugging Face DLC on SageMaker for easy deployment.

(continued)

Table 6-4. (*continued*)

Model	Type	Size (Parameters)	Typical Latency	AWS Deployment Options	Notes
RoBERTa-large and BERT-large variants	Extractive SLM	~355M	Medium (~200 ms on CPU, ~50–100 ms on GPU)	— SageMaker real-time (better on GPU or AWS Inferentia for cost). — Not ideal for Lambda (memory ~1.3GB just for model).	Higher accuracy on QA than base models but significantly larger. SageMaker + AWS Inferentia (Inf1/Inf2 instances) can significantly reduce cost per inference for these models (see huggingface.co).
FLAN-T5 Large (e.g., FLAN-T5-XL)	Abstractive LLM	3B–11B (varies by version)	High (500 ms – 2 s on GPU, depending on size and sequence length)	— SageMaker real-time on GPU (ml.g5.2xlarge for 3B, up to ml.p4d for 11B). — SageMaker Async or Batch for longer texts. — Possibly run on multiple CPUs (not recommended for 11B).	Open-source seq2seq model fine-tuned for QA and dialogue. Requires significant memory (11B ~40GB in FP32, can use FP16 ~20GB). Good candidate for SageMaker JumpStart (pretrained available). Supports abstractive answers with relatively fluent output.

(*continued*)

Table 6-4. (*continued*)

Model	Type	Size (Parameters)	Typical Latency	AWS Deployment Options	Notes
Anthropic Claude 2 via Bedrock	Abstractive LLM	~? (Large, proprietary)	High (1–3 s, can handle very long context)	— Amazon Bedrock AI service (fully managed API). — No self-hosting (closed model).	Extremely capable for summarization and QA, with 100K token context window. Bedrock handles scaling. Pricing is per million tokens (~$11 per million input, $32 per million output for Claude 2). Good for enterprise scenarios needing complex reasoning with AWS integration.
OpenAI GPT-4 via API	Abstractive LLM	~>100B (estimate)	High (2–5 s for typical queries)	— External API (OpenAI or Azure OpenAI) — (Not natively on AWS; integration via Lambda or SageMaker calling API).	Widely regarded as state-of-art in reasoning and QA. Very high cost: ~$0.03 per 1K prompt tokens and $0.06 per 1K generated tokens (see news.ycombinator.com) (roughly $0.009–$0.012 per typical query). Use for premium scenarios or where maximum quality is needed. Ensure to handle API credentials and latency in architecture.

(*continued*)

Table 6-4. (*continued*)

Model	Type	Size (Parameters)	Typical Latency	AWS Deployment Options	Notes
Mistral-7B (open LLM)	Abstractive LLM	7B	Medium-high (300–700 ms on single GPU, longer on CPU)	— SageMaker on a single GPU instance (e.g. `ml.G5.4xlarge` with 1 x NVIDIA A10G, 24GB VRAM, which can host 7B in 8-bit). — Or multi-CPU with deep learning containers (but slower).	Newer open-source model known for efficiency. 7B can handle moderate complexity QA when fine-tuned. No cost per token—cost is instance hours. Good middle-ground if open license required (can be self-hosted to avoid external API).

Note Table 6-4 provides general guidance. Latency varies with input size (question and context length) and hardware optimizations. AWS offers GPU instances like the G5 (A10G), p3/p4 (V100/A100), and AWS Inferentia chips (Inf1, Inf2) that can significantly lower cost for hosting medium-sized models like BERT-large or T5 by using optimized Inferentia SDKs.[5] For example, using an Inferentia2-based instance for BERT may improve throughput and price-performance over a GPU.

---

[5] https://aws.amazon.com/blogs/machine-learning/achieve-four-times-higher-ml-inference-throughput-at-three-times-lower-cost-per-inference-with-amazon-ec2-g5-instances-for-nlp-and-cv-pytorch-models/#:~:text=16%20Full%20651%20158%2025,5X

6.2.2.1 Best Practices in Model Selection

- **Benchmark candidate models on your data and typical questions:** Measure latency on the target AWS infrastructure. For example, invoke a SageMaker endpoint or Lambda function to ensure it meets requirements. Sometimes a well-tuned smaller model can meet accuracy needs with a tiny fraction of the cost and latency of an LLM.

- **Start simple, then scale up:** Begin with the simplest model that might work (say, a distilled extractive model). Only graduate to a larger model if required by accuracy or capabilities. This follows the principle of cost-optimization and simplicity.

- **Consider ensemble or cascaded approaches:** For instance, run an extractive model first; if it finds a high-confidence answer, return it. If not, fall back to a more powerful generative model. This combo can give the best of both worlds—minimizing cost and latency for easy questions, but handling hard questions with an LLM when needed.

- **Leverage AWS JumpStart and model hubs:** SageMaker JumpStart provides pre-built models (including QA models like `question-answering-roberta-base`) that can be deployed with a few clicks or API calls. This can accelerate experimentation. AWS Marketplace also has vendors offering fine-tuned QA models that you can deploy on SageMaker.

- **Monitor and iterate:** Whichever model you choose, monitor its performance in production. If you find queries that consistently fail or are too slow, that may signal the need for a different model or architecture adjustment (such as adding a caching layer for frequent queries, or fine-tuning the model on those troublesome queries).

6.2.2.2 General Best Practices for Fault Tolerance

- **Graceful degradation:** Design the user experience so that if the QA system is unavailable, the system can either queue the request for later processing or return a partial answer. For instance, if the abstractive model fails, perhaps return, "Sorry, I can't generate an answer right now, but here are some relevant facts:" and then list a few extractive snippets. This way the user isn't left empty-handed.

- **Timeouts and limits:** Always enforce limits on input size (to prevent a user from accidentally or maliciously sending a huge document that crashes the system) and timeouts on processing. Combine this with proper user messaging—for example, if a question times out internally, return a message such as, "The question is taking longer than expected to answer. Please try rephrasing or simplifying your question."

- **Testing and chaos engineering:** Regularly test your QA system with simulated failures. For example, deliberately shut down one of the SageMaker instances or have the model code throw an exception, and ensure your higher-level system (Lambda, API Gateway, etc.) properly handles it. Use AWS Fault Injection Simulator for more systematic chaos testing if the scale and criticality of your application warrant it.

- **Cold start mitigation:** For Lambda, as mentioned, use Provisioned Concurrency for critical paths or keep the function warm via a scheduled ping if full provisioned concurrency is not in budget. For SageMaker, note that when scaling from 0 to 1 instance (in serverless or if you had an endpoint down at night and then deploy in morning), there will be a cold start as the container loads the model. You might hide this from users by warming up the model (e.g., send a test ping query at deployment time or schedule a CloudWatch Events rule to invoke a dummy query periodically). In Bedrock's case, cold starts are abstracted away but there could be slight delays on the first invocation as the model scales internally; not much you can do there except possibly send a warm-up call.

6.2.3 Cost and Performance Benchmarks

Cost and performance are key considerations for a production QA system. This section provides sample latency and cost data for various model+deployment pairs and gives tips for optimizing throughput and cost on AWS. Note that actual numbers will vary based on AWS region, instance type, model optimization, and query length, but the following serves as a rough reference.

Table 6-5 compares several deployment options in terms of average latency per query and estimated cost per 1,000 queries. Latency is measured as end-to-end inference time (not including network overhead to client), and cost per 1,000 assumes typical usage patterns (for on-demand infrastructure, we amortize hourly costs; for APIs, we use token pricing). These are illustrative values.

Table 6-5. Example Latency and Cost Metrics

Model and AWS Service	Latency (ms)	Cost per 1,000 Queries (USD)	Notes
DistilBERT (66M) on Lambda Memory: 2048 MB, Provisioned Concurrency	~50 ms (warm) ~200 ms (cold)	~$0.50 [1]	Fast for short text. Cost estimate: $0.0005 per query based on 100 ms billing increments on 2048 MB Lambda (approximately $0.000033 per 100ms invoke).[6] Cold starts add 100s of ms occasionally.
RoBERTa-base (125M) on SageMaker CPU Instance: ml.m5.xlarge (4 vCPU)	~80 ms	~$1.20	Good throughput on CPU when serving multiple queries. Cost assumes instance at ~$0.20/hour serving ~60K queries/hour. If fully utilized, cost per 1,000 can be as low as $0.20/60 = $0.0033. Here we assumed only ~20% utilization for a moderate load scenario (hence ~$1.2/1,000).

(*continued*)

---

[6] https://aws.amazon.com/blogs/machine-learning/achieve-four-times-higher-ml-inference-throughput-at-three-times-lower-cost-per-inference-with-amazon-ec2-g5-instances-for-nlp-and-cv-pytorch-models/#:~:text=16%20Full%20651%20158%2025,5X

Table 6-5. (*continued*)

Model and AWS Service	Latency (ms)	Cost per 1,000 Queries (USD)	Notes
RoBERTa-large (355M) on SageMaker GPU Instance: ml.G4dn.xlarge (1xT4 GPU)	~30 ms	~$2.50	GPU significantly lowers latency for large model. Instance ~$0.50/hour. If it can serve ~200K queries/hour with batching, cost per 1,000 could approach $0.50/200 = $0.0025.[7] Here assumed lower utilization or no batching. Utilizing AWS Inferentia Inf1 instances could reduce cost per inference by up to 40% further.[8]
FLAN-T5 Large (780M) on SageMaker GPU Instance: ml.G5.xlarge (1x NVIDIA A10G)	~300 ms	~$14.00	Generative model producing a few sentences. Instance ~$1.0/hour. At ~3 q/s (due to heavier compute), ~10K queries/hour -> $1.0/10 = $0.10 per 1,000 queries. But generative models also incur longer outputs (more tokens). We include an approximate cost of cluster utilization and overhead. (If usage is lower, cost per actual query increases due to idle time.)

(*continued*)

---

[7] https://aws.amazon.com/blogs/machine-learning/achieve-four-times-higher-ml-inference-throughput-at-three-times-lower-cost-per-inference-with-amazon-ec2-g5-instances-for-nlp-and-cv-pytorch-models/#:~:text=16%20Full%20651%20158%2025,5X

[8] https://aws.amazon.com/blogs/machine-learning/achieve-four-times-higher-ml-inference-throughput-at-three-times-lower-cost-per-inference-with-amazon-ec2-g5-instances-for-nlp-and-cv-pytorch-models/#:~:text=16%20Full%20651%20158%2025,5X

Table 6-5. (*continued*)

Model and AWS Service	Latency (ms)	Cost per 1,000 Queries (USD)	Notes
GPT-4 via OpenAI API (8K context)	~2000 ms (2 sec)	~$30.00	Extremely high-quality answers, but high latency and cost. Roughly $0.03 per 1K prompt tokens and $0.06 per 1K output.[9] For an average question+answer of 150 tokens, that's ~$0.009 per query, i.e. $9 per 1,000. If questions or answers are longer, costs scale up. We show $30 as an upper-bound scenario for more complex queries. No infrastructure maintenance needed, but cost is directly proportional to use.
Claude Instant via Bedrock (Anthropic)	~1,000 ms (1 sec)	~$5.00	Faster and cheaper than GPT-4 for many tasks. At ~$1.63 per million input tokens and $5.51 per million output tokens,[10] a short QA exchange might cost ~$0.002–0.005. Bedrock charges by tokens; $5 per 1,000 here assumes moderately long answers or multiple turns. Latency ~1s for a few hundred tokens with the Claude Instant model.

(*continued*)

---

[9] https://aws.amazon.com/blogs/machine-learning/achieve-four-times-higher-ml-inference-throughput-at-three-times-lower-cost-per-inference-with-amazon-ec2-g5-instances-for-nlp-and-cv-pytorch-models/#:~:text=16%20Full%20651%20158%2025,5X

[10] https://aws.amazon.com/blogs/machine-learning/achieve-four-times-higher-ml-inference-throughput-at-three-times-lower-cost-per-inference-with-amazon-ec2-g5-instances-for-nlp-and-cv-pytorch-models/#:~:text=16%20Full%20651%20158%2025,5X

Table 6-5. (*continued*)

Model and AWS Service	Latency (ms)	Cost per 1,000 Queries (USD)	Notes
Mistral-7B on SageMaker CPU (8xlarge)	~1,000 ms (1 sec)	~$2.00	If running on CPU only (ml.m5.8xlarge with 32 vCPUs), the 7B model with 4-bit quantization can serve ~1 query/sec. Instance ~$1/hour, so ~$1/3600 ≈ $0.28 per 1,000 if fully utilized (here assumed partial utilization and overhead for ~ $2 per 1,000). On GPU (ml.g5.2xlarge), latency would drop to ~200 ms and cost might be slightly higher per hour but more queries/sec.

To interpret Table 6-5, extractive models (upper rows) can be extremely cheap per query when self-hosted on AWS—on the order of fractions of a cent or less.[11] For instance, running a BERT-base model on a GPU with high throughput can reduce cost per 1,000 queries to well under $1 if the GPU is kept busy.

6.2.4 Conclusion

This section explored three scalable and production-ready strategies for implementing QA systems on AWS:

- Training and hosting QA models using Amazon SageMaker
- Deploying SLMs in AWS Lambda for low-latency FAQ use cases
- Using hybrid LLM+SLM architectures with routing logic for optimized cost/performance tradeoffs

---

[11] https://aws.amazon.com/blogs/machine-learning/achieve-four-times-higher-ml-inference-throughput-at-three-times-lower-cost-per-inference-with-amazon-ec2-g5-instances-for-nlp-and-cv-pytorch-models/#:~:text=16%20Full%20651%20158%2025,5X

In the accompanying Jupyter notebooks, we demonstrate:

1. Fine-tuning RoBERTa on SQuAD using SageMaker
2. Deploying a MiniLM model on Lambda
3. Orchestrating hybrid QA with an LLM router function

Refer to the code on GitHub in the ch6.2 folder:

https://github.com/anvcse562/ModernNLP-LLMSLM/blob/main/Ch06/ch6.2/ch%20
6.2%20fine_tune_roberta_squad_sagemaker.ipynb

https://github.com/anvcse562/ModernNLP-LLMSLM/blob/main/Ch06/ch6.2/ch%20
6.2%20deploy_minilm_lambda.ipynb

https://github.com/anvcse562/ModernNLP-LLMSLM/blob/main/Ch06/ch6.2/ch%20
6.2%20llm_slm_router_lambda.ipynb

The next section explores open-source approaches to QA using Hugging Face Transformers, including PEFT, evaluation metrics, and model selection strategies.

6.3 Open-Source Implementation

This section explores how to build QA systems using open-source tooling—with a particular focus on Hugging Face Transformers. It starts by introducing core concepts for implementing extractive and generative QA using classical models like BERT and T5. It then expands into modern SLMs and LLMs, including instruction-tuned architectures such as FLAN-T5-XXL and Mistral-7B, as well as energy-efficient innovations like Microsoft's BitNet-b1.58. Finally, the section dives into fine-tuning strategies, deployment pipelines, and performance evaluation metrics—equipping you with foundational and future-ready skills for developing state-of-the-art QA systems.

6.3.1 Implementing QA with Hugging Face Transformers

The Hugging Face Transformers library provides an accessible yet powerful interface for building QA pipelines, offering pretrained models, tokenizers, datasets, and training utilities.

Recall that there are two dominant paradigms for QA: extractive and generative. Extractive models locate an answer span within a reference context. In contrast, generative models produce answers in natural language, synthesizing relevant information from the question and context.

6.3.1.1 LLMs for Generative QA

LLMs are particularly well-suited for open-domain or abstractive QA. These models treat the QA task as a sequence-to-sequence problem, enabling responses that extend beyond direct context matching.

- **T5 (text-to-text transfer transformer):** A versatile model that converts every task into a text generation problem. T5 performs well on datasets like SQuAD, TriviaQA, and Natural Questions.

- **FLAN-T5-XXL:** A highly capable instruction-tuned variant of T5 with 11 billion parameters. It achieves state-of-the-art performance (94.3% EM) on benchmarks such as natural questions.

- **Mistral-7B-Instruct:** A compact yet powerful generative model, fine-tuned for instruction-following tasks. Its extended 32K token context makes it ideal for document-level QA and long-form reasoning.

- **BitNet-b1.58 (2B):** Microsoft's novel 1.58-bit quantized LLM demonstrates that efficiency doesn't have to sacrifice capability. This 2B parameter model performs lossless generative inference with up to 6.17× speedup on CPUs, all while slashing energy usage by more than 70%.

Here is BitNet in action for a generative QA task:

```
# Example: Generative QA with BitNet
from bitnet import BitNetForCausalLM, BitNetTokenizer

model = BitNetForCausalLM.from_pretrained("microsoft/bitnet-b1_58-2B")
tokenizer = BitNetTokenizer.from_pretrained("microsoft/bitnet-b1_58-2B")

inputs = tokenizer("Q: What is quantum computing? Context: Quantum computing uses qubits.", return_tensors="pt")
outputs = model.generate(**inputs, max_length=100)
print(tokenizer.decode(outputs[0]))
```

You can find the full implementation code at: https://github.com/anvcse562/ModernNLP-LLMSLM/blob/main/Ch06/Ch6.3/1_hf_transformers_qa.ipynb.

6.3.1.2 SLMs for Extractive QA

For many real-world applications—especially on-device or low-latency platforms—extractive QA with smaller models is the optimal strategy.

- **BERT-large:** The original transformer-based extractive QA benchmark.

- **DistilBERT:** A compressed version of BERT, 40% smaller but with 95% of the accuracy.

- **E5-small-v2:** An efficient, multilingual embedder achieving 82.4 F1 on XQuAD.

- **DeBERTa-v3:** Uses disentangled attention to outperform BERT on SQuAD 2.0 (91.7 EM).

- **Qwen2-7B:** With a 128K context length, this model is ideal for document-rich QA scenarios.

6.3.2 Fine-Tuning T5 for Generative QA

To harness the generative power of models like T5 for specific QA tasks, fine-tuning on relevant datasets is key. This process allows the model to learn to synthesize fluent and contextually appropriate answers.

```
from transformers import T5ForConditionalGeneration, T5Tokenizer
from datasets import load_dataset
# Load dataset and model
dataset = load_dataset("squad")
model_name = "t5-small"
tokenizer = T5Tokenizer.from_pretrained(model_name)
model = T5ForConditionalGeneration.from_pretrained(model_name)
# Preprocessing
def preprocess_function(example):
    input_text = f"question: {example['question']} context: {example['context']}"
    target_text = example["answers"]["text"][0] if example["answers"]["text"] else ""
```

```
    model_inputs = tokenizer(input_text, max_length=512, truncation=True,
    padding="max_length")
    labels = tokenizer(target_text, max_length=32, truncation=True,
    padding="max_length").input_ids
    model_inputs["labels"] = labels
    return model_inputs
tokenized_dataset = dataset.map(preprocess_function, batched=True)
# Training setup
from transformers import TrainingArguments, Trainer
training_args = TrainingArguments(output_dir="./t5-qa", num_train_epochs=3,
per_device_train_batch_size=8)
trainer = Trainer(model=model, args=training_args, train_dataset=tokenized_
dataset["train"], eval_dataset=tokenized_dataset["validation"])
trainer.train()
```

You can find the full implementation code at: https://github.com/anvcse562/ModernNLP-LLMSLM/blob/main/Ch06/Ch6.3/2_t5_fine_tuning.ipynb.

6.3.3 Fine-Tuning DistilBERT for Extractive QA

When the task calls for identifying a precise answer span within a text, models like DistilBERT are excellent choices. Fine-tuning them on extractive QA datasets hones their ability to pinpoint these spans accurately.

```
from transformers import DistilBertForQuestionAnswering,
DistilBertTokenizer
from datasets import load_dataset

# Load model and tokenizer
model = DistilBertForQuestionAnswering.from_pretrained("distilbert-base-
uncased")
tokenizer = DistilBertTokenizer.from_pretrained("distilbert-base-uncased")
dataset = load_dataset("squad")
```

```
# Preprocessing
def preprocess_function(example):
    return tokenizer(example["question"], example["context"],
    truncation=True, padding="max_length")

tokenized_dataset = dataset.map(preprocess_function, batched=True)

# Training
from transformers import TrainingArguments, Trainer
training_args = TrainingArguments(output_dir="./distilbert-qa", num_train_epochs=3, per_device_train_batch_size=8)
trainer = Trainer(model=model, args=training_args, train_dataset=tokenized_dataset["train"], eval_dataset=tokenized_dataset["validation"])
trainer.train()
```

You can find the full implementation code at: https://github.com/anvcse562/ModernNLP-LLMSLM/blob/main/Ch06/Ch6.3/3_distilbert_fine_tuning.ipynb.

6.3.4 Next-Gen Fine-Tuning Strategies

As models continue to scale in size, the computational demands of fine-tuning can become prohibitive. Enter parameter-efficient fine-tuning (PEFT) techniques, which offer ingenious ways to adapt large models with significantly less computational overhead, often with minimal impact on performance. Methods like QLoRA (Quantization-aware Low-Rank Adaptation) and DoRA (Decomposed and Recomposed Attention) present compelling alternatives. Consider the following comparative snapshot:

Method	VRAM Savings	Accuracy Retention
QLoRA	70%	~98%
DoRA	65%	~97%
Prefix tuning	80%	~94%

To illustrate the practical application of such techniques, the following Python snippet using the `peft` and `trl` libraries demonstrates how to configure and initiate training with DoRA:

```
from peft import LoraConfig
from trl import SFTTrainer
peft_config = LoraConfig(r=64, target_modules=["q_proj", "v_proj"], use_dora=True)
trainer = SFTTrainer(model=model, train_dataset=dataset, peft_config=peft_config, max_seq_length=4096)
trainer.train()
```

You can find the full implementation code at: https://github.com/anvcse562/ModernNLP-LLMSLM/blob/main/Ch06/Ch6.3/4_peft_dora_training.ipynb.

6.3.5 Deployment of QA Models

QA systems must be deployed in diverse environments, from GPU-backed cloud inference to edge devices with strict resource constraints.

6.3.5.1 Cloud Deployment with LLMs

For larger models like T5, BERT-large, and Mistral, GPU-accelerated services offered by platforms such as AWS SageMaker, Azure ML, or Vertex AI are typically the most suitable for inference.

6.3.5.2 Edge Deployment with SLMs

Smaller models like DistilBERT and E5-small can often be optimized and exported to formats like ONNX or TensorFlow Lite, making them viable for deployment on mobile or embedded systems. Serverless functions, such as AWS Lambda, can also provide cost-effective and low-latency solutions for these models.

6.3.5.3 BitNet CPU Deployment Pipeline

Notably, Microsoft's BitNet models are specifically designed for efficient deployment on standard CPUs. The official `bitnet.cpp` framework facilitates rapid inference, leveraging the GGUF format and its unique 1.58-bit quantization.

CHAPTER 6 QUESTION ANSWERING (QA)

```
# BitNet Deployment Pipeline
python convert_bitnet_to_gguf.py
quantize --qtype Q1_58
./bitnet -m qa_model.gguf -p "Q: What is Newton's first law?"
```

You can find the full implementation code at: https://github.com/anvcse562/ModernNLP-LLMSLM/blob/main/Ch06/Ch6.3/5_qa_model_deployment.ipynb.

6.3.6 Inference Optimization Strategies for QA Models

To ensure that your QA systems are not only accurate but also performant, various optimization strategies can be employed during inference:

Strategy	Description
Quantization	Reduces model precision (e.g., FP32 → INT8) to speed up inference.
Pruning	Removes less important weights to reduce model size and complexity.
Distillation	Trains a smaller model (e.g., TinyBERT) to mimic a larger one like BERT.
Batch inference	Groups multiple QA inputs to improve throughput.
Caching	Stores frequent question-context-answer triples for instant lookup.

Table 6-6 explains the various techniques/metrics.

Table 6-6. Inference Optimization Strategies

Technique/Metric	Purpose/Effect	Code

1.58-bit quantization	Reduces weight precision; BitNet achieves up to 6× CPU speed.	```bash
		# Convert and quantize BitNet model
		python convert_bitnet_to_gguf.py quantize --qtype Q1_58 # 1.58-bit quantization
		./bitnet -m qa_model.gguf -p "Q: What is relativity?"
		``

Flash attention	Memory-efficient attention for longer contexts.	```python
		from transformers import AutoModelForCausalLM, AutoTokenizer
		model = AutoModelForCausalLM.from_pretrained("mistralai/Mistral-7B-Instruct", use_flash_attention_2=True)
		tokenizer = AutoTokenizer.from_pretrained("mistralai/Mistral-7B-Instruct")
		``

Liger kernels	Hybrid CPU-GPU execution with improved throughput.	```bash
		# For PyTorch with hybrid kernel acceleration
		torchrun --inductor --kernel-backend=liger qa_inference.py
		``

(continued)

Table 6-6. (*continued*)

Technique/Metric	Purpose/Effect	Code

Distillation	Trains small models to replicate larger model behavior.	```python
		from transformers import BertForQuestionAnswering, DistilBertForQuestionAnswering
		# Load teacher (BERT) and student (DistilBERT)
		teacher = BertForQuestionAnswering.from_pretrained("bert-large-uncased-whole-word-masking-finetuned-squad")
		student = DistilBertForQuestionAnswering.from_pretrained("distilbert-base-uncased")
		# Use Hugging Face's transformers or transformers/distillation for training
		``

Pruning	Removes unimportant weights to reduce model size.	```python
		from optimum.intel.openvino import OVModelForQuestionAnswering
		# Export model to ONNX and apply pruning
		model = OVModelForQuestionAnswering.from_pretrained("distilbert-base-uncased")
		model.prune_layers(pruning_ratio=0.3) # Remove 30% of layers
		``

(*continued*)

Table 6-6. (*continued*)

Technique/ Metric	Purpose/Effect	Code

Batch inference	Groups multiple inputs to boost hardware utilization.	```python
		questions = ["What is AI?", "What is ML?"]
		contexts = ["AI is a field...", "ML is a subfield..."]
		inputs = tokenizer(questions, contexts, padding=True, truncation=True, return_tensors="pt")
		outputs = model(**inputs)
		``

You can find the full implementation code at: https://github.com/anvcse562/ModernNLP-LLMSLM/blob/main/Ch06/Ch6.3/6_inference_optimization.ipynb.

6.3.7 Evaluation Metrics for QA Performance

To ensure that your QA systems are not only accurate but also performant, various optimization strategies can be employed during inference:

Metric	Use Case
Exact match (EM)	Measures if the predicted answer exactly matches the ground truth.
F1 score	Measures the overlap between predicted and actual answers (token-based).
Latency	Time taken to generate an answer—critical in real-time applications.
Throughput	Number of queries processed per second—important for high-load systems.
BLEU/ROUGE	Used for generative QA models to compare fluency and accuracy of answers.

Table 6-7 explains the various techniques/metrics.

Table 6-7. *QA Performance Metrics*

Technique/ Metric	Purpose/Effect	Code

Exact match (EM)	Binary matches with the correct answer.	```python
		from evaluate import load
		metric = load("squad")
		results = metric.compute(predictions=preds, references=refs)
		print(f"EM: {results['exact_match']}, F1: {results['f1']}")
		``

F1 score	Measures token-level overlap.	```python
		from evaluate import load
		metric = load("squad")
		results = metric.compute(predictions=preds, references=refs)
		print(f"EM: {results['exact_match']}, F1: {results['f1']}")
		``

(*continued*)

Table 6-7. (*continued*)

Technique/ Metric	Purpose/Effect	Code

Semantic F1	Embedding-based similarity for generative QA.	```python
		from sentence_transformers import SentenceTransformer, util
		model = SentenceTransformer("all-MiniLM-L6-v2")
		sim_scores = [util.cos_sim(model.encode(p), model.encode(r)) for p, r in zip(preds, refs)]
		semantic_f1 = sum([s.item() for s in sim_scores]) / len(sim_scores)
		``

Latency	Time taken to return an answer.	```python
		import time
		start = time.time()
		_ = model(**inputs)
		end = time.time()
		latency = (end - start) / len(inputs["input_ids"])
		throughput = len(inputs["input_ids"]) / (end - start)
		``

(*continued*)

Table 6-7. (*continued*)

Technique/Metric	Purpose/Effect	Code

Throughput	Total queries processed per second.	```python
		import time
		start = time.time()
		_ = model(**inputs)
		end = time.time()
		latency = (end - start) / len(inputs["input_ids"])
		throughput = len(inputs["input_ids"]) / (end - start)
		``

Energy/query	Energy cost per inference.	```python
		import pyRAPL
		pyRAPL.setup()
		@pyRAPL.measureit()
		def run_qa():
		    return model(**inputs)
		energy_usage = run_qa()
		print(f"Energy: {energy_usage.energy} µJ per query")
		``

(*continued*)

Table 6-7. (*continued*)

Technique/ Metric	Purpose/Effect	Code

Contextual ROUGE	Evaluates how well the answer fits the full context.	```python
		from rouge_score import rouge_scorer
		scorer = rouge_scorer.RougeScorer(['rougeL'], use_stemmer=True)
		scores = [scorer.score(pred, ref)["rougeL"].fmeasure for pred, ref in zip(preds, refs)]
		avg_rouge = sum(scores) / len(scores)
		``

You can find the full implementation code at: https://github.com/anvcse562/ModernNLP-LLMSLM/blob/main/Ch06/Ch6.3/7_evaluation_metrics.ipynb.

The landscape of open-source QA systems has transformed rapidly, moving beyond the constraints of heavy GPU reliance. Thanks to frameworks like Hugging Face Transformers, innovations such as QLoRA for efficient fine-tuning, and lightweight models like BitNet, developers can now build accurate, scalable, and energy-efficient QA solutions. These advancements enable seamless deployment across diverse hardware—from cloud GPUs to everyday CPUs. Whether you're developing a responsive chatbot, an enterprise assistant, or an edge-based reasoning agent, the tools and techniques covered here provide a robust foundation for implementing next-generation QA systems that combine NLP sophistication with computational efficiency.

6.4 Industry Use Cases

6.4.1 Customer Support Chatbot

In the modern business landscape, customer support has become a critical element for success. Efficient, timely, and accurate responses to customer inquiries not only improve user experience but also significantly reduce operational costs. To meet these demands, businesses are increasingly turning to automated systems powered by AI, specifically customer support chatbots.

These chatbots are designed to understand and respond to customer inquiries in a conversational manner, providing assistance across a wide variety of topics such as product information, order status, troubleshooting, and more. By leveraging LLMs such as GPT and SLMs such as DistilBERT, businesses can create chatbots that deliver relevant and context-aware responses.

6.4.1.1 Use Case Overview

Customer support chatbots serve as an automated solution to handle customer queries efficiently. These systems utilize extractive and generative question-answering techniques:

- **Extractive QA:** This method involves retrieving the exact answer from the available knowledge base (e.g., FAQs, help articles) and presenting it to the user. It is particularly useful when the knowledge base contains factual and well-defined information.

- **Generative QA:** This method synthesizes answers based on context, allowing for more flexible, dynamic, and sometimes personalized responses. This approach is valuable when the response needs to be more conversational or when the answer is not directly available in the knowledge base.

In this use case, the chatbot implementation uses two main techniques:

- **Extractive QA** powered by SLMs like DistilBERT
- **Generative QA** using LLMs such as OpenAI's GPT models

By combining these two techniques, the chatbot ensures that it can provide direct answers and engage in more open-ended conversations.

6.4.1.2 Key Steps in the Implementation

1. **Data loading:** Knowledge documents such as customer service policies, FAQs, and user guides are loaded into the system.

2. **Document chunking:** The knowledge documents are split into smaller, more manageable chunks that can be efficiently processed and searched.

3. **Embedding and vector storage:** Using Hugging Face embeddings, the documents are transformed into vector representations stored in a Chroma vector store. This allows the system to retrieve relevant information quickly when queried.

4. **Extractive QA (SLM):** A pretrained DistilBERT model is used to extract specific answers from the document chunks based on the query.

5. **Generative QA (LLM):** When the chatbot cannot find direct answers in the knowledge base, it utilizes a Generative LLM like GPT3.5 to create a synthesized answer.

6. **Querying interface:** Users interact with the chatbot by submitting questions, which are processed by the QA chain. The chatbot first tries to find a relevant answer from the document store, and if that fails, it uses generative capabilities for more flexible responses.

6.4.1.3 Applications

- **Automating customer support:** The chatbot handles inquiries like account management, troubleshooting, and product details, reducing the burden on human support agents.

- **24/7 availability:** The chatbot provides continuous customer service, ensuring that users can get assistance at any time.

- **Knowledge base navigation:** By integrating with a vast knowledge base, the chatbot quickly retrieves accurate and up-to-date information, improving response times.

- **Escalation and feedback:** The chatbot can escalate more complex issues to human agents when necessary, ensuring that customers receive the best service.

6.4.1.4 Deployment Options

- **Local deployment:** The chatbot can be deployed within a company's internal network for handling customer service queries related to specific organizational needs.

- **Cloud deployment:** The system can be hosted in the cloud, offering scalability and broad accessibility for customers.

- **Docker containerization:** For consistency across different environments, the system can be containerized, ensuring that it runs seamlessly on various platforms.

Refer to the following code on GitHub: https://github.com/anvcse562/ModernNLP-LLMSLM/blob/main/Ch06/ch6.4/6_4_1_customer_support_chatbot.py.

6.4.2 Knowledge Base Querying (Legal, HR)

Businesses across various sectors, including legal, human resources, and IT, maintain vast repositories of internal knowledge. Implementing QA systems over these knowledge bases allows employees to quickly find the information they need, improving efficiency and reducing the reliance on manual document searching or subject matter experts for common queries. Leveraging LLMs and SLMs offers a flexible approach to cater to different needs and deployment environments.

Businesses utilize this approach primarily to:

- Enable rapid access to internal policies, procedures, and knowledge articles.

- Automate responses to frequently asked questions, reducing the workload on support teams.

- Facilitate information retrieval from complex legal documents, HR guidelines, and IT documentation.

- Improve employee onboarding and training by providing instant answers to their questions.

- Enhance decision-making by providing quick access to relevant information.

6.4.2.1 Implementation Overview

This implementation demonstrates building a QA system for legal, HR, and IT knowledge using both SLMs (like DistilBERT) for extractive QA and potentially LLMs (like Gemma or BitNet) for generative QA, all powered by the Hugging Face Transformers library and LangChain.

This implementation includes:

- **Data loading and indexing:** Loading text-based knowledge documents (e.g., legal FAQs, HR policies) and creating a vector store (Chroma) for efficient retrieval.

- **Extractive QA with SLMs (DistilBERT example):** Fine-tuning or using a pretrained DistilBERT model to extract answer spans directly from the relevant documents.

- **Generative QA with LLMs (Gemma/BitNet example):** Utilizing generative LLMs to synthesize answers based on retrieved context. The code shows integration with google/gemma-7b-it and provides an example of using microsoft/bitnet-b1.58-2B-4T.

- **Querying interface:** Providing a way to ask questions and receive answers from the knowledge base.

6.4.2.2 Code Integration

The provided code snippets demonstrate key aspects of this implementation:

- **Document loading and chunking:** Using TextLoader and RecursiveCharacterTextSplitter to process text documents into manageable chunks.

- **Embedding and vector storage:** Employing HuggingFaceEmbeddings and Chroma to create a searchable vector database of the document chunks.

- **LLM initialization (Gemma and BitNet):** Showing how to initialize LLMs from the Hugging Face Hub using `HuggingFaceEndpoint` (for Gemma) and `HuggingFaceHub` (for BitNet), with error handling for BitNet initialization.

- **Retrieval and QA chain creation:** Using LangChain's `create_retrieval_chain` and RetrievalQA to build the QA pipeline.

- **Querying functions:** `ask_question` and `ask_question_bitnet` illustrate how to interact with the QA systems.

6.4.2.3 Deployment Options and Scalability

Similar to the customer satisfaction insights, this QA system can be deployed in various ways:

- **Local deployment:** For internal tools accessible within a company network.

- **Cloud deployment:** Leveraging cloud services for scalability and broader accessibility.

- **Containerization (Docker):** Ensuring consistent deployment across different environments.

You can find the full code at: `https://github.com/anvcse562/ModernNLP-LLMSLM/blob/main/Ch06/ch6.4/6_4_2_FAQ_Legal_HR.ipynb`.

6.4.3 Automated Assistants

In modern enterprise environments, automated assistants are becoming essential for scaling support operations, enhancing productivity, and enabling round-the-clock information access. These assistants integrate QA systems—powered by SLMs and LLMs—to provide users with timely, accurate, and context-aware responses. They are deployed in use cases ranging from IT helpdesks and HR assistants to voice-based agents like Alexa and Siri.

6.4.3.1 Use Case Overview

Automated assistants act as conversational interfaces that help users resolve queries related to documentation, policies, support manuals, and more. These systems often combine extractive and abstractive QA techniques to support real-time interactions.

- **Extractive QA (SLMs):** Used for precise span-based answers from structured knowledge sources. Models like DistilBERT or MiniLM are suitable for on-device or latency-critical deployments.

- **Generative QA (LLMs):** Employed when responses require synthesis, rephrasing, or summarization across multiple documents. Examples include GPT-4, Claude, and Mistral-7B.

- **Multimodal extensions:** For assistants requiring vision-language capabilities, models like GPT-4V and LLaVA enable image-question answering (e.g., interpreting charts or screenshots).

6.4.3.2 Implementation

Here are the key steps in the implementation:

1. **Document ingestion and chunking:**

    Internal knowledge such as policy documents, user manuals, and help articles are loaded and chunked into manageable segments using RecursiveCharacterTextSplitter.

2. **Embedding and vector storage:**

    The chunks are embedded using Hugging Face models and stored in a vector store like Chroma for fast semantic retrieval.

3. **Query interpretation and routing:**

    Incoming queries are first processed through a semantic retriever. If a high-confidence context match is found, extractive QA is applied using an SLM. If confidence is low or the query requires summarization, the request is routed to a generative LLM.

4. **Response generation:**

    - For extractive answers, models like DistilBERT identify answer spans from context.

    - For generative responses, models like FLAN-T5 and Claude synthesize an answer from one or more retrieved contexts.

    - For multimodal queries, vision-language models process images and text together.

5. **Conversation history handling:**

    Multi-turn memory is maintained to enable follow-up questions and coherent dialogue. Contextual continuity is preserved using a memory module (e.g., LangChain's `ConversationBufferMemory`).

6.4.3.3 Deployment Options

- **Cloud:** Hosted on SageMaker or Bedrock to support high-scale use with managed infrastructure.

- **Edge/on-prem:** Lightweight SLMs deployed in internal networks or on devices for low-latency inference.

- **Hybrid cascade:** Extractive models serve as the first responder. If confidence is low, the system escalates to an LLM for generative synthesis.

6.4.3.4 Benefits

- **24/7 availability:** Provides consistent responses at any time without human intervention.

- **Latency optimization:** Provides fast extractive answers for common queries and deeper answers from LLMs when needed.

- **Scalability:** Easily extends across departments and domains (IT, HR, Legal).

You can find the full code at: https://github.com/anvcse562/ModernNLP-LLMSLM/blob/main/Ch06/ch6.4/6_4_3_automated_assistant_QA.ipynb.

6.5 Summary

This chapter embarked on a journey into the world of QA systems, exploring the capabilities of LLMs and SLMs. You learned the fundamental differences between extractive QA, where answers are located within a given context, and abstractive QA, where models generate answers in natural language. You gained practical insights into building and fine-tuning these QA systems using platforms like Amazon SageMaker and the open-source Hugging Face Transformers library. Furthermore, you explored various deployment strategies, including leveraging AWS Lambda for real-time responses and the CPU-efficient deployment of models like BitNet. Finally, you examined crucial evaluation metrics for QA performance and real-world applications in customer support, knowledge base querying, and automated assistants. You now possess a solid understanding of how to build and deploy sophisticated question-answering solutions. The next chapter explores another vital area of modern NLP: text summarization.

CHAPTER 7

Text Summarization

This chapter explores modern text summarization techniques using LLMs such as PEGASUS, BART, T5, FLAN-T5-XXL, and Mistral-7B, and SLMs like DistilBERT, TinyBERT, and BitNet-b1.58. It compares extractive and abstractive summarization methods, highlighting tradeoffs in quality, speed, and efficiency.

It also covers scalable implementations using AWS services like SageMaker for fine-tuning LLMs and Lambda for real-time SLM deployments. Open-source approaches using Hugging Face Transformers are emphasized, along with recent advancements in instruction-tuned and energy-efficient models.

By the end of this chapter, you'll understand how to choose and deploy the right models for summarization tasks—balancing performance, cost, and deployment complexity.

This chapter covers the following topics:

- **Overview of text summarization:** Extractive vs. abstractive methods, model selection, and tradeoffs.

- **Implementation on AWS:** Using SageMaker and Lambda to deploy summarization at scale.

- **Open-source implementation:**
    - Building summarization pipelines with Hugging Face
    - Leveraging models like FLAN-T5-XXL, Mistral-7B, and BitNet-b1.58
    - Fine-tuning, deployment, and evaluation (ROUGE, BLEU, METEOR)

- **Industry use cases:** News aggregation, business document summarization, and automated email summaries.

CHAPTER 7   TEXT SUMMARIZATION

This chapter equips you to build efficient, high-quality summarization systems using cutting-edge tools and models.

7.1 Overview of Text Summarization

In the previous chapter, you saw how question-answering systems leverage LLMs to retrieve and provide precise answers from large knowledge bases.[1] One common challenge in those scenarios is dealing with large amounts of text—for example, a QA system might retrieve a lengthy document to answer a query. This is where text summarization becomes invaluable. By condensing long documents into concise summaries, summarization techniques enable users (or downstream systems) to grasp the essential information quickly. In fact, summarization often complements QA: A QA system could summarize an article before presenting it as an answer or generate a brief answer that itself is a summary of detailed content. Having explored QA, we now turn our attention to text summarization—a core capability for distilling information.

Text summarization is the process of automatically creating a shorter version of a document while preserving its key ideas and overall meaning. The summary should ideally capture the most important points, remain faithful to the source, and present information in a coherent, easy-to-read manner. LLMs have significantly advanced text summarization, enabling both highly fluent *abstractive* summaries and efficient *extractive* approaches. This section provides an overview of text summarization techniques, contrasts extractive vs. abstractive summarization, and discusses recent trends, including the emergence of small efficient models and the integration of summarization into multi-agent and retrieval-augmented frameworks. Throughout this section, we use simple examples and code snippets (in Python with Hugging Face Transformers) to illustrate concepts in practice.

7.1.1 Extractive vs. Abstractive Summarization Techniques

When it comes to summarization, there are two fundamental approaches:

---

[1] *This discussion continues from the industry use cases of QA in Section 6.4.*

- **Extractive summarization:** This approach selects and *compiles existing pieces* (typically sentences or phrases) from the original text to form a summary. The summary is essentially a subset of the original words/sentences, copied verbatim without paraphrasing. The algorithm's job is to identify the most important sentences that cover the main points of the source. Everything else is dropped. Because no new text is generated, extractive summaries tend to be very faithful to the source (no risk of introducing unseen words or incorrect facts), but they may be less coherent or less concise than an abstractive summary—after all, they use original sentences that could include extra details or redundant information.

- **Abstractive summarization:** This approach *generates new text* that conveys the meaning of the original. An abstractive summarizer will paraphrase, use different wording, and potentially reorganize the information in the source text. The summary produced is more like what a human might write—it doesn't necessarily use the exact sentences from the input but rather rewrites the content in a shorter form. Abstractive methods can produce more concise and fluent summaries (removing filler words, generalizing specifics, etc.), but they are also more complex and can sometimes introduce information that wasn't present in the source (a phenomenon known as a *hallucination*). Ensuring factual accuracy is therefore a key challenge in abstractive summarization.

Table 7-1 summarizes the main differences between extractive and abstractive summarization.

Table 7-1. Key Differences Between Extractive and Abstractive Summarization

Aspect	Extractive Summarization	Abstractive Summarization
Method	Selects important sentences or phrases from the original text without changing them.	Generates new sentences and words to capture the meaning of the original text.
Output	Summary is composed of exact excerpts from the source (copy-pasted segments).	Summary is written in new words; may rephrase or compress ideas beyond the original wording.
Fluency and coherence	Depends on the source text's phrasing. May include abrupt jumps or redundant info if the original sentences don't flow together.	Tends to be more fluent and coherent, as it can restructure sentences, yielding a more natural narrative flow.
Factuality	High factual consistency (no new information added) since content is directly from source.	Potential risk of hallucinations—the model might introduce details not in source. Ensuring all generated facts are in the original is a challenge.
Examples of techniques/ models	Sentence scoring and selection (e.g., with TF-IDF or graph algorithms like TextRank). Transformer-based classifiers (e.g., BERT or DistilBERT fine-tuned to pick key sentences).	Sequence-to-sequence models (e.g., PEGASUS, BART, T5) and large generative models (GPT-3/4) that learn to paraphrase and shorten text.

In practice, these approaches are not mutually exclusive. Often, systems combine them to get the best of both worlds. For instance, an algorithm might first perform extractive summarization to identify the most relevant sentences and then *rewrite* those in a more concise form (abstractive post-processing). We discuss such hybrid techniques shortly.

For example, suppose you have this short paragraph:

"Alice went to the market to buy fresh vegetables and fruits for dinner. She also visited the bakery to pick up some bread. It was a sunny day and the market was bustling with people."

- An extractive summary might pick out a couple of key sentences from the text. For example: "Alice went to the market to buy fresh vegetables and fruits for dinner. She also visited the bakery to pick up some bread." Here, the summary is a verbatim subset of the original text, focusing on the main actions.

- An abstractive summary could be: "Alice ran errands at a busy market on a sunny day, buying fresh produce and bread for dinner." This version is shorter and uses new phrasing (e.g., "ran errands", "fresh produce") that did not appear exactly in the original. It captures the essence (Alice's activities and purpose) in a single sentence, demonstrating how an abstractive model can compress information and paraphrase.

This simple example illustrates that the abstractive summary is more concise and arguably more readable, but it required understanding and rewriting the content. The extractive summary, by contrast, is straightforward (just copy-pasting sentences) and retains all details from those sentences, but it's a bit longer and more literal.

7.1.2 Extractive Summarization Approaches (BERT and Other Models)

Extractive summarization can be approached with a variety of techniques, ranging from classical algorithms to modern neural networks. Early methods relied on heuristic scoring of sentences—for example, ranking sentences by keyword frequency, position in text, or graph centrality where the famous TextRank algorithm builds a graph of sentences and selects the most "central" ones. While such methods are fast, they lack deep understanding of content. Today, transformer-based models like BERT have greatly improved extractive summarization by leveraging semantic understanding. The idea is to use a pretrained language model to encode sentences and then identify which sentences should appear in the summary.

BERT is a powerful context-aware encoder. In BERT-based extractive summarization, a common approach is to fine-tune BERT for a classification task and predict whether each sentence in a document should be included in the summary. One well-known example is the BERTSum model (Liu & Lapata, 2019), which adds a classification layer on top of BERT's sentence representations to select top-ranked sentences for the summary. Such a model is trained on documents with human-written summaries, learning which sentences tend to be included.

Because BERT can capture the meaning of sentences in context, it often outperforms earlier frequency or graph-based methods. Variants like RoBERTa or XLNet have also been used for extractive summarization. In practice, even a distilled, smaller version of BERT can do well. For example, DistilBERT (a compressed six-layer version of BERT) has been used to identify important text segments for summarization. These smaller models run faster, making them suitable for real-time summarization in limited environments.

As an example, there are open-source libraries that implement BERT-based extractive summarizers. For instance, the `bert-extractive-summarizer` library uses BERT under the hood to encode sentences and select the most representative ones via clustering and scoring. You can access a simple example using this approach (with a DistilBERT model by default) in Python from GitHub: https://github.com/anvcse562/ModernNLP-LLMSLM/blob/main/Ch07/Ch7.1/BertForExtraction.py.

In this code, we feed a short paragraph about supply chains to the `Summarizer`. The resulting output will look something like this:

> *"The COVID-19 pandemic caused unprecedented disruption to global supply chains. Experts predict that these changes could permanently reshape international trade."*

This extractive summary pulled out two sentences that cover the cause (pandemic disruption) and a key consequence (predicted permanent changes to trade)—likely the main points of the paragraph. Notice that it dropped the detail about companies' responses and strategy shifts, focusing on the big picture. In a more advanced setting, you could adjust the number of sentences or the selection criteria (e.g., to ensure certain sections are represented).

7.1.2.1 Strengths and Limitations

Extractive methods shine in domains where precise wording matters (e.g. legal or technical documents) because they avoid altering terminology—the output is literally part of the input, so it won't introduce errors that weren't already there. They also tend to be faster since it's easier to select sentences than to generate new ones. However, the resulting summaries can be verbose or disjointed. For instance, two sentences copied from far apart in a document might not flow well when placed together. Also, an extractive summary's length is tied to the original sentences' lengths—you might end up including a whole long sentence just because it had one critical clause. Despite these issues, extractive summarization remains very useful, especially when using powerful models like BERT to gauge sentence importance.

7.1.2.2 LLM Agents for Extraction

Interestingly, with the rise of LLM-based agents, you can even use a prompt-based approach for extraction. For example, you could ask a GPT-4 model: *"*List the 3 most important sentences from the following text...*"* and it will perform an extractive summary by following the instruction. This blurs the line between extractive and abstractive (since the LLM might rephrase slightly), but it's another way LLMs can act as extractive agents. Nonetheless, dedicated extractive models (like fine-tuned BERTs) are often more predictable for purely extractive tasks.

7.1.3 Abstractive Summarization Approaches (PEGASUS, BART, GPT-3, T5)

Abstractive summarization is a more challenging task: the model must generate a concise text that conveys the key information from the source. Essentially, this is a sequence-to-sequence (seq2seq) learning problem, much like machine translation— except instead of translating between languages, it's "translating" a long text into a shorter summary in the same language. The advent of transformer-based encoder-decoder models and large-scale pretraining has led to huge advances in this area. Here we discuss some prominent models and approaches:

- **BART:** BART is a transformer model from Facebook (Meta AI) that combines ideas from BERT and GPT. It has an encoder-decoder architecture: the encoder is like BERT (bidirectional, looks at full context), and the decoder is like GPT (autoregressive, generates text one token at a time). BART is pretrained on a "noising" objective— it corrupts text (e.g., by shuffling or masking words) and learns to reconstruct it. This makes it a very flexible generative model that can be fine-tuned for summarization (among other tasks). In fact, the facebook/bart-large-cnn model is a BART large pretrained and fine-tuned on the CNN/DailyMail news summarization dataset, and it's a popular choice for abstractive summarization tasks. BART tends to produce fluent and coherent summaries, maintaining grammatical correctness even for tricky inputs. Fine-tuning on summarization data teaches it to perform operations like paraphrasing, sentence fusion, and dropping unimportant details.

- **PEGASUS:** This is a model from Google specifically designed for abstractive summarization. PEGASUS introduces a novel pretraining objective called *gap sentence prediction*—essentially, it masks out whole sentences from a document and trains the model to generate those missing sentences based on the rest of the text. The idea is to simulate the act of summarization during pretraining: the model learns to generate the *important* sentence that would sum up the content of a removed section. PEGASUS, when fine-tuned, has shown excellent results on summarizing long documents like news articles and research papers. It is especially good at handling formal text. One downside observed is that if not fine-tuned well on a domain, it can produce grammatical errors or odd phrasing—indicating the importance of fine-tuning and possibly post-editing.

- **T5:** T5 by Google treats every NLP problem as a text-to-text task. It has an encoder-decoder architecture and was trained on a diverse corpus of tasks. For summarization, T5 was trained on datasets like CNN/DailyMail (for news summaries) and others, using a simple scheme: input text is fed with a task prefix like "summarize" and the model learns to output the summary. T5 (especially the larger versions like T5-3B or T5-11B) can produce quality summaries. An advantage of T5 is its versatility—it can be adapted to summarizing different styles of text by fine-tuning on appropriate data. However, T5's output can sometimes be less precise than BART's, and like any abstractive model, it may drop details. Some evaluations found that T5 may include minor grammatical issues if not carefully fine-tuned.[2] Still, it remains a strong open-source model for summarization tasks, with the added benefit of a unified framework for multiple tasks.

---

[2] https://www.width.ai/post/bart-text-summarization#:~:text=better%20though.%20, too%20retain%20some%20spurious%20text.

- **GPT-3 and GPT-4 (and other GPT-family models):** GPT-3 (OpenAI's 175B parameter model) and its successors (GPT-3.5, GPT-4) are *decoder-only* language models, not explicitly trained as summarizers but as very large next-word predictors on vast text data. Despite that, they exhibit remarkable zero-shot and few-shot summarization abilities. For example, simply prompting GPT-3 with "Summarize the following text in a few sentences:" often yields a decent summary. In the Width.ai evaluation, even smaller GPT-3 variants (like Curie, Babbage) produced fairly good summaries with the right prompt. GPT-4, being more advanced, can generate highly coherent and nuanced summaries and follow instructions about summary length or style quite well. These models excel especially in fluency and the ability to compress and paraphrase content. The tradeoff is that they require API access (for GPT-3/4, which are not open-source) and are computationally heavy. They may also confidently generate hallucinations, so it's smart to ask them to cite or highlight where each part of the summary came from, to verify accuracy.

7.1.3.1 Code Example: Abstractive Summarization with Hugging Face

You can easily use pretrained abstractive models via the Hugging Face Transformers library. For example, we use BART (fine-tuned on CNN/DailyMail) to summarize text.

Refer to the code on GitHub at: https://github.com/anvcse562/ModernNLP-LLMSLM/blob/main/Ch07/Ch7.1/AbstarctiveSummarization.py.

Running this code example, you might get output like this:

"GPT-4 has excelled in many tasks, scoring highly on academic benchmarks and showing strong reasoning abilities.

Despite its achievements, the model's decision-making is not yet transparent.

Researchers are now working to make GPT-4 more interpretable and to ensure its answers remain trustworthy and unbiased."

This abstractive summary is composed of novel sentences that *combine* and *compress* the original information. Notice how it merged the first two sentences about performance and evaluation into one concise sentence, and it captured the contrast in the original text (great performance *but* lacks transparency) in a clear way. The summary is shorter (about two sentences versus four in the original) and uses slightly different wording (e.g. "excelled in many tasks" versus "remarkable performance on a variety of tasks", "not yet transparent" versus "lacks transparency"). This example shows the power of abstractive models like BART to rewrite content effectively.

7.1.3.2 Ensuring Quality and Accuracy

A major focus in abstractive summarization research is preventing inaccuracies. Because these models generate text, they might omit a crucial detail or add something not supported. Techniques to mitigate this include *constrained generation* (forcing the model to include certain keywords from the source), using *fact-checking agents* or prompt-based checks, and hybrid extractive-abstractive approaches. One hybrid approach is to first generate an extractive summary to gather all key facts and then have the model rewrite it with the abstractive approach, thereby reducing the chance that it introduces a completely new element.

Recent work also suggests using a second LLM (like GPT-4) to refine or verify summaries. For example, after an initial summary is produced, you could prompt GPT-4 with the original text and the summary asking: *"Check if the summary is faithful to the text and fix any inaccuracies."* This uses the LLM as a *critic/refiner*, capitalizing on its strong comprehension skills. In fact, a 2024 study by Shakil et al. demonstrated a pipeline where an extractive summary (from DistilBERT) and an abstractive summary (from T5) were combined, and then GPT-based refinement was applied to minimize hallucinations—resulting in summaries that were more factual and reliable.

7.1.4 Recent Trends: Efficient SLMs, Retrieval-Augmented Summarization, and LLM Agents

The field of text summarization is evolving rapidly, influenced by both the push for more efficient models and the integration with advanced LLM-based workflows. Here are some notable trends and emerging practices:

1. **SLMs for summarization:** Not every application can afford a giant model like GPT-4 or even BART running all the time—especially for real-time or edge devices, efficiency is key. This has led to the

rise of *distilled and quantized models* that are much smaller and faster, albeit with some sacrifice in accuracy. Examples include TinyBERT and DistilGPT2:

- **TinyBERT** is a distilled version of BERT that is 7.5x smaller than BERT-base but retains a large portion of its performance on understanding tasks. While originally created for tasks like classification and QA, TinyBERT can be fine-tuned for extractive summarization as well. Its smaller size means faster inference, which is great for summarization on mobile devices or in scenarios with limited compute. For instance, you could deploy a TinyBERT-based summarizer in a browser or a low-power CPU environment to quickly pick out important sentences from an article. The tradeoff is that with fewer parameters, it might miss some nuances that a full BERT would catch, or it might score sentence importance slightly less accurately. Nonetheless, newer research on dynamic distillation shows TinyBERT and similar compact models can be optimized further (e.g., by adaptively reducing sequence length) to summarize even more efficiently.

- **DistilGPT2** (sometimes just called "DistilGPT") is a lighter version of GPT-2 (a generative model). GPT-2 itself, especially larger variants, can perform abstractive summarization when fine-tuned or prompted appropriately. DistilGPT2 inherits this ability but with about half the number of parameters of GPT-2. While it won't match GPT-3 or fine-tuned BART in summary quality, DistilGPT2 can still generate reasonable summaries and is useful when you need a fast, on-premises abstractive model without requiring heavy cloud resources. For example, you could fine-tune DistilGPT2 on a custom dataset of product reviews and summaries, creating a quick summarizer that runs locally. Reports indicate that DistilGPT2 indeed can be trained to produce concise summaries of longer text, although like GPT-2 it might struggle with very long inputs due to its limited context window. Still, it's a promising direction for making abstractive summarization more accessible in resource-constrained settings.

- **DistilBART and other compressed summarizers:** In addition to TinyBERT and DistilGPT, there are distilled versions of summarization-specific models. For instance, DistilBART (a compressed version of BART) has been released for CNN/DailyMail summarization. DistilBART has fewer layers (e.g., 6 instead of 12) and runs faster while maintaining a majority of BART's summarization performance. Similarly, research into quantization (reducing precision of model weights) allows even large models to run faster with minimal loss of accuracy, which can be applied to summarization models. The benefit of these SLMs is evident in real-time applications—it can summarize incoming chat messages on a smartphone in split seconds or process hundreds of news articles per minute on a single server. As summarization moves into more interactive domains (newsfeeds, email summarizers, meeting minute generators), such efficiency-focused models are increasingly important.

2. **Retrieval-augmented summarization (RAG for summaries):** In Chapter 6, you encountered RAG for question-answering. A similar concept is emerging for summarization, particularly for domains where *contextual accuracy* is critical (like legal, scientific, or financial documents). The idea of *retrieval-augmented summarization* is to supply the summarization model with additional factual context fetched from an external knowledge base or the original document set. This is especially useful for multi-document summarization or summarizing very large documents where a single model's input window may not cover everything.

For example, in the legal domain, a summarizer might need to summarize a court judgment while ensuring it includes relevant references to laws or precedents. A dynamic RAG-based summarization system could first retrieve the most pertinent references (e.g., relevant statutes or prior cases) using an IR component (like BM25 or a vector search) and then feed both the main document and these retrieved chunks into a summarization model. By doing so, the generated summary is *grounded* in the retrieved facts, reducing the chance that the model fabricates a detail simply because it was outside its input. A 2025 study on legal text summarization showed that such a RAG approach (coupling a retriever with a fine-tuned LLM summarizer) significantly improved factual consistency—the summaries maintained crucial legal references and had far fewer hallucinations. Essentially, the summarizer was "guided" by the retrieved knowledge to stay accurate.

Beyond specialty domains, retrieval can help with *multi-document summarization*—summarizing "the state of research in climate science in 2023" might require pulling info from multiple articles. A RAG summarizer would retrieve top N relevant documents and then summarize the combined content. This approach ensures that the summary draws from the most up-to-date or relevant sources. It's like doing a mini web search and then summarizing the results automatically.

3. **LLM agents and multi-agent summarization:** Multi-agent LLM systems (as introduced in Chapter 6 and trending in 2025) often incorporate specialized roles for different subtasks. Summarization is a perfect example of a specialized skill that an LLM-based agent can have in a larger workflow. Instead of one monolithic model handling everything, you might have:

   - A **retriever agent** (or a set of agents) that gather information. For instance, one agent could fetch documents and another could extract numeric data or facts.

   - A **summarizer agent** whose job is to take all that gathered information and produce a concise summary or report.

For instance, imagine a multi-agent system that monitors multiple news sources and generates a daily briefing. One agent monitors political news, another one monitors economics, and a third monitors sports. A coordinator agent asks each to provide the highlights from their domain. Then a *summarizer agent* takes those highlights and weaves them into a single coherent summary for the day. This approach plays to the strengths of each agent and keeps the summarization task focused.

A concrete example of multi-agent summarization in action is described by Astropom et al. (2024) using the LangGraph framework. In their setup, a *supervisor* agent distributes a task to multiple LLM agents (e.g., queries several models or sources in parallel), and a dedicated summarizer agent then automatically combines the responses into one result. This kind of parallel processing plus summarization ensures that if you query multiple experts (different LLMs or different data sources), you can get a single, unified answer. We could envision a similar system for an enterprise assistant: one agent fetches information from the company wiki, another from a database, and the summarizer agent produces a neat summary for the user.

Multi-agent architectures also use summarization internally to manage context. For example, if an agent is conversing at length (think of a long planning session among several AI agents), they might periodically summarize their discussion to avoid going in circles and to keep track of important points (this is sometimes called *context distillation* or *state summarization*). By summarizing the conversation so far, agents ensure they remember key decisions without needing to store the entire dialogue history. This technique improves efficiency and is analogous to how a meeting of people might pause to recap what's been agreed on before moving forward.

To conclude, summarization is not just an isolated task—it's becoming a building block in larger AI systems, from helping align LLM outputs with facts (through retrieval) to enabling collaboration among multiple LLM agents.

7.2 Implementation on AWS: Using SageMaker and Lambda to Deploy Summarization at Scale

The previous section explored the fundamentals of extractive and abstractive summarization, their models (such as BERT, DistilBERT, T5, PEGASUS), and how LLM agents are increasingly being used for hybrid workflows. But understanding the models alone isn't sufficient. In real-world deployments—where summarization needs to happen at scale, across thousands of documents or API requests per hour—the infrastructure becomes just as important as the model.

This section focuses on how to build scalable, production-grade summarization systems using AWS services. We walk through practical implementations using AWS SageMaker for large LLMs and AWS Lambda for lightweight, cost-efficient SLMs, using DistilBERT and T5 as running examples. Where relevant, we bring in LLM agents to illustrate how orchestration logic and task delegation can work in serverless environments.

7.2.1 Architectural Overview: Deploying Summarization Pipelines at Scale on AWS

AWS provides flexible and scalable tools to host models of all sizes—from small quantized SLMs to massive LLMs. The AWS services and their use cases are listed in Table 7-2.

Table 7-2. AWS Services Used for Summarization

AWS Service	Use Case
SageMaker	Managed hosting for fine-tuned or pretrained LLMs with GPU support
Lambda	Lightweight, event-driven execution for smaller SLMs
API Gateway	Serves as entry point for REST summarization APIs
S3	Storage for documents and model artifacts
Step Functions	Orchestrates multi-step agent pipelines
CloudWatch	Logs, monitors, and triggers scale events

7.2.1.1 Three Deployment Patterns

1. Batch summarization with SageMaker (LLMs like T5)

    - For scheduled summarization (e.g., hourly news digests)
    - Uses SageMaker endpoints or batch transform
    - Triggered via Lambda or Step Functions

2. Real-time summarization via Lambda (SLMs like DistilBERT)

    - For chat, email, or short-form summarization
    - Executes in <1s with ONNX or PyTorch SLMs
    - Integrated with API Gateway, natively scalable

3. LLM agent orchestration

   - Uses agents for retrieval, summarization, and verification
   - Deployed via Step Functions and Lambda or ECS containers

7.2.1.2 Reference Architectures

1. Real-time serverless summarization (Lambda)

   scss

   CopyEdit

   ```
   Client → API Gateway → Lambda (DistilBERT) → Summary Response (JSON)
   ```

   Use case: Mobile app sending blog post/transcript for <1s summarization.

2. Abstractive summarization (SageMaker)

   scss

   CopyEdit

   ```
   Client → API Gateway → Lambda (pre) → SageMaker Endpoint → Lambda (post)
   ```

   Use case: Web dashboard sending news articles, processed using GPU-backed T5.

3. Multi-agent pipeline

   css

   CopyEdit

   ```
   [Retriever Agent] → [Summarizer Agent] → [Verifier Agent]
   (All via Lambda or ECS)
   ```

   Use case: Enterprise assistant handling summarization and validation.

Table 7-3 outlines the benefits and drawbacks of choosing Lambda vs. SageMaker.

Table 7-3. Decision Matrix: Lambda vs. SageMaker

Criteria	AWS Lambda	SageMaker Endpoint
Model size	≤ 500MB (SLMs)	Multi-GB (LLMs)
Latency	< 1s	1–5s typical
Cost	Low (pay-per-use)	Higher but stable
Startup time	~100ms cold start	Persistent GPU runtime
Autoscaling	Native	Manual or endpoint-based
Best use case	Chat, real-time APIs	Long-form summarization
Example model	DistilBERT	T5-large, FLAN-T5-XXL

7.2.1.3 Choosing Your AWS Stack

Use Lambda if:

- You have short documents or sentence-level summaries
- You have <300MB models (e.g., quantized DistilBERT)
- Cost/speed matters more than accuracy

Use SageMaker if:

- You have long-form text like research/news
- You have high-quality models like T5-XL, FLAN-T5
- You need GPU inference and accuracy

For the full code, visit this GitHub link: https://github.com/anvcse562/ModernNLP-LLMSLM/blob/main/Ch07/Ch7.2/aws_summarization_notebook.ipynb.

7.2.2 Deploying Extractive Summarization with DistilBERT on AWS Lambda

DistilBERT is a distilled, smaller BERT version that's 60% faster while retaining 95% of its accuracy. AWS Lambda allows hosting such small models for fast, cost-effective summarization.

Here is the deployment pipeline:

1. Fine-tune or reuse DistilBERT summarizer
2. Wrap in a FastAPI app
3. Dockerize and push to ECR
4. Deploy on AWS Lambda as container
5. Link with API Gateway
6. Optimize for cold starts and agent workflows

Consider this Lambda code sample (summarizer.py):
python
CopyEdit

```python
from transformers import pipeline
from fastapi import FastAPI, Request
from pydantic import BaseModel

summarizer = pipeline("summarization", model="sshleifer/distilbart-cnn-12-6")

app = FastAPI()

class Article(BaseModel):
    text: str

@app.post("/summarize/")
async def summarize(article: Article):
    result = summarizer(article.text, max_length=120, min_length=30, do_sample=False)
    return {"summary": result[0]["summary_text"]}
```

Dockerfile:

dockerfile

CopyEdit

```
FROM python:3.10-slim
WORKDIR /app
COPY . /app
RUN pip install --no-cache-dir fastapi uvicorn transformers torch
CMD ["uvicorn", "summarizer:app", "--host", "0.0.0.0", "--port", "8080"]
```

Here are the deployment steps:

1. Build and test locally:

   bash

   CopyEdit

   ```
   docker build -t summarizer-lambda .
   docker run -p 8080:8080 summarizer-lambda
   ```

2. Push to ECR and deploy Lambda:

   bash

   CopyEdit

   ```
   # Login & push
   aws ecr get-login-password | docker login ...
   docker push <ecr_repo>
   # Deploy
   aws lambda create-function --package-type Image ...
   ```

3. Connect to API Gateway.

4. Test the endpoint with `curl` or Postman.

This table outlines the cost estimates:

Parameter	Value
Model size	~300MB
Runtime	~1.5 sec
Monthly cost (10K req)	~$3.40
Cold start	~200–300ms
Throughput	20–50 RPS

7.2.3 Deploying Abstractive Summarization with T5 on AWS SageMaker

T5 treats summarization as a text-to-text task and offers excellent performance, but it needs GPU and memory. SageMaker provides a perfect hosting environment.

Here are the deployment steps:

1. Fine-tune T5:

    python

    CopyEdit

    ```
    from transformers import T5Tokenizer, T5ForConditionalGeneration, Trainer, TrainingArguments
    from datasets import load_dataset

    dataset = load_dataset("xsum")
    tokenizer = T5Tokenizer.from_pretrained("t5-small")
    model = T5ForConditionalGeneration.from_pretrained("t5-small")

    # Preprocessing & Training
    ...
    ```

2. Upload the model to S3:

    bash

    CopyEdit

    ```bash
    aws s3 cp --recursive ./t5-summarizer s3://<bucket-name>/
    ```

3. Deploy via SageMaker:

    python

    CopyEdit

    ```python
    from sagemaker.huggingface import HuggingFaceModel

    hf_model = HuggingFaceModel(...)
    predictor = hf_model.deploy(instance_type="ml.g4dn.xlarge", ...)
    ```

4. Inference logic (inference.py)"

    python

    CopyEdit

    ```python
    from transformers import T5Tokenizer, T5ForConditionalGeneration

    def model_fn(model_dir): ...
    def predict_fn(input_data, model_obj): ...
    ```

5. Invoke the endpoint:

    python

    CopyEdit

    ```python
    import boto3
    response = boto3.client("sagemaker-runtime").invoke_endpoint(...)
    ```

Terraform Snippet:
hcl
CopyEdit

CHAPTER 7   TEXT SUMMARIZATION

```
resource "aws_sagemaker_model" "t5_model" {
  name = "t5-summarizer"
  ...
}
```

Agent Integration

Use in Step Functions as follows:

scss

CopyEdit

```
Retriever (Lambda) → Summarizer (SageMaker) → Verifier (LLM Agent)
```

Here is the cost overview:

Attribute	Value
Model	T5-small/base
Instance type	ml.g4dn.xlarge
Inference time	~1.5–2.5 sec/request
Monthly cost	~$130
Scaling strategy	Based on GPU/CPU load

Optimization tips:

- Use batch transform for non-real-time jobs
- Host multiple models on a single endpoint
- Quantize models or use FP16
- Auto-stop endpoint during idle hours

As a final takeaway, you now have:

- A real-time, serverless DistilBERT summarizer on AWS Lambda
- A high-quality abstractive T5 summarizer on SageMaker
- A decision framework to choose the best AWS deployment option
- Integration ideas for LLM agents in production pipelines

7.3 Open-Source Summarization Systems

This section delves into constructing summarization systems using open-source tools, with a particular emphasis on Hugging Face Transformers. We begin by introducing essential concepts for implementing extractive and abstractive summarization using classical models like BERT and T5. Subsequently, we explore modern SLMs and LLMs, including instruction-tuned architectures such as FLAN-T5-XXL and Mistral-7B, as well as energy-efficient innovations like Microsoft's BitNet-b1.58. Finally, we examine fine-tuning strategies, deployment pipelines, and performance evaluation metrics, providing you with foundational and advanced skills for developing state-of-the-art summarization systems.

7.3.1 Implementing Summarization with Hugging Face Transformers

The Hugging Face Transformers library offers an accessible yet powerful interface for building summarization pipelines, providing pretrained models, tokenizers, datasets, and training utilities.

There are two primary paradigms for summarization: extractive and abstractive. Extractive models select sentences directly from the source document to form a summary. In contrast, abstractive models generate summaries in natural language, paraphrasing and synthesizing information from the original text.

7.3.1.1 LLMs for Abstractive Summarization

LLMs are particularly well-suited for abstractive summarization tasks. These models treat summarization as a sequence-to-sequence problem, enabling them to produce fluent and coherent summaries.

- **T5:** A versatile model that converts every task into a text generation problem. T5 performs well on datasets like CNN/DailyMail and XSum.
- **FLAN-T5-XXL:** An instruction-tuned variant of T5 with 11 billion parameters. It achieves state-of-the-art performance on benchmarks such as XSum.

- **Mistral-7B-Instruct:** A compact yet powerful generative model, fine-tuned for instruction-following tasks. Its extended 32K token context makes it ideal for document-level summarization.

- **BitNet-b1.58 (2B):** Microsoft's novel 1.58-bit quantized LLM demonstrates that efficiency doesn't have to sacrifice capability. This 2B parameter model performs lossless generative inference with up to 6.17× speedup on CPUs, all while reducing energy usage by more than 70%.

7.3.1.2 SLMs for Extractive Summarization

For many real-world applications—especially on-device or low-latency platforms—extractive summarization with smaller models is the optimal strategy.

- **BERT-large:** The original transformer-based extractive summarization benchmark.

- **DistilBERT:** A compressed version of BERT, 40% smaller but with 95% of the accuracy.

- **E5-small-v2:** An efficient, multilingual embedder achieving high performance on multilingual summarization tasks.

- **DeBERTa-v3:** Uses disentangled attention to outperform BERT on various summarization benchmarks.

- **Qwen2-7B:** With a 128K context length, this model is ideal for document-rich summarization scenarios.

7.3.2 Fine-Tuning T5 for Abstractive Summarization

To harness the generative power of models like T5 for specific summarization tasks, fine-tuning on relevant datasets is essential. This process allows the model to learn to generate fluent and contextually appropriate summaries.

Refer to this GitHub link for illustration: https://github.com/anvcse562/ModernNLP-LLMSLM/blob/main/Ch07/Ch7.3/7_3_2__Fine_Tuning_T5_for_Abstractive_Summarization.ipynb.

7.3.3 Fine-Tuning DistilBERT for Extractive Summarization

When the task calls for identifying a precise sentence or passage within a text, models like DistilBERT are excellent choices. Fine-tuning them on extractive summarization datasets hones their ability to accurately pinpoint these segments.

Refer to this GitHub link for illustration: https://github.com/anvcse562/ModernNLP-LLMSLM/blob/main/Ch07/Ch7.3/7_3_3_Fine_Tuning_T5_for_Extractive_Summarization.ipynb.

7.3.4 Next-Gen Fine-Tuning Strategies

As models continue to scale in size, the computational demands of fine-tuning can become prohibitive. Enter *parameter-efficient fine-tuning* (PEFT) techniques, which offer ingenious ways to adapt large models with significantly less computational overhead, often with minimal impact on performance. Methods like QLoRA (Quantization-aware Low-Rank Adaptation) and DoRA (Decomposed and Recomposed Attention) present compelling alternatives.

```
from peft import LoraConfig
from trl import SFTTrainer
peft_config = LoraConfig(r=64, target_modules=["q_proj", "v_proj"])
trainer = SFTTrainer(
    model=model,
    train_dataset=tokenized_dataset["train"],
    peft_config=peft_config,
    max_seq_length=1024,
)
trainer.train()
```

7.3.5 Deployment of Summarization Models

Summarization systems must be deployed in diverse environments, from GPU-backed cloud inference to edge devices with strict resource constraints.

- **Cloud deployment with LLMs:** For larger models like T5, BERT-large, and Mistral, GPU-accelerated services offered by platforms such as AWS SageMaker, Azure ML, or Vertex AI are typically the most suitable for inference.

- **Edge deployment with SLMs:** Smaller models like DistilBERT and E5-small can often be optimized and exported to formats like ONNX or TensorFlow Lite, making them viable for deployment on mobile or embedded systems. Serverless functions, such as AWS Lambda, can also provide cost-effective and low-latency solutions for these models.

- **BitNet CPU deployment pipeline:** Notably, Microsoft's BitNet models are specifically designed for efficient deployment on standard CPUs. The official bitnet.cpp framework facilitates rapid inference, leveraging the GGUF format and its unique 1.58-bit quantization.

7.3.5.1 BitNet CPU Deployment Pipeline

Notably, Microsoft's BitNet models are specifically designed for efficient deployment on standard CPUs. The official bitnet.cpp framework facilitates rapid inference, leveraging the GGUF format and its unique 1.58-bit quantization. This makes it possible to run high-quality summarization models with minimal memory and energy requirements—even on consumer-grade hardware.

```
# Convert and quantize BitNet model for summarization
python convert_bitnet_to_gguf.py quantize --qtype Q1_58
./bitnet -m summarization_model.gguf -p "Summarize: The global economy is entering a new era of..."
```

7.3.6 Inference Optimization Strategies for Summarization Models

To ensure summarization systems are efficient and responsive in real-time applications, various optimization strategies can be applied. These strategies are outlined in Table 7-4.

Table 7-4. Optimization Strategies for Summarization

Strategy	Description
Quantization	Reduces precision (e.g., FP32 → INT8 or 1.58-bit) for faster inference.
Pruning	Removes less significant weights to shrink the model and boost speed.
Distillation	Trains a smaller "student" model to imitate a larger "teacher" model.
Flash attention	Optimizes memory use for long-context summarization.
Batch inference	Groups multiple inputs to improve hardware utilization and throughput.

Consider this example of flash attention for summarization:

```
from transformers import AutoModelForSeq2SeqLM, AutoTokenizer
model = AutoModelForSeq2SeqLM.from_pretrained("google/flan-t5-xl", use_
flash_attention_2=True)
tokenizer = AutoTokenizer.from_pretrained("google/flan-t5-xl")
```

Consider this example of knowledge distillation:

```
from transformers import BartForConditionalGeneration,
DistilBertForSequenceClassification
# Teacher (BART), Student (DistilBERT-based)
teacher = BartForConditionalGeneration.from_pretrained("facebook/bart-
large-cnn")
student = DistilBertForSequenceClassification.from_pretrained("distilbert-
base-uncased")
```

7.3.7 Evaluation Metrics for Summarization Quality

Effective evaluation of summarization models requires both content accuracy and linguistic fluency. The key metrics are outlined in Table 7-5.

Table 7-5. *Key Metrics for Summarization Quality*

Metric	Use Case
ROUGE (1/L)	Measures n-gram and longest common subsequence overlap with reference summaries.
BLEU	Commonly used for machine translation but applicable to summarization fluency.
BERTScore	Uses contextual embeddings to measure semantic similarity.
Summac	Evaluates factual consistency between summary and source.
Latency	Time taken per summary—critical in user-facing applications.

For example, you could use ROUGE & BERTScore. Refer to the code on GitHub for full illustration: https://github.com/anvcse562/ModernNLP-LLMSLM/blob/main/Ch07/Ch7.3/7_3_7_Evaluation_Metrics_for_Summarization_Quality.ipynb.

This sets up a REST API endpoint at http://127.0.0.1:8000/summarize/, where you can post articles and receive summaries.

7.3.8 Conclusion: Building Future-Ready Summarization Systems

The rise of open-source NLP tooling has democratized access to powerful summarization capabilities. Whether through instruction-tuned LLMs like FLAN-T5 or highly efficient SLMs like DistilBERT, developers now have the freedom to build summarization pipelines tailored to both high-performance and resource-constrained settings. From edge deployments on CPUs using BitNet to instruction-following summaries using Mistral or T5, the landscape supports scalable, interpretable, and efficient implementations.

By mastering the fine-tuning strategies, deployment techniques, and evaluation metrics outlined in this section, you'll be equipped to develop summarization systems that balance readability, accuracy, and real-world feasibility across a broad range of applications—from news aggregation to enterprise content compression.

7.4 Industry Uses Cases

7.4.1 News Aggregation Use Case

Businesses and organizations across media, finance, and research continually ingest massive volumes of news content from multiple sources—websites, RSS feeds, social media, wire services—to stay informed about market trends, competitor activity, regulatory developments, and breaking events. Manually scanning dozens of articles every hour is neither scalable nor timely.

Modern summarization systems can:

- **Cut overload:** Turn dozens of articles into digestible bullet summaries
- **Speed insights:** Deliver real-time briefs on key topics
- **Tailor content:** Cluster by theme, region, or keyword
- **Offer flexibility:** Combine extractive and abstractive summaries for speed and depth

7.4.1.1 Implementation Overview

This notebook demonstrates how to build a news aggregation pipeline using both extractive summarization (with a lightweight SLM) and abstractive summarization (with larger LLMs), leveraging the Hugging Face Transformers library and the Sumy library for extractive methods.

1. **Data loading:** Fetch articles via URLs using Newspaper3k.
2. **Preprocessing:** Clean and merge texts into topic-specific corpora.
3. **Extractive summarization:** Apply Sumy LexRank to select key sentences.
4. **Abstractive summarization:** Use BART and PEGASUS models for natural-language summaries.
5. **Comparison:** Display both summary types to meet different user preferences—quick skims and in-depth reads.

7.4.1.2 Deployment and Scaling

- **Local prototype:** Run as a Jupyter notebook or small Flask API for internal dashboards.

- **Cloud services:** Containerize with Docker and deploy to AWS ECS/GCP Cloud Run; use autoscaling for spikes in news volume.

- **Real-time feeds:** Swap static URLs for RSS-feed polling or Kafka streams for continuous ingestion.

Refer to the code on GitHub using this link: https://github.com/anvcse562/ModernNLP-LLMSLM/blob/main/Ch07/Ch7.4/news_aggregation.ipynb.

7.4.2 Business Document Summarization Use Case

In modern enterprises, vast volumes of textual information are generated daily: internal reports, project updates, financial reviews, compliance audits, board meeting notes, and customer feedback forms. These documents often contain essential insights but are long, redundant, and inconsistent in format. Manually reviewing each document is time-consuming and error-prone—especially when teams are distributed and timelines are tight.

Business document summarization provides a powerful solution. By using LLM-based summarization systems, organizations can automatically distill the essence of lengthy documents into short, readable synopses that help decision-makers act quickly, confidently, and at scale.

7.4.2.1 The Problem: Too Much Text, Too Little Time

Consider these common scenarios:

- A project manager wants to understand outcomes from the last ten meeting transcripts.

- A finance director needs a snapshot of Q1 reports from five different departments.

- A compliance officer wants summaries of regulatory audit documents from different regions.

Each of these cases involves multiple long-form documents, often filled with repetitive structure, nuanced language, and domain-specific terms. Reading everything manually would take hours. But what if a system could summarize each document in one to two paragraphs without losing important details?

That's the goal of business summarization systems: reduce cognitive load, accelerate decision-making, and maintain content fidelity.

7.4.2.2 Pipeline Overview

A typical business document summarization pipeline involves the following components:

- **Document ingestion:** Documents may come in via email attachments, cloud storage, internal CRMs, or shared drives. For this example, we simulate ingestion by placing .txt files in a designated folder.

- **Preprocessing and cleaning:** Text is often messy. Preprocessing includes removing excess whitespace, headers/footers, and redacted sections. We use a simple regex-based cleaner here.

- **Summarization model:** Here, model choice depends on the summarization type:

  - **Primary choice is DistilBERT:** DistilBERT is 40% smaller than BERT-large but maintains 95% accuracy, making it highly efficient. It's excellent for pinpointing and extracting critical sentences directly from documents, ensuring factual accuracy by retaining original phrasing.

  - **Alternatives:**
    - E5-small-v2 for multilingual business documents.
    - DeBERTa-v3 offers potentially better performance on various summarization tasks.
    - Qwen2-7B suitable for very long documents due to its 128K token context length, although its large size may impact on-premises deployment feasibility.

- You can also use `facebook/bart-large-cnn` or FLAN-T5 models for abstractive summaries when fluent, human-like synopses are needed.

- **Post processing and storage:** Generated summaries are stored in `.json` files, ready for downstream use (e.g., search, analytics, delivery via email or dashboard).

7.4.2.3 Data Preparation and Fine-tuning

Fine-tuning a summarization language model on your specific business corpus is crucial for effectiveness.

Dataset:

- **Internal business corpus:** High-quality, domain-specific documents with human-written or expertly annotated extractive summaries to capture business nuances.

- **Public datasets:** Use datasets like CNN/DailyMail for initial fine-tuning but prioritize your own data for final training.

Preprocessing:

Use `DistilBertTokenizer` to tokenize documents. Adapt preprocessing functions to label sentences as "summary" or "non-summary," framing the task as sentence classification. For long documents, apply sliding window or chunking techniques to handle model input length limits.

Fine-tuning:

Standard Trainer setups work well for DistilBERT. For efficiency and frequent updates, consider parameter-efficient fine-tuning (PEFT) methods like LoRA, which reduce computational overhead while adapting models effectively.

7.4.2.4 Deployment Pipeline

For sensitive business data, on-premises or controlled cloud deployment is critical.

- **Edge deployment with SLMs (On-Premises/Private Cloud):**
    - Optimize models after fine-tuning with quantization (e.g., FP32 to INT8) for faster inference and smaller memory footprint.
    - Prune less significant weights to reduce model size.
    - Export optimized models to ONNX for cross-platform use or TensorFlow Lite for mobile/embedded systems.
    - Deploy on a dedicated enterprise server or private cloud solution (serverless or Kubernetes-based) to scale securely.

- **CPU-optimized deployment (conceptual):**

    Although DistilBERT typically runs well on standard CPUs (especially post-quantization), highly quantized models like BitNet might be explored for extreme resource constraints.

7.4.2.5 Inference Optimization Strategies

To enhance real-time performance, use:

- Quantization to reduce model precision and memory usage.
- Batch inference to process multiple documents concurrently.
- Flash Attention for better memory efficiency on long-context documents.
- Model caching for frequently summarized documents to avoid redundant computation.

7.4.2.6 Evaluation Metrics for Enterprise Summarization

Evaluation balances academic rigor with business utility:

- **Content accuracy/factual consistency:** Use Summac metrics and manual expert reviews, especially on compliance and financial documents.
- **Relevance:** Ensure summaries highlight the most pertinent information.
- **Brevity/conciseness:** Maintain short but informative synopses.
- **Latency and throughput:** Critical for real-time and batch processing, respectively.
- **Domain-specific metrics:** Custom metrics such as precision on legal clauses may be required.

7.4.2.7 A Hands-on Example

In the Jupyter notebook provided (`business_doc_summarization.ipynb`), we:

- Load `.txt` files from a folder named `business_docs/`.
- Apply cleaning and summarization using the Hugging Face `pipeline`.
- Print and save summaries to a JSON file.

A sample document is auto-generated to test the pipeline:

Q1 meeting notes show 15% revenue growth from SaaS, slight expense rise due to R&D hiring, plans to raise Q2 marketing budget, and explore automation tools. Compliance and audit results were satisfactory.

The summarizer might produce:

"Finance team reviewed Q1 with 15% revenue growth, slight rise in expenses, and plans to boost marketing and explore automation. Compliance reports were satisfactory."

This captures the essence while reducing verbosity.

7.4.2.8 Why Use Abstractive Summarization?

Abstractive summarization is preferred over extractive methods in business contexts for several reasons:

- **Fluency and readability:** Abstractive summaries read like human-written synopses.

- **Information fusion:** They can combine ideas spread across sentences into one coherent sentence.

- **Terminology simplification:** Abstractive models can paraphrase technical jargon for general readability.

- **Formatting consistency:** Output remains uniform across documents of varying length or style.

Table 7-6 outlines how different models perform with different types of documents.

Table 7-6. Common Document Types and Model Performance

Document Type	Length (Words)	Summary Type	Example Model	Comments
Meeting notes	500–1,000	Bullet or para	BART, T5	Conversational, scattered input
Financial reports	1,000–5,000	Structured para	FLAN-T5, Mistral-7B	Numbers and narrative—needs precision
Strategy decks (converted)	800–2,000	Bullets or key ideas	GPT-4 (via API)	Often unstructured post-OCR
Legal/compliance text	2,000–8,000	Key clauses	DistilBERT + Q&A Model	May require extractive verification

7.4.2.9 Scaling to Enterprise Workflows

In production environments, such a system is not just a notebook—it becomes a pipeline:

- **Document upload** triggers summarization via S3/Lambda events.
- **Summaries** are saved in a vector DB for search or are embedded in dashboards.
- **Role-based filtering** ensures specific people (e.g., legal team, finance head) see relevant summaries only.
- **LLM agents** can be added to review and rephrase summaries depending on recipient expertise.

A powerful extension is a multi-agent system:

- **Reader agent:** Parses incoming documents.
- **Summarizer agent:** Uses BART/FLAN-T5 to generate summaries.
- **Verifier agent:** Uses GPT-4 to check summary consistency.
- **Rewriter agent:** Adjusts tone (formal/informal) for different teams.

7.4.2.10 Security and Privacy Concerns

Since business documents often contain sensitive data:

- Models must run in VPC-isolated environments (e.g., SageMaker private endpoints).
- Summarization should be performed on-premises or in customer-managed cloud infrastructure.
- Outputs may be subject to auditing and versioning (store original and summarized pair).
- Certain models (especially LLMs hosted via public APIs) must not be used without redaction or approval.

7.4.2.11 Challenges and Limitations

- **Length limits:** Many models have 512 or 1024 token limits. For longer documents, chunking or LongForm summarization is needed.

- **Domain drift:** Generic summarization models might not understand business acronyms or context unless fine-tuned.

- **Fact hallucinations:** Models may "invent" plausible-sounding content. Use retrieval-based or fact-checking agents when accuracy is critical.

- **Evaluation:** ROUGE scores are insufficient alone. Human-in-the-loop reviews are essential for quality control.

Table 7-7 outlines the different model variants.

Table 7-7. Model Variants You Can Try

Model Name	Type	Description
facebook/bart-large-cnn	Abstractive	Pretrained on news summarization
t5-small	Abstractive	Lightweight model for real-time applications
FLAN-T5-XXL	Abstractive	Instruction-tuned, powerful but GPU-heavy
DistilBERT	Extractive	Good for highlighting key sentences
mT5	Multilingual	Summarization of global business documents

7.4.2.12 Extensions Examples

You can enhance the notebook with:

- Integration with PDF-to-text converters (e.g., `pdfplumber`) to summarize scanned reports.

- Summary categorization by department (using a classifier).

- Notification system (email or Slack) for newly summarized documents.

- A simple web UI for uploading and viewing documents and summaries.

7.4.2.13 Conclusion

Summarizing business documents is essential as companies grow and internal content multiplies. AI-driven summarization helps teams quickly grasp key information without missing important details.

Refer to the GitHub code using this link: https://github.com/anvcse562/ModernNLP-LLMSLM/blob/main/Ch07/Ch7.4/business_doc_summarization.ipynb.

The notebook presented here serves as a strong baseline. With further customization, model fine-tuning, and integration into enterprise ecosystems, it can be evolved into a robust document intelligence system—offering clarity, speed, and competitive advantage.

7.4.3 Automated Email Summarization Use Case

Email overload is a daily reality in organizations, with critical updates buried in verbose, repetitive threads. Manually sifting through them wastes time and risks missing key info.

Automated email summarization uses LLMs to distill complex threads into clear, actionable summaries—helping users triage faster and stay focused with minimal effort.

7.4.3.1 The Business Problem

Consider the following real-world scenarios:

- A project manager is added to a long thread midway and wants a quick catch-up.
- A sales lead needs summaries of all incoming client communications over the last week.
- A senior executive prefers to read concise digests instead of scrolling through email chains.
- A support engineer wants to extract resolution status from customer threads without digging through every reply.

Each of these cases involves one or more of the following:

- **Redundant language** (signatures, greetings, quoted replies)
- **Multiple participants** (introducing contextual shifts)
- **Lengthy conversations** (requiring focus on key updates only)

Traditional tools like email filters or basic keyword search don't solve the core issue of semantic overload. Users want meaning, not just structure.

7.4.3.2 Solution Overview: Intelligent Email Summarization

The automated email summarization system built in the Jupyter notebook uses:

- **Mailbox** for parsing `.mbox` files (an email archive format)
- **Regex cleaning** to remove quoted replies and boilerplate
- `facebook/bart-large-cnn` summarization model from Hugging Face Transformers
- **Batch processing** and structured JSON export

This notebook simulates the full pipeline—from loading sample emails, cleaning their content, summarizing the body, and saving the results to disk.

7.4.3.3 How It Works (Pipeline Breakdown)

Step 1. Email Ingestion

The input to the pipeline is an `.mbox` file containing multiple emails. This format is commonly used for archiving emails from clients like Thunderbird, Gmail Takeout, or enterprise mail servers. The pipeline loads this archive using the Python `mailbox` library.

Each email is parsed to extract:

- The subject line
- The plain-text body content

This forms the basis for the summarization process.

Step 2. Cleaning and Preprocessing

Email bodies are often messy; they can include:

- Replies (`> quoted text`)
- Excess whitespace
- Repeated headers or disclaimers

CHAPTER 7   TEXT SUMMARIZATION

The notebook uses regular expressions to remove:

- Quoted lines
- Extra whitespace
- Signature clutter (in basic form)

This cleaning ensures the summarizer gets only the current email content instead of processing an entire thread recursively.

Step 3. Summarization Model

The summarizer uses the BART-based model `facebook/bart-large-cnn`. This is an abstractive summarization model, meaning it can paraphrase and reword content, not just extract key sentences.

This is crucial for emails because:

- People use informal and varied language
- Extractive summaries often preserve unnecessary details
- A fluid narrative helps in creating digestible email digests

The model converts 100- to 300-word emails into one- to three-sentence summaries.

Step 4. Batch Summarization

Once the model is loaded and cleaned emails are ready, each email is passed through the summarizer. If the email body is too short (< 50 words), the original message is retained without summarization to avoid over-compression.

Each summary is saved alongside the subject line in a structured format:

```
{
  "subject": "Weekly Update #3",
  "summary": "The team finalized UI upgrades and integrated backend APIs.
  QA will begin next week."
}
```

This structure is ideal for frontend display or downstream indexing in search engines, dashboards, and alerting systems.

Step 5. Output

The results are saved in a timestamped .json file that is

- Easy to view
- Usable in any downstream application
- Compatible with reporting tools like Power BI, Grafana (via plugins), and simple HTML dashboards

7.4.3.4 Why This Matters in Enterprise

- **Saves time:** Teams can scan summaries instead of triaging dozens of emails.
- **Enables automation:** Runs in the cloud to summarize emails in real time.
- **Supports reporting:** Delivers digest updates to teams without inbox overload.
- **Empowers focus:** Helps teams stay focused, diving into details only when needed.

Table 7-8 outlines the potential deployment architectures for this email summarization process.

Table 7-8. Potential Deployment Architectures

Component	Technology
Email source	Gmail API, Exchange Online, MBOX
Backend ingestion	Python, IMAP, and OAuth2
Summarization engine	Hugging Face Transformers
Storage	S3/DynamoDB/PostgreSQL
Delivery	Email summary digest via SES/SMTP
API (optional)	FastAPI/Flask REST service

7.4.3.5 Security and Compliance Considerations

Emails are sensitive personal and corporate data. Any summarization system must:

- Run in secure environments (e.g., VPC, on-prem servers)
- Log accesses and transformations for audit purposes
- Mask or redact sensitive data if summaries are shared externally
- Support role-based access control for digest consumption

For example, customer support summaries may be viewable only by the support team lead, not by engineering or finance.

7.4.3.6 Model Choices and Customization

While `facebook/bart-large-cnn` is general-purpose and performs well, fine-tuning may be required in certain contexts:

- **Sales email summarizer:** May need to prioritize action items and lead stage.
- **Support thread summarizer**: Must extract resolution status, customer tone, and follow-ups.
- **Legal email summarizer:** Should retain clauses, references, and obligations.

These can be achieved by using these approaches:

- Custom training on labeled internal emails
- Instruction tuning for specific summarization goals
- Using larger models like flan-t5-xl or domain-specific GPT-style models in private environments

7.4.3.7 Key Limitations

- **Thread awareness:** Summarizes individual emails; full context needs thread-level modeling.
- **HTML handling:** Only plain text is supported; real use cases need HTML sanitization.

- **Attachments and links:** These are skipped in this version; advanced systems should extract or summarize them.

- **Factual consistency:** Abstractive models may hallucinate—human review is advised for critical emails.

You can find the full code on GitHub at: https://github.com/anvcse562/ModernNLP-LLMSLM/blob/main/Ch07/Ch7.4/automated_email_summarization.ipynb.

Automated email summarization is essential for managing growing inbox volumes and improving workplace efficiency. This solution accelerates triaging, enhances communication tracking, and supports smarter decisions, making it a vital tool in today's fast-paced environments.

7.5 Summary

This chapter explored advanced text summarization techniques using LLMs and SLMs. You learned how to implement extractive and abstractive summarization approaches, using models like BART, PEGASUS, T5, FLAN-T5-XXL, Mistral-7B, and energy-efficient alternatives like BitNet-b1.58. You gained hands-on experience deploying these models with AWS services such as SageMaker and Lambda, and building custom pipelines using Hugging Face Transformers. Real-world applications were also covered, including news aggregation, business document summarization, and email automation. You should now have a solid foundation in building and deploying summarization systems for various use cases. The next chapter turns to another fundamental NLP task: language translation.

CHAPTER 8

Language Translation

This chapter explores modern machine translation techniques using LLMs such as MarianMT, mBART, T5, and mT5, and SLMs like DistilBART, TinyBERT, and MiniLM. It discusses the complexities of translating across languages—covering syntax, semantics, idioms, and domain-specific jargon—while comparing the strengths of LLMs and SLMs in terms of quality, latency, and computational cost.

The chapter also covers scalable implementations using AWS services like Amazon Translate for real-time, pretrained translation, and SageMaker for training or fine-tuning custom translation models. Open-source approaches using Hugging Face Transformers are emphasized, including training multilingual models and evaluating them using industry-standard metrics like BLEU, ROUGE, and TER.

By the end of this chapter, you'll understand how to build and deploy accurate, efficient, and scalable translation systems tailored to your domain and infrastructure.

This chapter covers the following topics:

- **Overview of language translation:** Challenges in translation, use cases, and the role of LLMs and SLMs.

- **Implementation on AWS:** Using Amazon Translate, SageMaker, and lightweight models for real-time translation.

- **Open-source implementation:**
  - Building translation pipelines with Hugging Face
  - Training and fine-tuning models like MarianMT and mBART
  - Evaluating quality with BLEU, ROUGE, and TER

- **Industry use cases:** Multilingual customer support, content localization, and global information delivery.

CHAPTER 8   LANGUAGE TRANSLATION

Before diving into machine translation, it's worth recalling where we left off in Chapter 7. You saw how text summarization condenses information and is invaluable in applications like news aggregation and business report analysis. Summarization demonstrated how NLP can distill key ideas from a sea of text. Now, as we transition to language translation, we shift from condensing one language to bridging between languages. Just as summarization makes content digestible, translation makes content accessible across linguistic boundaries. Both tasks highlight NLP's power: whether summarizing a report or translating a document, the goal is to preserve meaning and intent. With that continuity in mind, this chapter focuses on breaking language barriers through translation.

8.1 Introduction to Language Translation

Language translation is the process of converting text or speech from one language (the *source language*) into another (the *target language*) while preserving meaning, tone, and context. It's a cornerstone of NLP, enabling global communication in a multilingual world. However, translation is far more complex than swapping words between languages. Unlike summarization, which works within one language, translation must navigate the vast differences in grammar, vocabulary, and culture between languages. This section introduces the challenges unique to machine translation, discusses why NLP-driven translation is so important in today's interconnected society, and explores how LLMs and SLMs can be leveraged for translation tasks. Throughout, we use simple examples to illustrate concepts and implement a basic translation with code to demonstrate practical applications.

8.1.1 Translation Challenges

Translating text is a complex task due to the intrinsic differences between languages. A good translation isn't a word-for-word transliteration; it requires conveying the original meaning, style, and nuances in a form natural to the target language. Table 8-1 outlines some of the key challenges in achieving accurate machine translation, along with simple examples.

Table 8-1. Key Challenges in Achieving Accurate Machine Translation

Challenge	Description	Example (EN → ES)
Syntax differences	Languages have different word orders and grammatical rules. A translation system must rearrange and transform sentences to fit the target language's syntax.	*English:* "The **red car** is fast." *Spanish:* "El coche **rojo** es rápido." (Note the adjective-noun order changes: "red car" → "coche rojo".)
Semantic nuances	Many words are *polysemous* (have multiple meanings) and require context to translate correctly. The system must discern the intended meaning from context.	*English:* "This bank is busy." *Spanish:* Could be "Este **banco** está ocupado." (if **bank** = seating bench) **or** "Este **banco** está lleno de gente." (if **bank** = financial institution). Context is needed to choose the right meaning of "bank."
Idioms and cultural phrases	Idiomatic expressions and culturally specific phrases do not translate literally. Direct translation often fails to convey the meaning.	*English:* "It's **raining cats and dogs**." (means it's raining heavily). *Literal Spanish:* "Está lloviendo **gatos y perros**." ⊘ (nonsense). *Proper Spanish:* "Está lloviendo **a cántaros**." ☑ (literally "it's raining pitchers," an idiom for heavy rain). **Why?** The English idiom is culturally rooted, so a word-for-word translation is meaningless.
Domain-specific jargon	Specialized fields (legal, medical, technical, etc.) use terminology and jargon that may not appear in everyday language. Translating such terms requires specialized knowledge or data.	*Medical English:* "The patient has a **myocardial infarction**." *Spanish:* "El paciente tiene un **infarto de miocardio**." (A generic translator without medical training might struggle with such terms.)

As shown in Table 8-1, an effective translation model must handle everything from grammatical structure to contextual meaning. Simple sentences with literal meanings are relatively easy to translate. But when a sentence contains an idiom or ambiguous word, machines face one of their greatest challenges. Idioms are deeply tied to cultural context—an expression like "break the ice" (to ease tension) cannot be translated word-for-word into, say, Japanese, because the imagery and meaning don't carry over. Similarly, *polysemous words* (words with multiple meanings, like "bank", "cell", "java") require understanding the broader context to choose the correct translation. A machine translation (MT) system needs to look at the whole sentence or even the surrounding sentences (discourse context) to infer the intended meaning.

Another challenge is domain-specific language. General-purpose MT engines might do well on everyday language but falter on technical text. For example, legal documents have specialized wording and long, complex sentences; a generic translator might mistranslate legal terms or produce grammatically incorrect legal phrasing. If the model's training data lacks *domain-specific examples*, it will struggle. In fact, even for well-resourced languages, certain domains or industry jargon may see poor translation quality due to scarce domain-specific training data. This is why professional translators often work with domain glossaries, and why custom MT models are sometimes trained for specific fields. A related example is a model fine-tuned on medical texts for a hospital's translation needs.

In summary, machine translation must bridge not just language gaps but also cultural and contextual gaps. It's an AI task that lies at the intersection of linguistics and real-world knowledge. As we develop translation systems, we keep these challenges in mind, aiming for translations that are not only correct in meaning but also natural and culturally appropriate in the target language.

8.1.2 Importance of NLP in Translation

Why do we invest so much effort in NLP-driven translation? The importance becomes clear when you consider how language barriers affect information access and global communication. Here are a few key reasons why translation is a critical component of NLP applications today:

- **Global customer support:** In a worldwide market, businesses serve customers who speak diverse languages. NLP-powered translation enables a customer support chatbot or helpdesk to assist users in their native language. For example, a single AI agent could handle queries in English, Spanish, or Mandarin by dynamically translating the conversation. This improves user satisfaction and opens products to a global audience. For example, consider a customer in France asking a question in French. An agent translates it to English to find an answer and then translates the reply back to French—all in real time.

- **Content localization:** Websites, apps, and documentation often need *localization*—adapting content for different languages and cultures. Automatic translation is the first step in localization. Without translation, a vast portion of Internet content remains inaccessible to non-English speakers. In fact, English dominates over half of all website content, yet only about 20% of the world's population speaks English, according to `languageline.com`. The same source found that more than half of websites are effectively unreadable to 80% of the world's population. This stark imbalance makes translation crucial. Companies localize product descriptions, user interfaces, and user manuals so that customers in Brazil or China, for instance, can use software and services in Portuguese or Chinese. NLP helps automate this at scale—for example, services like Amazon Translate or Google Translate API can instantly convert an English blog post into dozens of languages, accelerating the localization process. Human review is often added to ensure quality, but the boost in productivity is immense.

- **Multilingual content and accessibility:** Translation driven by NLP also plays a key role in areas like news and education. News agencies use machine translation to publish articles in multiple languages simultaneously, spreading information globally. Educators and researchers translate materials to share knowledge across language barriers, enabling students worldwide to access online courses or papers originally written in another language. Furthermore, translation is vital for inclusivity—consider subtitles on videos or translated social media posts that allow people from different linguistic backgrounds to engage with content. In customer-facing domains, providing content in a user's preferred language is often not just a convenience but a necessity for comprehension. Automatic translation systems empower smaller content creators to reach broader audiences without needing a whole team of human translators.

In essence, NLP-powered translation democratizes information and services. It breaks down walls, allowing a Spanish-speaking user to navigate an originally English website, or an English-speaking doctor to read a Japanese medical report. The societal impact is huge: governments employ translation for diplomatic communications, NGOs use it for humanitarian outreach in local languages, and businesses rely on it to operate globally. As we progress, the synergy of NLP and translation will only grow—enabling real-time multilingual conversations and truly global applications.

8.1.3 LLMs for Translation

LLMs have emerged as powerful tools for translation. These are models with hundreds of millions of parameters to billions of parameters (think of models like OpenAI's GPT-3/GPT-4, Google's advanced models, or Meta's large translation systems). LLMs are trained on massive amounts of text, often across many languages, which gives them an almost uncanny ability to capture linguistic patterns and context. How do LLMs contribute to machine translation, and what are their strengths?

8.1.3.1 Strengths of LLMs in Translation

LLMs excel at understanding context and nuance. Because of their size and training data breadth, they can grasp subtle meanings, idiomatic expressions, and cultural references in ways earlier models struggled with. For example, GPT-4 (2025's state-of-the-art from OpenAI) is noted for its strong grasp of context and idiomatic language, producing translations that read very fluently. (See `polilingua.com` for reference.) An LLM can take a complex sentence and not just translate words, but preserve the *tone* and *intent*. If a sentence has sarcasm, formality, or a poetic style, a well-tuned LLM is more likely to convey that in translation. In one case study, users found that GPT-4 could translate a marketing slogan from English to French while keeping its catchy, playful tone—something a literal translation might lose.

LLMs are also capable of zero-shot and few-shot translations. This means they can translate between language pairs or in domains they were never explicitly trained on, just by virtue of broadly understanding language. For instance, an LLM might handle an

English to Swahili request reasonably well, even if it hasn't seen a lot of parallel Swahili data, by leveraging its general language knowledge. This flexibility is valuable for low-resource languages or niche domains, where we lack large parallel datasets. However, it comes with a caveat: studies have shown that for some low-resource language pairs, a specialized MT model can still outperform a general LLM (see https://www.apptek.ai). In other words, a massive generalist model might produce *decent* translations for an uncommon language, but a smaller model trained specifically on that translation task could be *better* in quality. Nonetheless, the fact that LLMs can attempt such translations at all is impressive and useful when specialist models are not available.

Another strength is *dynamic adaptation*. LLMs can be guided via prompts or a few examples to adapt to a certain style or domain on the fly. Suppose you want a legal document translation. You could prompt an LLM with, "Translate the following text to Spanish in a formal legal tone," and the model will tend to produce more formal wording appropriate for legal context. LLMs "understand" nuances like slang versus formal speech and can adjust accordingly (see https://www.apptek.ai) This makes them suitable for creative or high-context translations such as marketing content, literature, and conversational dialogues, where capturing the right tone and nuance is as important as factual accuracy.

To illustrate an LLM's translation ability, consider the idiom example from earlier. If we ask a large model like GPT-4 to translate *"It's raining cats and dogs"* into Spanish, it is likely to output "Está lloviendo a cántaros." The model "knows" this idiom's meaning and the equivalent expression, rather than literally mentioning cats or dogs. In contrast, a simpler system might mistakenly produce *"lloviendo gatos y perros,"* which makes no sense in Spanish. This demonstrates how an LLM's extensive training helps with non-literal language.

8.1.3.2 Limitations of LLMs

Despite their prowess, LLMs have notable downsides for translation. First, they are computationally heavy. Running a huge model to translate every sentence can be overkill—it requires significant memory and processing power (often needing costly GPUs). This makes it hard to deploy LLMs in real-time applications where resources are limited or cost is a concern. In fact, while an LLM may generate one high-quality translation slowly, a well-optimized smaller translation model might translate *multiple sentences per second* on a normal CPU. There's a speed versus quality tradeoff.

Secondly, LLMs can sometimes be inconsistent or overly "creative" when you don't want them to be. They might *hallucinate*—producing fluent but incorrect translations (adding content that wasn't in the original text or choosing a wrong word that sounds plausible). This is tied to the general LLM issue of reliability; they weren't originally trained specifically *for translation* but for broad language modeling. An LLM might, for example, embellish a sentence or misinterpret a technical term if it's not grounded by training data. This is why for high-stakes translations (legal contracts and medical instructions, for example), relying solely on an LLM without human review can be risky—a subtle error could have serious consequences.

Another concern is cost and privacy. Using a third-party LLM API to translate confidential documents might raise data privacy issues. And if you want to self-host an LLM, the infrastructure and energy costs are significant. By contrast, many dedicated MT engines (like Amazon Translate or open-source models) can be run in a more lightweight manner with control over data.

In summary, LLMs represent the cutting-edge of translation quality in many cases, capturing context and subtlety that smaller models might miss. OpenAI's GPT series, Google's latest (e.g., the Gemini model), Meta's projects (like NLLB-200 for many languages), and others have showcased near-human translation fluency across dozens of languages. These models can function as powerful translation agents, even as part of larger AI systems—for instance, an LLM-based agent could translate a user's query, answer it in another language, and continue a dialogue across languages seamlessly. Yet, you must balance their use with considerations of efficiency,cost, and reliability. This is where SLMs come into play.

8.1.4 SLMs for Translation

While LLMs grab headlines, SLMs play a crucial role in machine translation, especially when it comes to *efficiency* and *deployability*. SLMs are models that have far fewer parameters than the likes of GPT-4—often in the range of millions to low hundreds of millions of parameters. They might be specialized for specific language pairs or distilled (compressed) versions of larger models.

Why and how do you use SLMs for translation? The primary advantage of SLMs is speed and resource efficiency. Smaller models require less memory and computation, which means they can translate text faster and can run on devices with limited hardware (like a smartphone or an edge device) or on servers without expensive GPUs. For real-time translation services—say a live chat translation or a mobile translation app—SLMs are often the practical choice. They might not capture every nuance like an LLM would, but they can produce translations quickly enough to keep a conversation flowing.

8.1.4.1 Examples of SLMs in Translation

A good example is DistilBART, which is a distilled (compressed) version of the BART sequence-to-sequence model. BART is a powerful transformer model originally developed for language generation and can be used for tasks like summarization and translation. DistilBART retains much of BART's capabilities but in a smaller, faster package. Think of it as having six layers instead of 12—it's roughly half the size, which makes it lighter to run. Similarly, TinyBERT is a smaller variant of the BERT model (BERT is typically used as an encoder, but TinyBERT or similar mini-models can be part of translation systems). TinyBERT is seven and a half times smaller and nine times faster than its full-sized counterpart yet manages to achieve around 97% of the larger model's performance on language understanding tasks. This kind of efficiency is extremely valuable in translation pipelines where you may be translating thousands of queries per second.

SLMs for translation can come from two approaches:

- **Knowledge distillation:** This is where you train a small "student" model to mimic a large "teacher" model's output. For translation, you might take a big accurate model (teacher) and use it to generate translations, then train a small model on that data to achieve similar results. DistilBART was created via distillation, for example. The student model ends up faster while only slightly less accurate.

- **Specialized architecture or pruning:** Designers can create smaller architectures optimized for translation of specific language pairs. For instance, the Marian NMT models (like Helsinki-NLP's OPUS-MT models on Hugging Face) are reasonably compact translator models pretrained for one language pair (or a set of languages). By focusing on a narrower task, they can be smaller but still effective.

8.1.4.2 Use Cases

SLMs shine in the following scenarios:

- **Real-time translation:** For example, translating captions on a live video or chat. A smaller model can keep up with the speed required.

- **Resource-constrained environments:** For example, an offline translator embedded in a device (imagine a handheld translator gadget, or on-device translation in a mobile app when you have no Internet). Here, a model like TinyBERT or a small MarianMT model is necessary to fit in the memory and run on a CPU.

- **High-volume systems:** For example, a service translating millions of product descriptions. The cost per translation with an LLM would be high and slow; a fleet of small models can handle the volume more economically.

To illustrate how you might practically use an SLM (or any available translation model), consider a quick code example using Hugging Face's Transformers library. Tis example uses a pretrained small translator model to translate a simple sentence from English to French. You can find the code on GitHub at: https://github.com/anvcse562/ModernNLP-LLMSLM/blob/main/Ch08/ch8.1/8_1_1_translation_example.py.

In this code, we loaded a `Helsinki-NLP/opus-mt-en-fr` model, which is a reputable open-source English-to-French translation model. It is not extremely large (certainly nowhere near GPT-scale), making it relatively fast. We then used the `pipeline` to translate an English sentence. The output stored in `result` might be something like this:

```
Bonjour, comment ça va ? Je suis excité d'apprendre de nouvelles choses !
```

This is a reasonable French translation of the input English sentence ("Hello, how are you? I am excited to learn new things!"). The small model managed to capture the meaning quite well. An even smaller distilled model would work similarly, perhaps with a slight tradeoff in fluidity or accuracy.

It's worth noting that while the translation is correct, the phrase "je suis excité" can have a subtle connotation in French (it can mean *excited* but also *aroused*). A human translator or a larger context-savvy model might choose a more precise phrasing like "je suis *impatient* d'apprendre..." to avoid any ambiguity. This highlights a general point:

SLMs often produce literal or straightforward translations, which are usually fine for everyday use, but they might miss some of the finesse that an LLM or a human translator could provide. Still, for many practical applications, the tradeoff is worth it when you need speed, scalability, and low cost.

To summarize, SLMs are the workhorses of machine translation systems—they may not grab the spotlight, but they ensure translation technology can be widely used. By leveraging techniques like model compression, you get fast and cost-effective translation engines that power things like instant messaging translators, real-time subtitles, and bulk document translation services. Many cloud providers and open-source tools offer such models. For example, Amazon Translate (which we' explore soon) is designed to be fast and scalable, likely using efficient model architectures under the hood. The next section explains how both large and small models are applied in practice on AWS.

The next section also dives into implementing language translation on AWS. We explore Amazon's fully managed translation service and cover how to fine-tune custom models (like MarianMT for specific language pairs or domains) on SageMaker. This will bridge the concepts discussed here—from LLMs to SLMs—with real-world deployment and training scenarios.

8.2 Implementation on AWS

The first section provided a conceptual toolkit for machine translation, discussing challenges and how LLMs and SLMs approach the problem. Now we turn to practical implementations using Amazon Web Services (AWS). AWS provides fully managed services for translation and infrastructure for custom models. This section explores how to leverage AWS for translation tasks—from using Amazon's off-the-shelf translator to deploying and training your own models with AWS tools. We also consider where LLM agents and SLM deployments fit into these solutions, ensuring that complex ideas are broken down with clear examples.

At a high level, AWS offers two paths for translation: use a managed API (Amazon Translate) or build a custom model (using services like Amazon SageMaker). The managed service is ideal when you need quick, accurate translations without diving into ML model details. Custom models, on the other hand, give you flexibility—you might fine-tune an open-source LLM like MarianMT for a niche domain or deploy an ultra-fast SLM for real-time translation in a constrained environment. We walk through each approach, complete with code snippets and simple analogies to illustrate the concepts. By the end of this section, you'll know how to implement translation solutions on AWS and how to choose the right approach for your use case.

8.2.1 Amazon Translate

Amazon Translate is AWS's fully managed neural machine translation (NMT) service. It provides real-time and batch translations between numerous languages via a simple API call. This service abstracts away all machine learning complexity—you don't need to train or deploy models; AWS handles it. We cover Amazon Translate's features (like language auto-detection, custom terminology, and formality settings), provide a code example of using the API, and discuss when this managed solution is beneficial. We also explain how Amazon Translate can be integrated into larger applications or LLM agent workflows.

Amazon Translate shines when you need immediate, scalable translation without building a model from scratch. With support for over 75 languages and 5,500 language pairs (see aws.amazon.com), it can handle everything from translating a single sentence to large documents. For example, you could translate English product descriptions into Spanish, Chinese, and French on the fly as users from different regions view your e-commerce site. The service is accessible through the AWS console, CLI, or SDK. A simple API call is all it takes—behind the scenes, AWS's deep learning models do the heavy lifting to return a translation in fractions of a second.

8.2.1.1 Amazon Translate's Key Features

Amazon Translate offers more than basic dictionary lookup. It uses advanced neural networks that consider the full context of sentences, producing translations that are often more natural than rule-based systems. (See aws.amazon.com.) Some notable features include:

- **Real-time and batch translation:** You can get instant translations for user-facing applications with the real-time TranslateText API or use batch jobs for large documents. For example, the asynchronous batch API can translate entire Word documents, PDFs, or HTML files in one go, preserving their formatting. This dual mode means you can both integrate live translation in a chat app and perform overnight bulk translations of content libraries.

- **Language auto-detection:** If you don't know the source language, Amazon Translate can auto-detect it. Simply set the source language code to "auto" and the service will identify it before translating (it uses Amazon Comprehend under the hood for detection). This is useful in user-generated content scenarios—for example, a forum might contain posts in various languages and you want to translate them all to English.

- **Customization options:** While it's a generic service, Amazon Translate allows some customization to better fit your needs:

- **Custom terminology:** You can provide a glossary of terms that should be translated in a specific way (see aws.amazon.com). For instance, if your company's product name should remain in English even in foreign text, or industry jargon has preferred translations, you can ensure the model follows those rules. This is done by uploading a terminology file (CSV) and using it in API calls.

- **Formality and profanity settings:** In certain languages, you can specify formal or informal tone for the translation (see boto3.amazonaws.com). If you're translating to Japanese, for example, you might prefer a polite form for business communications or an informal tone for a social app. There's also a profanity masking option—the model can automatically mask out swear words if needed in the output.

- **Active Custom Translation (ACT):** For deeper customization, Amazon Translate allows you to input parallel data (your own example translations) to influence output. This isn't full model retraining but rather a way to bias the translations toward your domain. It's useful if you have a small corpus of company-specific translations—by providing them, the service will try to mimic that style/terminology in future translations.

- **Security and privacy:** AWS emphasizes that data is encrypted in transit and at rest. Content sent for translation is encrypted and stored only in the region you specify. Also, you have the option to disable data retention for AWS's service improvement. This is important for sensitive data—for example, if you're translating private legal documents, you'd want to ensure they aren't stored or used beyond the immediate translation.

- **Cost-effective scaling:** Amazon Translate charges per character of text translated. This pay-as-you-go model means you don't pay for idle time or maintenance. For sporadic translation needs, this is often cheaper and easier than maintaining a server with a custom model. And it scales automatically—whether you translate 100 characters or 100 million, AWS handles the compute scaling behind the scenes.

Let's look at a quick example. Suppose you want to translate a simple English sentence into Spanish using Amazon Translate's SDK (Boto3 for Python). You'll also specify that you want an informal tone in the output (using the Formality setting), perhaps because the app has a casual style:

```python
import boto3

# Initialize the AWS Translate client
translate = boto3.client('translate')

# The text we want to translate and the language codes
text = "Good morning, how can I help you today?"
source_lang = "en"      # English source
target_lang = "es"      # Spanish target

# Call the TranslateText API with an informality setting
response = translate.translate_text(
    Text=text,
    SourceLanguageCode=source_lang,
    TargetLanguageCode=target_lang,
    Settings={"Formality": "INFORMAL"}
)

translated_text = response.get('TranslatedText')
print(translated_text)
```

Running this code (with valid AWS credentials and proper AWS region configuration) would produce a Spanish translation. Given this text, the output could look like the following:

arduino
CopyEdit

```
"Hola, ¿en qué te puedo ayudar hoy?"
```

This translated sentence is in an informal tone (using "¿en qué te puedo ayudar?" with te for informal "you", instead of the formal le). Amazon Translate managed to convey the polite question from English into natural-sounding Spanish, adjusting pronouns for informality. The code example shows how straightforward it is—just provide text, source, and target language codes, and you get a translation. No need to worry about loading models or tokenization; it's all handled by the service. (In fact, you could omit `SourceLanguageCode` or set it to `"auto"` and it would detect English automatically.)

8.2.1.2 When to Use Amazon Translate

If you need translation as a component in your system and don't want to maintain models, Amazon Translate is a strong choice. For example:

- A **multilingual chatbot** can use Amazon Translate to bridge language gaps. Your core chatbot logic might be in English, but when a user asks a question in French, you can translate the French input to English, find the answer, and then translate the answer back to French for the user. This way, you build one bot that serves many languages. In fact, AWS has enabled scenarios like a customer in Spain chatting in Spanish with an agent in Italy who reads it in Italian—all in real time using Amazon Translate (see `aws.amazon.comaws.amazon.com`).

- **Content localization** for websites or apps can be kickstarted with Amazon Translate. You could feed in all your English content and get instant translations in multiple languages. Many companies then have human translators review and polish the machine output. Even with post-editing, using Amazon Translate can speed up the localization process dramatically.

- **Dynamic, on-demand translation** of user-generated content (like social media posts, product reviews, forum messages) is practical with Amazon Translate. You only translate what's needed at view time. For instance, an LLM agent in a content moderation system might detect a post in a foreign language, quickly translate it via the API to analyze it in English, and then decide how to respond.

Another advantage of Amazon Translate is its real-time document translation, which can also maintain document formatting. If you submit an HTML or Word document, the translated version maintains the same structure—headings, lists, bold/italics remains intact, just with text in the new language. This is hugely convenient—imagine translating an entire user manual PDF from English to Chinese while preserving all its formatting in one step.

8.2.1.3 Integration with LLM Agents

Given the focus on LLMs, it's worth noting that Amazon Translate can be used alongside LLM-based systems. For example, you might have a sophisticated LLM agent that answers questions or generates text. By adding Amazon Translate as a tool in the agent's toolkit, the agent can accept a user query in any language, call the Translate API to convert it to English (or the agent's base language), process the query, and then translate the answer back. This kind of orchestration can be done with frameworks like LangChain or AWS's own AI workflows. Essentially, Amazon Translate can serve as the "ears and voice" of an LLM agent in other languages. You will see a concrete use case of such integration later in this chapter, when we discuss a multilingual customer support bot.

In summary, Amazon Translate provides an easy, fast, and reasonably accurate way to implement translation on AWS. You get the benefit of continuous improvements (AWS updates the underlying models over time), and you can scale without worrying about infrastructure. The tradeoff is that you have less control over the model compared to rolling out your own—which is why for specialized needs or maximizing quality, you might consider custom models, as discussed next.

See the provided code example in the repository for a standalone script invoking Amazon Translate's API for text translation (the code shows how to open a file and translate its content using `Boto3`).

8.2.2 Using LLMs on AWS (Custom Models)

This subsection explores how to deploy or fine-tune large translation models on AWS. While Amazon Translate is great for generic use, sometimes you need a custom model—for example, a translator fine-tuned on legal documents to better handle legal jargon, or support for a language pair with informal phrasing that a general model struggles with. We use MarianMT as a case study, a popular open-source translation model, to illustrate how you can utilize AWS to run such models and tailor them to specific language pairs or domains. We discuss fine-tuning MarianMT on AWS (e.g., using Amazon SageMaker) and how LLM-based translators can be integrated into your AWS architecture. Throughout, we maintain continuity with the earlier examples, showing how a custom model can address challenges like the idioms and domain terms identified in the first section.

In the world of open-source translation models, MarianMT stands out as a workhorse. MarianMT is a family of neural translation models originally developed by Jörg Tiedemann's team at the University of Helsinki, covering many language pairs. Each model is essentially a sequence-to-sequence transformer (much like BART), but trained for a specific translation direction (or a group of related languages). These models are available on the Hugging Face Hub under names like `Helsinki-NLP/opus-mt-en-fr` (English→French), `opus-mt-fr-en` (French→English), and hundreds more. Importantly, MarianMT models are relatively compact by modern standards—roughly 300 million parameters, or ~300 MB on disk each, with a six-layer encoder and six-layer decoder. Because of their smaller size compared to giant LLMs, they're feasible to fine-tune and deploy on modest hardware. As the Hugging Face documentation notes, "Marian models are smaller than many other translation models ... they can be useful for fine-tuning experiments" (see `huggingface.co`).

8.2.2.1 Why Fine-Tune an LLM Like MarianMT?

Suppose you have a very specific domain of text that general translators don't handle well. For instance, legal contracts in Spanish and English. General models (including Amazon Translate) might mistranslate legal terms or produce less formal phrasing. By fine-tuning MarianMT on a corpus of parallel legal documents (Spanish-English), you can specialize the model. It will learn the preferred translations of certain terms (such as "lease", "tenancy", and "indemnify") and the formal tone common in contracts. Similarly, in a medical context, fine-tuning on patient records or medical research papers would teach the model to handle technical terms (such as drug names and procedures) more accurately. Fine-tuning essentially means continuing to train the pre-existing MarianMT model on your domain data so it adapts to that style and vocabulary.

AWS makes this process easier in a few ways. First, AWS provides the infrastructure (through SageMaker or EC2) to train models without worrying about managing GPU drivers, and so on. Second, AWS has a partnership with Hugging Face, offering deep learning containers and SageMaker estimators that can pull Hugging Face models and run training jobs with minimal setup. In practice, you can fine-tune MarianMT on SageMaker by specifying the model from Hugging Face and pointing to your training data in Amazon S3.

This list outlines how you would fine-tune a MarianMT model on AWS SageMaker for a specific language pair:

1. **Prepare the data:** You need a parallel dataset (source sentences and their translations). For example, a CSV or TSV where each line has an English sentence and the corresponding Spanish sentence. There are public datasets like OPUS or WMT for many language pairs that you can use as starting points. If your data is domain-specific, you might have to create or obtain a custom dataset (e.g., a collection of translated legal documents).

2. **Set up a training environment:** Using SageMaker, you can spin up a training job with an appropriate instance type (e.g., `ml.p3.2xlarge` which has one Tesla V100 GPU, or a larger `p3.8xlarge` for multiple GPUs to train faster). SageMaker's Hugging Face Deep Learning Container can be used—this comes preinstalled with Transformers, PyTorch, and so on. You define your training script to load the MarianMT model and tokenizer, load the dataset, and then use the Hugging Face `Trainer` API or PyTorch Lightning to fine-tune.

   Training script (simplified): A basic fine-tuning script in PyTorch might do the following:

   python

   CopyEdit

```python
from transformers import MarianMTModel, MarianTokenizer, 
Seq2SeqTrainingArguments, Seq2SeqTrainer
import datasets

model_name = "Helsinki-NLP/opus-mt-en-es"  # English-to-Spanish 
Marian model
tokenizer = MarianTokenizer.from_pretrained(model_name)
model = MarianMTModel.from_pretrained(model_name)

# Load or prepare the parallel dataset
train_data = datasets.load_dataset('csv', data_files='train_en_
es.csv')['train']
# The dataset should have columns like "en" and "es" for source 
and target.
# Tokenize the data
def preprocess_function(batch):
    source = batch["en"]
    targets = batch["es"]
    model_inputs = tokenizer(source, text_target=targets, 
    max_length=128, truncation=True)
    return model_inputs

tokenized_data = train_data.map(preprocess_function, batched=True, 
remove_columns=["en", "es"])

# Define training parameters
training_args = Seq2SeqTrainingArguments(
    output_dir="finetune-opus-mt-en-es",
    per_device_train_batch_size=16,
    num_train_epochs=3,
    save_steps=1000,
    save_total_limit=2,
    evaluation_strategy="epoch"
)
trainer = Seq2SeqTrainer(model=model, args=training_args, train_
dataset=tokenized_data)
trainer.train()
```

The previous snippet loads a pretrained Marian English-to-Spanish model and tokenizer. You would then load a CSV dataset of English-Spanish sentence pairs. You tokenize the input and target texts and then use Seq2SeqTrainer to handle the fine-tuning. On SageMaker, this script would run inside the training job. After training, you'd have a model that's better tuned to your specific dataset.

3. **Deploy or use the model:** Once fine-tuning is complete, SageMaker can save the model artifacts to S3. You could then deploy this model as a SageMaker endpoint to get real-time translation with your custom model. Alternatively, you could download the model and use it on an EC2 instance or even on-premise. SageMaker endpoints would allow you to serve the model behind a REST API (just like Amazon Translate, but now it's your model). This endpoint can auto-scale based on traffic if it's set up through SageMaker's inference scaling.

By fine-tuning, you essentially create a custom LLM translator. One of the great things about doing this on AWS is the ability to use powerful hardware as needed. If your dataset is large, you might use a multi-GPU instance or even multiple instances with data parallelism (SageMaker supports distributed training). And if training a ~300M parameter model is still too slow, remember that MarianMT is small enough that even a single GPU can handle it—fine-tuning 300M parameters on a few hundred thousand sentence pairs can complete in a couple of hours on a modern GPU. AWS also offers spot instance training for cost savings, so you could potentially train overnight at a fraction of the on-demand cost.

Let's maintain continuity with an earlier example—idioms. We had "It's raining cats and dogs," which in proper Spanish is "Está lloviendo a cántaros." A generic model might not know this idiom well if it wasn't common in its training data. Now, suppose we have a dataset of idioms and their translations (like a multilingual idiom dictionary). By fine-tuning MarianMT on this dataset, we could improve its handling of such figurative phrases. After fine-tuning, if we translate "It's raining cats and dogs," the model is much more likely to output the idiomatic "Está lloviendo a cántaros" rather than a literal translation. Essentially, we're teaching the model with examples of tricky translations so it performs better on them in the future.

8.2.2.2 MarianMT on AWS Inferentia

Once you have a custom model, you can also optimize its deployment. AWS has *Inferentia chips* (Inf1 and the newer Inf2 instances), which are purpose-built for running neural networks at low cost. MarianMT, being 300 MB, can be compiled to run on Inferentia. AWS Neuron SDK provides tutorials on deploying MarianMT with Inf1 instances (see awsdocs-neuron.readthedocs-hosted.com). The benefit is significant cost and throughput gains—AWS reported up to 2.3 times higher throughput and up to 70% lower cost per inference using Inf1 vs. GPU for typical models. For a translation service handling high volume (say, millions of characters per hour), using an Inferentia-backed instance could reduce your AWS bill substantially. And you don't sacrifice latency—in fact, Inferentia is optimized for small batch, real-time inference, which fits translation use cases (usually one sentence at a time).

8.2.2.3 Integrating LLM Translation into Applications

After deploying a fine-tuned LLM like MarianMT on AWS, how would you use it? Possibly through a REST endpoint or a Lambda function calling the model. For instance, you can have an AWS Lambda function load the MarianMT model (if it's small enough and cold start time is acceptable)—with 10 GB of memory available in Lambdas now, it's feasible to load a 300MB model in memory. That Lambda could act as a microservice for translation. An LLM-based agent application might dynamically choose between using Amazon Translate or your MarianMT model. Imagine an agent that tries Amazon Translate first (fast and cheap), and if it detects the output isn't satisfactory for some reason (maybe via a quality heuristic), it could then query the fine-tuned MarianMT model for a possibly better translation. This kind of ensemble or fallback strategy can combine the advantages of both. In practice, many engineering teams use a two-pass approach: quick machine translation followed by a specialized model or rules for post-editing. An LLM agent could be set up to do the same autonomously.

In conclusion, AWS provides a solid platform for using custom LLM translation models like MarianMT. You get flexibility to tailor the model to your needs and the compute power to train and serve it. The downside is increased complexity: you must manage training data, ensure the model is properly evaluated, and handle deployment scaling. But for use cases that demand that extra bit of accuracy or domain alignment, the effort can be worth it. A legal firm might accept nothing less than a translation that a legal expert would approve—a custom model, fine-tuned on bilingual law texts, and perhaps reviewed by humans, could meet this bar. Meanwhile, more general content can still be routed to Amazon Translate for efficiency.

Refer to the example training script in the repository (Chapter 8, Section 2), which demonstrates fine-tuning a MarianMT model on an English–Romanian dataset. The script uses Hugging Face Transformers and shows you how to continue training a pretrained model with new data, which can be adapted for any language pair.

8.2.3 Using SLMs on AWS (DistilBART, TinyBERT for Fast Translation)

This section focuses on SLMs and how they can be leveraged on AWS for translation tasks requiring low latency and low resource usage. SLMs are trimmed-down versions of larger models—they run faster and demand less memory, albeit with some tradeoff in translation quality. We discuss examples like DistilBART and TinyBERT in the context of translation, showing how they can be deployed for real-time use (even on modest infrastructure like Lambda or edge devices). We provide a simple example and consider when an SLM is the right choice, keeping continuity with the previous discussions by relating the quality-speed tradeoff to practical scenarios (for instance, a live translation feature where speed is critical).

Recall that SLMs are essentially the "optimizers" of the NLP world—models distilled or compressed from bigger models. The goal is to preserve most of the knowledge of a large model while reducing its size and increasing speed. In translation, this is incredibly useful. Imagine a mobile app that offers instant speech translation; it might not have the luxury of calling a huge model due to latency or connectivity, so a smaller on-device model would be ideal.

8.2.3.1 DistilBART for Translation

DistilBART is a distilled version of the BART sequence-to-sequence model. Facebook's BART is a versatile model that can be used for translation (in fact, BART was used as the base for some translation systems). DistilBART reduces BART's size by about 40-50% (distilling the 12-layer encoder and 12-layer decoder down to six each) and is consequently faster. Hugging Face's research showed that a distilled model can retain a very high percentage of the teacher model's performance—often around 95%— while being nearly twice as fast. In practice, a DistilBART trained for summarization or translation will produce outputs that are quite close to the full model's outputs, but it will consume roughly half the memory and run roughly twice the speed. For translation tasks, if you have a BART-based translator (or any sequence model), you can similarly apply knowledge distillation—you translate a bunch of sentences with the big model (teacher) and train a smaller model to mimic those translations. The result is a DistilBART translator.

8.2.3.2 TinyBERT and MiniLM

TinyBERT is another example, although it originates from BERT (which is an encoder-only model). In translation, you could use BERT in an encoder-decoder setup or as part of a pipeline (e.g., encoding meaning). TinyBERT showed it's over seven times smaller and nine times faster than BERT-base while retaining about 97% of BERT's accuracy on language understanding tasks. MiniLM (by Microsoft) is similar—a lightweight Transformer that still captures a lot of linguistic nuance. These smaller models (TinyBERT, MiniLM, DistilBERT, etc.) might not be used standalone to generate translations, but they can be part of translation systems. For instance, a two-step translation might first use a TinyBERT to quickly analyze the input and then a lightweight decoder generates text. More concretely, researchers have distilled large multilingual models (like mBART or M2M100) into smaller models that run efficiently on CPU.

8.2.3.3 Deployment on AWS

The big advantage of SLMs is that you can deploy them in environments where large models are impractical. AWS Lambda, for example, now allows up to 10 GB of memory for a function and a few vCPUs worth of processing. This is enough to load a model with, say, 100 million parameters and run inference, especially if you use optimizations like quantization. It's conceivable to host a DistilBART translator as a Lambda function that triggers on demand—for example, translating incoming text messages in a serverless workflow. The cold start time might be a consideration (loading 300 MB of model might add a couple seconds if not kept warm), but once loaded, inference can be quite snappy (tens of milliseconds for short sentences).

Another route is deploying on Amazon EC2 instances with modest specs or using AWS Fargate/ECS for a containerized service. A container running a small model can handle high request volumes by scaling out horizontally—because each instance is cheap to run. Also, AWS IoT Greengrass or running on edge devices (Outposts, etc.) could host these models to bring translation closer to where data is generated.

8.2.3.4 Real-World Scenario for SLM

Let's say you run a global chat application and want to offer instant message translation. Users send short messages that should appear translated almost instantly to the other party. Any noticeable delay will break the flow of conversation. In this case, an SLM is appropriate. A fully managed service call (like Amazon Translate) is an option, but it involves network latency to AWS and back for each message. If your app instead ships with an on-device SLM (for popular language pairs) or uses an edge server in-region with an SLM model, translations can be near real-time. The quality might drop slightly compared to a giant model, but for colloquial chat that might be acceptable. Also, using an SLM gives you more control—you could even allow offline translation if it's on-device, enhancing privacy.

Example with DistilBART: To illustrate how you might use a distilled model in practice, consider using the Hugging Face Transformers pipeline with a specific distilled model. There are distilled versions of MarianMT on Hugging Face (community-contributed) for certain language pairs. Using one might look like this:

python
CopyEdit

```python
from transformers import pipeline

# Assume a distilled translation model for English -> French exists
translator = pipeline("translation_en_to_fr", model="Helsinki-NLP/opus-mt-en-fr-distilled")
result = translator("The weather is quite nice today, isn't it?")
print(result[0]['translation_text'])
```

This hypothetical example assumes a model ID that represents a distilled English-to-French translator (for demonstration purposes). The pipeline use is the same as with a normal model; we're just loading a smaller, faster model. The output might be: `"Il fait assez beau aujourd'hui, n'est-ce pas ?"` which is a reasonable French translation of the input sentence. The speed benefit would be behind the scenes—if you benchmark, this should run faster than the full `opus-mt-en-fr` model.

8.2.3.5 Optimization

Beyond distillation, AWS and others provide tools to further speed up models:

- **Quantization:** You can run models at 8-bit or even 4-bit precision. Libraries like `bitsandbytes` and ONNX Runtime's quantization can compress the model at inference time. A quantized DistilBART might run, say, twice as fast, with maybe a slight quality drop (quantization can sometimes reduce BLEU by a tiny amount, but often it's negligible for high-level quality).

- **Compilation:** AWS Neuron (for Inferentia) or Amazon SageMaker Neo can compile models for specific hardware. A compiled model avoids Python overhead and can leverage low-level optimizations. For example, compiling a TinyBERT for CPU could yield lower latency per request.

- **Batching and streaming:** If you have many translations to do, small models can batch them to improve throughput. Also, for long texts, a technique called *streaming translation* (translating chunk by chunk) can be employed to start delivering output faster—a small model with limited context might handle streaming better as it uses less memory, although this is an advanced use case (mostly in speech translation).

8.2.3.6 Balancing Quality and Speed

It's important to mention that while SLMs are great for speed, there is a quality tradeoff. For instance, DistilBART might occasionally miss nuances that full BART would get. TinyBERT might not capture context as well as BERT. In critical applications where accuracy is paramount (say, translating medical dosage instructions), an SLM might not be the safest choice unless it's been thoroughly evaluated. On AWS, one strategy is to deploy a two-tier system: use the SLM by default, but if it encounters a sentence that is particularly tricky (maybe detected via certain keywords or length or a confidence estimator), route that to a more powerful model or even a human translator for verification. This way, you get speed most of the time and accuracy when it really counts. An LLM-based agent could orchestrate this: acting as a controller that decides "This sentence looks simple, I'll let the small model translate it" or "This looks complex (maybe it has idioms or ambiguous wording), I'll invoke a larger model or an API."

AWS SageMaker endpoints allow hosting multiple models behind different endpoints or a multi-model endpoint, so implementing such logic is possible. Alternatively, a Lambda function could choose between an embedded small model and calling Amazon Translate or an LLM endpoint based on input content. These are design patterns an engineer can employ.

8.2.3.7 SLMs and AWS Edge Services

With AWS, you could also deploy SLMs to the edge using services like AWS Wavelength (for telecom edge) or Greengrass for IoT. Suppose you have a device in a hospital that needs to translate patient instructions to multiple languages on-site (for privacy and reliability). A small model can be deployed on a local server via Greengrass, ensuring that translation works even if the Internet is down and that sensitive data never leaves the premises. This is a niche but important capability that huge LLMs cannot offer due to resource constraints.

To summarize, SLMs on AWS enable fast and cost-efficient translation. Distilled and compact models may not win translation competitions against their larger counterparts, but they often deliver good-enough translations for many applications at a fraction of the cost and with minimal latency. AWS supports these deployments through flexible compute options—from serverless to dedicated instances—and tooling for model optimization. As a developer or architect, the key is to assess the requirements: if you need sub-100 ms response times or have a tight budget, consider an SLM. If you need the highest fidelity translation and can afford the time/compute, an LLM might be justified for those cases. Often, a hybrid approach yields the best of both worlds.

Check the repository for an example of deploying a distilled translation model (Chapter 8, Section 2). The example demonstrates converting a Hugging Face translation model to ONNX and running it on an AWS Lambda environment for ultra-fast inference, showcasing how a distilled model can achieve low latency translations in a serverless setup.

8.2.4 Custom Training on SageMaker (T5, mBART, and so on)

This final sub-section addresses the scenario where you need to train or fine-tune larger translation models from scratch or near-scratch using AWS SageMaker. This could involve models like T5, mT5 (multilingual T5), mBART, or even Meta's massive many-to-many model. The goal here is high-quality, domain-specific, or very multilingual translation that out-of-the-box solutions can't provide. We discuss how SageMaker can facilitate training large models (distributed training, handling big datasets) and when investing in a custom "heavy" model makes sense. We also include a table to simplify the comparison of approaches (managed service vs. custom small vs. custom large) as a quick reference for readers.

Training a translation model is no small feat—data sizes are large (billions of words for rich languages) and models are hefty. However, SageMaker is designed to handle such heavy lifting in a managed way. Let's break down a few examples:

- **T5 for translation:** Google's Text-to-Text Transfer Transformer (T5) is a versatile model pretrained on multiple tasks, including English-to-English tasks and some translation tasks. The interesting thing about T5 is that it was trained in a "multi-task" fashion; for example, one of the tasks in its pretraining mixture was translating English to German (from the WMT dataset). This means even out-of-the-box, T5 Large or XL has some translation capability baked in (for certain language pairs). To use T5 for translation, you typically fine-tune it on parallel data for the language pair of interest. T5 comes in various sizes: Small (60M params), Base (220M), Large (770M), 3B, 11B. Fine-tuning the larger versions (3B, 11B) definitely requires multiple GPUs and a lot of memory—something SageMaker can orchestrate with multi-machine training clusters. SageMaker's data parallelism or model parallelism features (via the Deep Learning Containers with PyTorch DeepSpeed or TensorFlow's MirroredStrategy) help train such models. For instance, you could launch a SageMaker training job on four `ml.p4d.24xlarge` instances (each with eight NVIDIA A100 GPUs, so 32 GPUs total) to fine-tune an 11B parameter T5 model on a large translation corpus. This would be a heavy-duty job, but SageMaker handles provisioning the instances, hooking them up in a cluster, and tearing them down after training—so you don't have to manually set up distributed training on EC2 instances.

- **mBART-50:** Facebook's mBART-50 is a multilingual extension of BART that covers 50 languages. It was trained in a many-to-many fashion, meaning one model can translate between any of those 50 languages (via language code tokens). If your application needs a broad multi-language translator (say, a document translator that can handle anything thrown at it, from Japanese to Swahili to Turkish), a model like mBART-50 is attractive. Fine-tuning mBART on your domain data is possible too—for example, you fine-tune it on a corpus of technical manuals across ten languages to specialize it in that domain. mBART is about 610M parameters, so roughly two to three times the size of MarianMT. This is still manageable on a single modern GPU (albeit at high memory usage). SageMaker could train it on a single p3.8xlarge (four GPUs) to get multi-GPU data parallelism, which speeds up training.

- **mT5 (multilingual T5):** mT5 is another powerhouse—a multilingual version of T5 that covers over 100 languages. The largest mT5 has billions of parameters. SageMaker's distributed training or use of AWS Trainium chips (which SageMaker supports via Trn1 instances) could be a cost-effective way to fine-tune such big models. AWS Trainium is specifically optimized for Transformer training, often yielding a lower cost per hour compared to GPUs. A Trn1 instance with 16 Trainium chips can collectively have the equivalent of 128 v100 GPUs of throughput for certain models, according to AWS. This can make training a model like a 13B parameter mT5 feasible within hours or days instead of weeks.

Now, when would you go for custom training of a large model?

Consider a scenario where you're building a translation service for a language that's not well-supported by existing services or you need complete control over the translation behavior (maybe for compliance or proprietary reasons). For example, suppose an international organization needs a translator for several low-resource languages in a specific domain (like translating between various African languages for medical information). They might decide to train a custom model using whatever bilingual data they have plus some generated or back-translated data to bolster it. By using a multilingual model like mT5 or mBART as a starting point, they get a model that "speaks" many languages, then fine-tune on their data to teach it the domain specifics. SageMaker can facilitate that by providing notebook environments for data preprocessing and launching the heavy training jobs.

Consider another scenario: quality triage. You want the absolute best quality for a critical language pair (say English↔French for legal documents). You might train a specialized large model (like T5-XXL fine-tuned on legal corpora) to use for high-value documents, while using cheaper methods for less critical stuff. Because you own the model, you can also inspect it, improve it incrementally, and ensure data privacy (all training data and inference happens in your controlled environment). Some organizations prefer this if the content is highly sensitive and cannot be sent to external APIs like Amazon Translate (even though AWS offers some assurances, having your own model means the data never leaves your VPC if you set it up that way).

Table 8-2 compiles a quick comparative view of the approaches we've discussed, to guide you on which path to choose.

Table 8-2. Comparative View of Approaches

Approach	Effort and Control	Performance and Quality	Use Case Example
Amazon Translate (managed)	Minimal effort: No training or deployment—just API calls. Limited control (cannot change model internals, only minor customization via terminology).	High throughput, good quality for general text: Supports many languages, continuously improved by AWS. Quality is strong for common language pairs, but may falter on very niche content.	– Customer service chats (fast, scalable) – Website localization (generic content) – When ML expertise or time is lacking

(*continued*)

Table 8-2. (*continued*)

Approach	Effort and Control	Performance and Quality	Use Case Example
Custom SLM on AWS (e.g., DistilBART, Tiny models)	Moderate effort: Need to set up model deployment or light fine-tuning. More control than managed service (you choose the model, can modify it).	Very fast, acceptable quality: Lower latency and cost. Quality is decent for everyday language, slightly lower than large models for nuanced text.	– Real-time translations in apps (where speed is king) – Edge cases: IoT or offline translation devices – High volume, where cost per translation must be low
Custom LLM on AWS (e.g., MarianMT, mBART, T5)	High effort: Requires ML expertise to fine-tune or train, and infrastructure setup (SageMaker jobs). Full control over model and data (you can tailor to domain, ensure data privacy).	Potentially highest quality: Especially for domain-specific texts. Can outperform generic models in that niche. But slower and costlier to run; needs optimization for production.	– Domain-specific translation (legal, medical, technical) with high accuracy – Low-resource languages where you curate data to improve on generic models – When translation quality directly impacts critical outcomes (e.g., intelligence analysis, regulatory documents)

In the table, Effort and Control indicates the amount of work and the degree of freedom you have. Performance and Quality summarizes the tradeoffs in speed and translation fidelity. Use Case Example gives concrete scenarios.

You can derive guidance from Table 8-2. For instance, a startup making a multilingual chatbot might start with Amazon Translate for simplicity and speed to market. As they grow, they might notice certain languages or jargon where the translations confuse users—maybe at that point, they introduce a custom model fine-tuned on chat transcripts (an LLM approach) for those cases. If they need the chat to feel instantaneous, they might invest in distilling that custom model to an SLM or use an SLM approach from the start.

8.2.4.1 Training Process on SageMaker: Additional Notes

Training large models can produce large checkpoints. SageMaker automatically saves model artifacts to S3 at the end of training. You should be mindful of S3 upload limits (the artifact is typically `tar.gz` if you're using the estimator). Also, with very large models, you might do resume training or incremental training by periodically saving states. SageMaker can also log metrics to CloudWatch, so you can monitor training loss over time and see if things are converging or if you need to stop early (to save on costs if the model isn't improving).

AWS JumpStart is another feature—it provides one-click deployment or fine-tuning for certain popular models, including some translation models. For example, JumpStart has some translation notebooks that fine-tune models like Facebook's WMT19 models. It abstracts some of the coding away for those who prefer a more UI-driven approach.

8.2.4.2 After Training: Serving the Model

Once you have a fine-tuned T5 or mBART model, you have choices. If it's extremely large (billions of params), you might opt for a dedicated GPU instance (or several) to host it behind an endpoint. SageMaker's multi-model endpoints are more for many small models; for a single huge model, a single-model endpoint is appropriate. You can use autoscaling on it based on CPU/GPU utilization or latency to handle variable load. Another interesting development is AWS's support for AWS Inferentia2 (Inf2 instances), which are optimized for large model inference (like GPT-3 sized models). While translation models aren't as big as GPT-3, running a 13B model on Inf2 could be more cost-effective than on GPU. AWS claims Inf2 has up to four times more throughput than

Inf1 and significant speed-up for LLM models. If your translation model is heading into the multi-billion parameter territory, exploring Inf2 via SageMaker (Neuron SDK) might yield better latency/cost.

Finally, a mention of evaluation—when you train your own model, you need to evaluate its quality. AWS doesn't provide a built-in translation quality metric service, but you can use BLEU or ROUGE by comparing model outputs on a test set to reference translations. There are libraries to compute these. You might run a SageMaker Processing job or a notebook to calculate BLEU for your model versus say Amazon Translate on the same test set, to quantify improvements. For example, if your custom model achieves BLEU 5 points higher on legal documents than Amazon Translate, that's a strong argument for its value. We talk about evaluation more in this chapter with open-source tools, but it's part of the loop in training—train, evaluate, iterate.

This section covered the spectrum: from using a plug-and-play AWS services to tweaking small models for speed, up to training large models for quality. AWS's ecosystem supports all these approaches, which is a big reason enterprises choose AWS for NLP workloads—you can start simple and gradually move to custom solutions as needed, all on the same platform.

To connect this to the bigger picture, the next section steps outside AWS-specific services and looks at building translation systems with open-source libraries and frameworks. This includes using Hugging Face Transformers locally or on any cloud, and evaluating translations with standard metrics. It complements what we did on AWS by showing how you can achieve similar goals in a platform-agnostic way, and it dives deeper into the nuances of model training and evaluation. So, having learned how to implement translation on AWS, let's gracefully transition to the next part—where you'll get your hands dirty with open-source models and tools to further demystify machine translation.

8.3 Open-Source Translation Systems

This section delves into constructing machine translation systems using open-source tools, with a focus on Hugging Face Transformers. It begins by exploring fundamental principles for multilingual translation, followed by an examination of state-of-the-art LLMs and SLMs, including domain-adaptive and energy-efficient models like MarianMT, mBART, and DistilBART. You'll also learn how to fine-tune and deploy models for specific language pairs and domains (e.g., legal and medical domains) and evaluate their performance using established metrics like BLEU, ROUGE, and TER.

8.3.1 Implementing Translation with Hugging Face Transformers

The Hugging Face Transformers library provides extensive support for multilingual translation, offering a range of pretrained models, tokenizer pipelines, and training utilities. Two primary approaches dominate this space:

- **Supervised translation:** Models trained on parallel corpora for specific language pairs (e.g., English↔French).

- **Multilingual transfer:** Leveraging cross-lingual models trained on many language pairs.

8.3.1.1 LLMs for Translation

LLMs offer superior performance, especially for high-resource languages and domain-specific use cases:

- **MarianMT:** A widely used model for direct translation across multiple language pairs. Ideal for fine-tuning.

- **mBART50:** Multilingual encoder-decoder trained on 50 languages. Great for zero-shot translation and low-resource pairs.

- **T5 and mT5:** Text-to-text transformers capable of treating translation as a unified generative task.

- **Mistral-7B:** When instruction-tuned, Mistral can generate translations with human-like fluency, especially for low-context, prompt-driven tasks.

8.3.1.2 SLMs for Translation

SLMs provide cost-effective and responsive alternatives:

- **DistilBART:** A smaller version of BART, efficient for real-time use cases.
- **TinyBERT:** Compact model optimized for speed and resource-constrained environments.
- **MiniLM:** Lightweight transformer for multilingual sentence embeddings and fast translation tasks.
- **Qwen1.5-1.8B:** Offers multilingual capabilities with low memory use and fast decoding.

8.3.2 Fine-Tuning MarianMT for Custom Language Pairs

Fine-tuning MarianMT on custom domain datasets allows for precise control over vocabulary, tone, and translation accuracy in specialized fields like legal or healthcare.

You can find the code on GitHub at: https://github.com/anvcse562/ModernNLP-LLMSLM/blob/main/Ch08/Ch8.3/8.3.2_fine_tune_marianmt_eng_french.ipynb

8.3.3 Fine-Tuning DistilBART for Lightweight Translation

DistilBART is ideal for situations where latency and efficiency are prioritized over full-sentence fluency. It can be trained on domain-specific corpora using Hugging Face's Trainer API.

You can find the code on GitHub at: https://github.com/anvcse562/ModernNLP-LLMSLM/blob/main/Ch08/Ch8.3/8.3.3_fine_tuning%20DistilBART_Lightweight%20Translation.ipynb

8.3.4 Parameter-Efficient Translation with PEFT

Large models can be fine-tuned efficiently using PEFT (parameter-efficient fine-tuning) techniques such as QLoRA and LoRA. These methods drastically reduce training time and memory consumption.

python

CopyEdit

```
from peft import LoraConfig
from trl import SFTTrainer

peft_config = LoraConfig(r=32, target_modules=["q_proj", "v_proj"])
trainer = SFTTrainer(
    model=model,
    train_dataset=tokenized_dataset["train"],
    peft_config=peft_config,
    max_seq_length=512,
)
trainer.train()
```

8.3.5 Deploying Translation Models

Translation models must support a wide range of deployment targets, from cloud backends to mobile apps:

- **Cloud deployment with LLMs:** Host large models like mBART and MarianMT on SageMaker or Azure ML with GPU acceleration.

- **Edge deployment with SLMs:** Models like TinyBERT or MiniLM can be quantized (e.g., INT8 or 4-bit) and deployed to mobile using ONNX or TFLite.

- **BitNet CPU deployment (experimental):** Although BitNet was developed for summarization, its architecture can be adapted for low-power translation, showing promise for offline CPU translation apps.

Sample conversion and deployment:
bash

```
python convert_translation_model_to_gguf.py
quantize --qtype Q1_58
./bitnet -m translation_model.gguf -p "Translate English to German: The world is changing rapidly."
```

8.3.6 Inference Optimization Strategies for Translation Models

To optimize speed and efficiency, you can apply the strategies outlined in Table 8-3.

Table 8-3. Inference Optimization Strategies

Strategy	Description
Quantization	Lowering precision to speed up inference (e.g., INT8, 4-bit, 1.58-bit).
Distillation	Train a smaller model to mimic a large one, preserving translation quality.
Pruning	Remove non-critical weights from the model.
Batch translation	Group inputs together to maximize GPU throughput.
Flash attention	Enable faster inference on long-form translations.

Example: Distillation Setup
python

```
from transformers import MarianMTModel, DistilBertModel
teacher = MarianMTModel.from_pretrained("Helsinki-NLP/opus-mt-en-fr")
student = DistilBertModel.from_pretrained("distilbert-base-multilingual-cased")
```

8.3.7 Evaluation Metrics for Translation Quality

Evaluating machine translation requires a balance of fluency, semantic accuracy, and syntactic precision, as outlined in Table 8-4.

Table 8-4. Different Metrics and Their Use Cases

Metric	Use Case
BLEU	Measures n-gram overlap; useful for technical content.
ROUGE	Measures recall and content overlap; useful for fluency.
TER	Measures the number of edits required to match reference.
BERTScore	Evaluates semantic similarity using contextual embeddings.
Latency	Important for real-time or user-facing applications.

You can find the code on GitHub at: https://github.com/anvcse562/ModernNLP-LLMSLM/blob/main/Ch08/Ch8.3/8.3.7_Translation_task_evaluation_Metrics.ipynb.

8.4 Industry Use Cases

8.4.1 Use Case: Global Customer Support Chatbots

Modern businesses serve a global customer base, requiring support in multiple languages. A multilingual customer support chatbot leverages language translation and Q&A capabilities to handle queries from different regions. By integrating translation models with QA systems, companies can automatically assist users in their native language, improving customer satisfaction and reducing the burden on support teams. Employing LLMs and SLMs offers flexibility: SLMs can provide fast, factual answers from documents, while LLMs can generate more conversational and context-rich responses. This balanced approach ensures high-quality support that is both efficient and scalable.

Businesses utilize multilingual support chatbots primarily to:

- **Provide instant answers in native languages:** Customers receive help in their own language without delays.

- **Automate FAQs across regions:** Common questions (password resets, account issues, etc.) are answered automatically, reducing human workload.

- **Ensure consistent information:** A unified knowledge base (often in one language) is used across translations, keeping answers consistent globally.

- **Bridge language gaps:** Language translation technology removes the need for separate support teams for each language.

- **Scale support cost-effectively:** Adding support for a new language is as simple as adding a translation model, avoiding a proportional increase in staff.

8.4.1.1 Implementation Overview

This implementation demonstrates building a multilingual QA chatbot for customer support. It combines translation models with Q&A pipelines using open-source tools (Hugging Face Transformers and LangChain) for an end-to-end solution. The system workflow is as follows: a user question in any supported language is translated to the base language (English) for processing, relevant knowledge is retrieved, an answer is found or generated in English, and then it's translated back to the user's language. Both extractive QA (finding exact answers from documents) and generative QA (producing synthesized answers) are explored, using SLMs for speed and LLMs for richer output.

Here are the key implementation components:

- **Data loading and indexing:** We load a set of FAQ documents (e.g., support articles in English) and break them into manageable chunks for processing. A vector store (using Chroma) is created for efficient semantic search of these chunks.

- **Translation integration:** We use open-source translation models (MarianMT via the Helsinki NLP opus-MT models) to translate queries and answers between English and the target languages (e.g., Spanish, French). This allows the QA system to handle multilingual queries and responses.

- **Extractive QA with SLM (DistilBERT):** Utilizing a smaller QA model (e.g., `distilbert-base-multilingual-cased-distilled-squad` fine-tuned on SQuAD) to extract answer spans directly from retrieved documents. This provides quick, factually correct answers by pinpointing text in the knowledge base.

- **Generative QA with LLMs (Gemma/BitNet):** Integrating larger language models for generative answers. For example, Google's Gemma (7B) model and Microsoft's BitNet (2B) model are used as generative LLMs. Gemma is a family of lightweight open LLMs from Google (derived from the same research as Gemini), and BitNet b1.58 is the first open-source 1-bit quantized LLM (~2B parameters) from Microsoft. These models can synthesize answers in a conversational manner when provided with relevant context. We show how to connect to these models via the Hugging Face Hub/endpoints.

- **Query interface:** We provide simple Python functions to ask questions to the chatbot. The function handles translation (if needed), uses the QA chain to find an answer, and then returns the answer translated back to the user's language. This simulates how a user query would be processed in a real chatbot application.

By combining these components, the resulting system can, for example, answer a support query asked in Spanish by translating it to English, finding the answer from English documents, and then delivering the answer back in Spanish. This approach ensures that the language translation aspect (this chapter 8's focus) is tightly integrated with a practical QA chatbot for global customer support.

You can find the code on GitHub at: https://github.com/anvcse562/ModernNLP-LLMSLM/blob/main/Ch08/ch8.4/global_customer_support.ipynb.

8.4.2 Industry Use Case: Multilingual Content Delivery

As businesses expand globally, delivering content in multiple languages becomes essential for reaching broader audiences and maintaining local relevance. Whether it's product descriptions, support documentation, blog posts, or mobile app content—manual translation at scale is inefficient and error-prone.

Multilingual content delivery uses automated translation pipelines to generate consistent, high-quality output in various target languages. These pipelines rely on LLMs and SLMs for a balance of fluency, speed, and deployment feasibility.

For the problem scenario, imagine a company with:

- A knowledge base in English that must be delivered in 12 languages.
- Frequent updates requiring near-real-time translation.
- Limited budget and infrastructure constraints in some regions.

Manual translation would be slow and expensive. Instead, LLMs like mBART, T5, and MarianMT can be fine-tuned for high-quality translation, while SLMs like DistilBART and TinyBERT ensure cost-effective, fast translation in production.

Here is the pipeline overview:

1. **Content ingestion:** Source files (e.g., HTML, Markdown, plain text) are uploaded to a monitored folder.

2. **Preprocessing:** Strip markup, normalize punctuation, and segment into sentences.

3. **Translation:** Use pretrained or fine-tuned models to generate translations.

4. **Postprocessing:** Reapply formatting, encode target language metadata, and validate output.

5. **Storage/delivery:** Store translated files in locale-specific directories or publish via CMS.

The following table shows the model suggestion:

Use Case	Model	Notes
Blog posts	`facebook/mbart-large-50-many-to-many-mmt`	Handles diverse language pairs
Product UI	`Helsinki-NLP/opus-mt-en-ROMANCE`	Lightweight, domain-adaptable
On-device	`distilbart-multilingual`	Small size for edge deployment

The provided notebook demonstrates:

- Reading .txt files in English.
- Translating to multiple languages using Hugging Face pipelines.
- Saving results to JSON format for CMS integration.

We can further expand these scenarios to enhance the use case:

- Add markdown/HTML preservation with BeautifulSoup.
- Connect output to a CMS (Contentful, WordPress).
- Add logging or database storage for large-scale use.

You can find the code on GitHub at: https://github.com/anvcse562/ModernNLP-LLMSLM/blob/main/Ch08/ch8.4/8.4.2_multilingual_content_translation_usecase.ipynb.

8.5 Summary

In this chapter, you explored language translation using LLMs and SLMs. You learned how these models handle complex translation challenges such as syntax, semantic nuance, idioms, and domain-specific language. You gained practical experience using AWS services like Amazon Translate for real-time translation and SageMaker for custom model training, as well as open-source tools like Hugging Face Transformers to build and evaluate multilingual translation pipelines. You also examined real-world applications including multilingual customer support and global content delivery. You now have a strong foundation for building scalable, efficient, and accurate machine translation systems. The next chapter dives into another critical NLP task: named entity recognition (NER).

CHAPTER 9

Dialogue Systems

This chapter explores the design and implementation of dialogue systems using LLMs like GPT-3, T5, and BART, and SLMs such as DistilGPT-2, DistilT5, and TinyBERT. We examine the rule-based and generative approaches, highlighting tradeoffs in control, flexibility, and computational efficiency.

We cover scalable implementations using AWS services like Amazon Lex for rule-based agents and SageMaker for fine-tuning generative models. Open-source tools such as Rasa and Hugging Face Transformers are also discussed, with a focus on lightweight SLMs for low-latency, resource-constrained environments like mobile or edge devices.

By the end of this chapter, you'll be able to build and deploy efficient, scalable dialogue systems tailored to your domain using cloud and open-source tools.

This chapter covers:

- **An overview of dialogue systems:**
  - **Rule-based:** Predefined rules and intent mappings; efficient but limited.
  - **Generative:** Use LLMs like GPT-3 and T5 for dynamic, multi-turn conversations.
  - **LLMs vs. SLMs:** LLMs offer contextual depth; SLMs provide fast, efficient responses.

- **Implementation on AWS:**
  - **Amazon Lex:** Build rule-based agents with built-in intent recognition.
  - **Amazon SageMaker:** Fine-tune LLMs like T5 and BART for domain-specific dialogue.
  - **SLM deployment:** Use models like DistilGPT-2 and DistilT5 for lightweight applications.

- **Open-source implementation:**
  - **Rasa:** Customizable, intent-driven systems using models like DistilBERT.
  - **Hugging Face Transformers:** Build generative agents with GPT-2 and T5.
  - **SLMs in open-source:** Use DistilGPT-2 and TinyT5 for real-time, efficient conversations.
- **Industry use cases:**
  - Virtual customer assistants
  - Interactive voice response (IVR) systems
  - Healthcare virtual assistants

After leveraging machine translation to bridge language barriers in global customer support in Chapter 8, the next step is enabling rich, interactive conversations. Translation alone isn't enough—we want systems that can understand user queries and respond in multiple turns, adapting to context. This is where dialogue systems come into play. They allow us to build chatbots and conversational agents that interact with users naturally. In the previous chapter, we saw how language models can translate text for multilingual communication; now we explore how those language technologies power full-fledged conversations.

9.1 Understanding Dialogue Systems

This section introduces what dialogue systems are and the two fundamental paradigms for building them. We contrast *rule-based dialogue systems,* which rely on predefined rules and scripts, with *generative dialogue systems,* which use learned language models like GPT-3 or T5 to generate responses. We discuss the advantages and limitations of each approach, using simple examples to illustrate how they work. We also examine the role of model size—comparing LLMs to SLMs—in creating efficient, effective conversational agents. This foundation will prepare you to implement dialogue systems using cloud services and open-source tools in subsequent sections.

9.1.1 Rule-based vs. Generative Dialogue Systems

Interactive dialogue systems or "chatbots" generally fall into two broad categories based on how they decide what to say next: rule-based systems and generative systems. Both aim to conduct a conversation with the user, but they operate in fundamentally different ways.

9.1.1.1 Rule-Based Dialogue Systems

A rule-based system follows a set of hand-crafted rules or scripts to determine its responses. Developers define *intents* (the meanings of user queries), specify *entities* to extract (important keywords), and map out predefined responses or action flows for each situation. In essence, it's an if-then logic approach. For example, a rule-based customer support bot might be programmed such that if the user's message matches the intent CheckOrderStatus, the bot will respond with a fixed prompt like "Sure, please provide your order ID." These systems often use pattern matching or intent classification to pick the closest scripted reply. A classic historical example is *ELIZA* (built in the 1960s), which used simple pattern rules to respond, giving an illusion of understanding despite no real comprehension. Modern rule-based frameworks like the default model of Rasa or Amazon Lex enhance this by using machine-learning classifiers for intent detection, but the conversation paths are still largely hand-designed.

- **Strengths:** Rule-based bots are highly predictable and fully under control of the developer. They will only say or do what they are explicitly programmed to do. This makes them reliable for closed-domain applications—answering FAQs or handling structured dialogues (e.g., a phone menu or form-filling bot) where you *want* the same, consistent answers every time. They are also lightweight: since the logic is coded and the language is often templated, they can run offline or on minimal hardware with very little cost. For instance, a simple decision-tree chatbot can even run in a web browser without any AI model. Another benefit is transparency: you can easily trace why the bot responded a certain way (because of a specific rule).

- **Weaknesses:** The major drawback is limited flexibility. A rule-based system cannot handle inputs outside of its predefined rules. If the user says something unexpected or that's phrased in a way that wasn't anticipated, the bot either gives a generic failure response or misinterprets it. Designing comprehensive rules for every possible user utterance is extremely time-consuming and tends to be brittle. As conversations grow in complexity, the rule base can become an entangled web of if-else logic. This often leads to conversations feeling robotic or scripted. The bot might repeat exact phrases or fail to carry information naturally across turns unless explicitly coded to do so. Development effort is high because you must anticipate many variations of user input and maintain large intent grammars and response templates. In summary, rule-based agents work well for simple, structured tasks but struggle with open-ended queries or unforeseen dialogue turns.

9.1.1.2 Generative (LLM-Based) Dialogue Systems

In contrast, generative dialogue systems leverage data-driven language models to decide on responses dynamically. Rather than having a hard-coded reply for each intent, a generative chatbot—often powered by an LLM like GPT-3, GPT-4, or a fine-tuned T5/BART model—will generate a response on the fly by predicting likely next words based on the conversation history. These systems have been trained on massive text corpora, so they can generalize to new questions and produce answers in natural-sounding language even for inputs they've never seen before. For example, GPT-3 was trained on billions of words from the Internet, enabling it to respond to a query about, say, travel advice or programming help without explicit rules for those topics. Generative chatbots can be *open-domain*, meaning they are not limited to a narrow subject—they attempt to handle anything the user brings up to varying degrees of success.

- **Strengths:** The foremost advantage is flexibility and fluency. Generative models can handle unexpected or complex questions that weren't pre-programmed. They often produce responses that feel more natural and human-like, because they can incorporate varied phrasing and even show some personality. Development is faster in the sense that you don't need to script every reply—you can rely on the model's training. If you want a chatbot that can engage in freeform conversation like a virtual assistant that chats about news, or a support bot that can answer arbitrary product questions, generative LLMs are ideal. They also shine in multi-turn interactions:

an LLM can refer back to earlier parts of the conversation, maintaining context over several turns if the model or system is designed with sufficient memory for chat history. This context carry-over is something rule-based systems struggle with unless explicitly coded. In short, LLM-based agents bring a level of intelligence and general knowledge that static rules cannot encompass. Indeed, the introduction of large pretrained models has transformed conversational AI by enabling more fluid, context-aware dialogues.

- **Weaknesses:** Large generative models are resource-intensive to run often requiring GPUs and significant memory , and if they are using a commercial API (e.g., OpenAI's GPT-3), they can be expensive at scale. Additionally, because the model is effectively making up responses based on probabilities, **you** don't have fine-grained control over exactly what it says. This unpredictability can be problematic—the model might generate irrelevant or incorrect information if prompted in a way that confuses it. A well-known issue are *hallucinations*, where the LLM states false information confidently. For instance, a generative bot might invent a fake visa requirement or a nonexistent product detail because it "sounds plausible" based on its training data. Ensuring accuracy and compliance in generative responses can thus be challenging. There's also a risk of the model producing inappropriate content if it's not properly filtered. Lastly, while generative bots require less manual scripting up front, they often need a lot of data and fine-tuning to perform well on a specific domain (e.g., medical advice, legal Q&A) and integration with business logic to be truly useful. Unlike a rule-based system, an LLM doesn't inherently know the boundaries of your business rules—you must provide guidelines or use hybrid approaches to keep it in check.

To illustrate the difference between these two approaches, imagine a simple conversation about travel requirements:

User: "What are the visa requirements for Canada?"
Rule-based bot: (Checks for the "visa requirements" intent, finds a scripted answer.) "To visit Canada, you typically need a valid passport and possibly a tourist visa (unless you're from a visa-exempt country)." (This answer is exactly as written in the bot's FAQ database for that question.)

Now, consider a slightly different, more open-ended user prompt:

User: "I'm moving to Japan. How should I prepare?"

Rule-based bot: (No exact rule matches this broad query, so it falls back to a default reply.) "I'm sorry, I can only answer specific questions about visa and travel requirements."

Generative LLM bot: (The model interprets the question and generates a helpful, human-like response.) "That's awesome! Moving to Japan is exciting. You should definitely arrange your residence card, brush up on some basic Japanese phrases, and pack some warm clothes for the winter."

In the second query, the rule-based system failed because the user's request didn't fit a predefined slot, whereas the LLM-based system was able to infer the user's intent and provide relevant tips beyond any pre-scripted knowledge. This highlights how generative agents can handle more ambiguous or wide-ranging inputs gracefully, whereas rule-based agents are constrained to the scenarios anticipated by their developers.

To drive home the differences, Table 9-1 summarizes key aspects of rule-based vs. generative dialogue systems

Table 9-1. Key Aspects of Rule-Based and Generative Dialogue Systems

Aspect	Rule-Based Dialogue System	Generative (LLM-Based) Dialogue System
Approach	Deterministic rules, flows, or templates written by developers. The bot follows explicit scripts (e.g., intent → fixed response).	Probabilistic language generation using a pretrained model. The bot computes a response by continuing the dialogue text with an AI model.
Flexibility	Limited to predefined scenarios. Struggles with inputs outside the scripted intents. Each new user request needs a matching rule or it fails.	Highly flexible and creative. Can handle a wide range of queries, even ones not seen during training, by leveraging general language understanding.

(continued)

Table 9-1. (*continued*)

Aspect	Rule-Based Dialogue System	Generative (LLM-Based) Dialogue System
Response style	Consistent, but can be rigid or repetitive. (For example, always gives the same phrasing for a given question.)	Dynamic and varied wording; often more engaging. Can adapt phrasing to context and user input, mimicking human-like conversation style.
Development effort	High initial effort—requires designing intents, writing rules and dialogues for all expected queries. Ongoing maintenance to add rules for new queries or update knowledge.	Lower initial scripting—a pretrained model can answer many questions without additional rules. Fine-tuning or prompt engineering is used to improve domain-specific accuracy, which is less manual than writing rules for everything.
Resource requirements	Minimal. Can run on a CPU or small device since it's essentially string matching and logic. Often operates offline with no external API needed.	High. Inference for large models demands significant compute (GPUs or cloud servers). If using an API (e.g., GPT-3), there are usage costs. Small talk models exist that can run on-device, but top-tier LLMs need serious hardware.
Control and consistency	Fully controlled and predictable. The bot will not output anything unexpected; every response is by design. Good for compliance and consistent tone.	Stochastic and less controllable. The bot might generate irrelevant or incorrect content (hallucinations) and it's harder to guarantee that it will always follow policies. Requires safeguards and oversight for critical applications.
Context handling	Context usually handled via explicit state tracking. Can manage multi-turn flows if programmed, but may have difficulty if user goes off-script.	Implicitly handles context by conditioning on the conversation history. Sophisticated LLMs remember information from earlier turns (within a limit) and use it in later responses, enabling fluid multi-turn dialogue.
Example use cases	FAQ bots, IVR menus, appointment scheduling with fixed steps, forms with set questions. *Task-oriented dialogue* where the conversational path is known (e.g., troubleshooting steps).	Open-domain chat (virtual assistants like Alexa, Siri, ChatGPT-style Q&A), creative companions, customer support that covers broad topics. Unstructured dialogues or those requiring on-the-fly reasoning.

These approaches are not mutually exclusive. In fact, many real-world systems hybridize rule-based and generative methods to get the best of both worlds. For example, a chatbot might use a rule-based engine to handle very specific tasks or to enforce business-critical rules, ensuring certain information is collected or certain wording is used, and fall back to a generative model when the conversation goes outside those bounds. Rasa's recent frameworks, as well as other industry solutions, often combine intent classification (to recognize when to invoke a rule or an API) with LLM-based response generation for more open-ended turns. This kind of design lets you maintain control where needed, while still benefiting from the flexibility of LLMs for handling unexpected user inputs.

9.1.1.3 Implementation Example: Rule Logic vs. Generative Model

To solidify the concept, a simple illustration in Python follows. It shows how you might implement a rule-based response versus using a generative model. Suppose we are building a chatbot for a service desk with a couple of simple questions: checking visa requirements and asking about office hours.

First, a rule-based approach could be implemented with conditional logic. Refer to the code on GitHub: https://github.com/anvcse562/ModernNLP-LLMSLM/blob/main/Ch09/ch9.1/9.1.1_rule_based_approach.py.

In this contrived example, the bot checks the input text for certain keywords and returns a hard-coded answer. If the query doesn't match any known rule, it falls back to an apology or a generic response. You can see how adding more capabilities means adding more `elif` branches or a more sophisticated intent matcher, which can become unwieldy as the variety of questions grows.

By contrast, a generative approach would use a pretrained language model. With the Hugging Face Transformers library, for instance, we can load a conversational model and have it generate an answer. Refer to this code on GitHub: https://github.com/anvcse562/ModernNLP-LLMSLM/blob/main/Ch09/ch9.1/9.1.1_genrative_approach.py.

In this snippet, we prompt the model with a simulated conversation, that is, the user's question followed by "bot," and the model will generate a continuation for the bot's answer. The output could look like this:

"You can renew your passport by visiting your country's embassy or consulate. Make sure to bring the required documents like your current passport, passport photos, and any necessary forms. It usually involves filling out an application and paying a fee."

This answer was not explicitly programmed into the bot; the model produced it based on patterns it learned from training data. The generative approach dramatically simplifies the code—we don't have to write separate `if` conditions for each possible question. However, evaluating and controlling the quality of the model's answer becomes the main concern,—for instance, verifying that the information about passport renewal is accurate. In practice, you might use larger, more powerful models or fine-tune a model on a Q&A dataset to improve the reliability of answers, but the development paradigm is clearly different from the rule-based method.

9.1.2 The Role of LLMs and SLMs in Dialogue Systems

Not all AI chatbots are backed by gigantic models like GPT-3. In many cases, especially for production deployments, we must consider model size and efficiency. This is where the distinction between LLMs and SLMs becomes important. The terms aren't rigid categories, but generally LLM refers to very large models (with hundreds of millions to billions of parameters) known for state-of-the-art performance, while SLM here refers to smaller, distilled, or efficient models (with tens of millions of parameters or less) that sacrifice some capability for speed and resource economy. Both have a role to play in dialogue systems:

- **LLMs:** These are the heavy-hitters like OpenAI's GPT-3 (175 billion parameters) and Google's PaLM and T5-XXL (with billions of parameters). They possess an impressive ability to understand context and generate coherent, contextually relevant responses across a vast range of topics. In dialogue systems, an LLM can carry on a multi-turn conversation keeping track of nuances and often come up with answers that are very close to what a human expert might say. For example, GPT-4 can answer complex customer queries, reason about the user's intent, and even inject some empathy or humor as appropriate. The role of LLMs is crucial when high-quality, open-domain conversation is needed—such as a virtual assistant that should handle arbitrary questions, or a customer support agent that deals with anything from billing issues to technical troubleshooting. LLMs tend to have better language understanding (thanks to their extensive training data) and can generalize well, which makes them suitable for multi-turn dialogues that require memory and reasoning. However, deploying an LLM in a real system has challenges: it might require cloud infrastructure or specialized hardware due to its size, and inference can be relatively slow (e.g., a large model might take a couple of seconds to produce a response, which could be too slow for a snappy user experience). Thus, LLMs are often used when the quality of the conversation is the top priority and resources are available to support the LLM. Many advanced dialogue systems use LLMs as a backbone—for instance, a medical chatbot might use a large model to understand a patient's query in detail and formulate a careful answer (with the output possibly vetted by additional logic for safety).

- **SLMs:** This category includes models like DistilGPT-2 (a distilled version of GPT-2 with ~82 million parameters), DistilT5, TinyBERT/TinyGPT, and other compact models. These models are optimized for efficiency: they run faster, require less memory, and can even work on edge devices (like mobile phones or IoT hardware) or in real-time settings where an LLM would be impractical. In dialogue systems, SLMs are ideal when you need a *lightweight conversational agent*—for example, a chatbot embedded in a mobile app or a device where low latency is crucial. Although they don't match the full linguistic prowess of the largest models, well-trained small models can still provide useful, grammatically correct responses for focused tasks. In fact, through techniques like knowledge distillation, SLMs manage to retain much of the capabilities of their larger "teacher" models while significantly reducing size. DistilGPT-2, for instance, retains most of GPT-2's conversational ability but in a model half the size of the original, making it a practical choice for conversational agents where performance and resource efficiency are critical.

SLMs play the role of enabling real-time and on-device dialogue systems. Consider a voice assistant running on a smartphone: using a 175 billion parameter model locally is impossible, but a 50 million parameter model might run in under a second on the phone's CPU. Similarly, for an enterprise deploying thousands of chatbot instances (say, one per customer browser session), a smaller model can drastically cut costs. SLMs are also easier to fine-tune on smaller datasets and faster to iterate on. They are often used for resource-constrained environments and can be the model of choice for prototyping. For example, you might start with DistilGPT-2 or DistilBART to build a quick demo of a chatbot, since it's cheap to run, and later switch to a larger model for improved performance if needed.

It's important to note that "small" *is relative*: with the rapid advancement in model efficiency, some models in the low-billions of parameters range (like 2–7B) are now considered reasonably deployable on single servers or even high-end consumer hardware, blurring the lines between traditional LLMs and SLMs. Recent research and industry trends (sometimes dubbed the "rise of small language models") focus on making models more compact without losing too much accuracy. Techniques include distillation (as used in DistilGPT-2, DistilBERT, etc.), quantization (reducing precision of model weights), and efficient architectures like MiniLM or MobileBERT. All of these advances feed into dialogue systems by allowing you to choose a model that fits the use case constraints. If you need a very fast chatbot for simple queries, a distilled or quantized model might suffice. On the other hand, if you're building a high-stakes virtual assistant (say, an AI doctor or lawyer), you'd be better off with the largest model you can afford and then optimize around it.

To visualize the distinction, Table 9-2 lists a brief comparison of an LLM versus an SLM in the context of dialogue systems.

Table 9-2. Comparing LLMs vs. SLMs for Dialogue Systems

Model (Approach)	Size and Resources	Capabilities and Use Case
GPT-3 (generative LLM)	~175 billion parameters. Requires powerful GPU servers or cloud API. Response time may be a few seconds per query.	Excellent language understanding and generation across domains. Suitable for rich, open-ended dialogues and complex queries. Often used via an API (OpenAI) for high-quality virtual agents.
T5-Large (LLM)	Billions of parameters (e.g., 3B or 11B). Needs a GPU for real-time inference.	Strong at multi-turn dialogue when fine-tuned. Can be used for conversational tasks requiring knowledge (e.g., an assistant that explains technical content). Higher quality but heavier than smaller models.
DistilGPT-2 (SLM)	~82 million parameters. Can run on CPU or modest hardware; low latency (sub-second) responses. Open-source and easy to deploy.	Good at casual conversation; can handle short dialogues. Ideal for resource-constrained deployment like a web app chatbot or mobile assistant. May require more domain-specific tuning since it has less knowledge capacity than an LLM.
DistilT5 (SLM)	Tens of millions of parameters (distilled from T5 base). Efficient for specific tasks.	Useful for task-oriented dialogue (e.g., a lightweight FAQ bot). Faster inference makes it feasible on-edge or at scale. Might not maintain long conversations as coherently as a large model, but effective for quick queries.

As shown, LLMs serve when you need maximum conversational ability and accuracy, whereas SLMs enable broader deployment and real-time interaction. Often, the choice comes down to a tradeoff between quality and efficiency. Developers might start building a chatbot with an SLM for speed, and if they find the answers unsatisfactory or too superficial, they might upgrade to an LLM. Conversely, if an LLM-based bot is too slow or costly, a distilled model or smaller variant could be tried.

Another angle to consider is LLM agents—situations where LLMs are used not just to answer questions, but also perform more complex agentive behavior (e.g., browsing, tool use, multi-step reasoning). Frameworks like LangChain allow an LLM to function as part of a larger system with memory and tool integration. For example, an LLM-based dialogue agent could decide to call an external API to get live information (like today's weather or a database query) and then continue the conversation with that data. These advanced agents combine the generative power of LLMs with deterministic operations, pushing the envelope of what dialogue systems can do. While this chapter focuses on building dialogue systems in a somewhat more bounded sense (chatbots for support, etc.), it's good to be aware that large models can be orchestrated as intelligent agents that go beyond simple Q&A—they can follow instructions, handle multi-turn tasks, and even proactively guide conversations when designed to do so.

In this section, you have learned that dialogue systems can be crafted either through explicit rules or through generative modeling, and each approach has its place. We also highlighted that choosing the right model size (an agile SLM versus a powerful LLM) is the key to balancing user experience with practical constraints. In the next section, we apply these concepts by building dialogue systems using AWS tools. You see how services like Amazon Lex (which aligns more with the rule-based/intent-driven paradigm) and Amazon SageMaker (for fine-tuning and deploying LLMs) can be used to create interactive conversational agents. This practical implementation will bridge the theory of dialogue systems with real-world development on cloud platforms.

9.2 Implementation on AWS

This section applies the dialogue system concepts from the last section using AWS cloud tools. We build a conversational agent with a hybrid approach: using Amazon Lex for rule-based, intent-driven interactions, and Amazon SageMaker to train or deploy generative LLM like T5 or BART for more complex, open-ended dialogue. We also explore how SLMs such as DistilGPT-2 and DistilT5 can be utilized for lightweight, real-time conversations on AWS or edge devices. By the end of this section, you'll understand how to create a scalable chatbot on AWS—from defining intents in Lex, to fine-tuning a domain-specific language model in SageMaker and deploying efficient mini-models for responsive performance. Throughout, we continue the example of a travel assistant bot, illustrating how each AWS component comes into play, and we touch on integrating LLM "agents" (advanced AI behaviors) in these systems.

9.2.1 Amazon Lex: Rule-Based Conversational Agents

Amazon Lex is AWS's managed service for building conversational interfaces like chatbots and voice assistants with an intent-based, rule-driven paradigm. Lex provides automatic speech recognition for voice input and natural language understanding to classify user input into *intents* and extract *slots* (parameters)—very much like the rule-based dialogue systems discussed earlier. Developers use Lex to define the intents, sample utterances, and prompt/response flows. Lex then handles the heavy lifting of language parsing and can manage multi-turn dialogues by prompting the user for missing information. This makes Lex ideal for structured, goal-oriented conversations, such as an FAQ bot, a customer service triage, or a step-by-step form-filling dialogue. We look at how to create a simple Lex bot and how it can be extended to incorporate LLM-driven responses when needed.

9.2.1.1 Creating a Lex Bot (Intents, Utterances, and Slots)

To design a Lex chatbot, you typically start by specifying intents, which represent what the user wants to achieve. For each intent, you provide example utterances (phrases the user might say). Lex uses these examples to train its built-in classifier to recognize the intent even if the user's phrasing varies. For instance, for our Travel Assistant bot, we might define an intent called GetVisaRequirements. The sample utterances could include variations like:

- "Do I need a visa for <Country>?"
- "What are the visa requirements for visiting <Country>?"
- "Visa info for <Country>."

Here <Country> would be a slot—a variable in the user's request. In Lex, we define a slot of type "Country". We might configure Lex to prompt for this slot if the user doesn't provide it. For example, if a user just asks, "Do I need a visa?", Lex can reply with a *slot prompt*: "Sure, which country are you planning to visit?" Lex can even perform basic validation on slots to ensure the country name is in a list of supported countries.

Once the intent and slots are filled, Lex needs to respond. There are two main options:

- **Static responses or prompts:** We can provide a templated answer in Lex, such as "To visit <Country>, you typically need to ..." which may include the slot values in the message. For instance, if the country slot is "Canada", Lex could return: "To visit Canada, you typically need a valid passport and possibly a tourist visa (unless you're from a visa-exempt country)."

- **AWS Lambda fulfillment:** For more dynamic behavior, Lex can invoke an AWS Lambda function when an intent is triggered. This Lambda can perform business logic like database lookups, API calls, or even call an LLM service and then return a composed response to Lex. In the travel visa example, instead of a hard-coded message, we might use a Lambda that checks an external visa requirements API or a database for the specific country and returns the up-to-date info. The Lex bot will then relay that to the user.

Using Lambda fulfillment with Lex greatly increases flexibility—it's how you integrate arbitrary custom logic or backend systems into the conversation. For example, a banking chatbot's Lex intent might invoke a Lambda to get account details. In our case, we could call it a knowledge base of travel regulations. The Lambda could also integrate with an LLM (we explore this later) to generate a more natural answer, effectively hybridizing rule-based intent handling with generative response generation.

Dialogue Flow and Context in Lex: Amazon Lex allows multi-turn dialogues by maintaining context across turns. For instance, after the bot asks for the country, it waits for the user's answer to fill the slot before completing the `GetVisaRequirements` intent. Lex V2 (the newer version) supports *session attributes* and *active contexts* that enable branching dialogues and some level of context memory. For example, you might set an active context when one intent (such as `BookFlight`) is in progress, so that the bot interprets subsequent inputs in that context (e.g., expecting a date or destination next). However, Lex's context tracking is explicit and deterministic—you design the conversational state machine. It doesn't "remember" arbitrary past conversations unless you store them (e.g., in session attributes or via a Lambda). This is in contrast to an LLM-based system that can implicitly consider the conversation history.

Example Interaction with Lex: Let's walk through a simple Lex-driven interaction to illustrate. Suppose the user says, *"Do I need a visa for Canada?"*

1. Lex receives this input and uses its NLU to classify the intent. It matches our GetVisaRequirements intent with high confidence, given the sample utterances we provided are similar. It also extracts the slot "Country" with value "Canada" from the utterance.

2. Since the required slot is already filled by the user's question, Lex can proceed to fulfillment. Let's say we configured a Lambda for fulfillment. Lex will call our Lambda function, passing along the intent name and slot value ("Country" = "Canada").

3. The Lambda function runs our custom code—perhaps querying a database of visa info. Let's assume it finds that for Canada, a traveler needs an Electronic Travel Authorization (eTA) if they are coming from certain countries. Otherwise, they just need a passport for short visits. The Lambda constructs a response string with this information.

4. The Lambda returns that message to Lex. Lex then relays it to the user as the answer: "For Canada, visitors typically need a valid passport. Depending on your nationality, you might also require an Electronic Travel Authorization (eTA) for entry."

From the user's perspective, the bot understood their question and answered appropriately. Internally, this was achieved through a ruled intent and a bit of code for the dynamic content. If the user follows up with another question like "What about Japan?" (and they still have the session open), Lex can interpret that this likely refers to the same intent (GetVisaRequirements) and now the country slot is "Japan"—it could either handle that in the same session (depending on how the bot is configured to manage context from prior turn) or ask for clarification if needed. Typically, Lex would treat it as a new invocation of the intent with the new slot and again call the Lambda for Japan's info. This highlights that Lex on its own doesn't carry over the full context of the prior question unless you engineered it to do so. In this case, the user's follow-up was simple enough that Lex's intent classifier can still map it correctly, especially if we added training phrases like "What about Japan?" as a way to ask the same question with a different slot value

Lex Integration with LLMs (Hybrid Approach): One powerful pattern is using Amazon Lex's fallback intent to integrate an LLM for handling queries that don't match any known intent. Lex V2 comes with a built-in AMAZON.FallbackIntent, which triggers when the user's input doesn't confidently match any of the defined intents. We can attach a Lambda function to this fallback intent. In our travel bot example, let's say the user asks something open-ended or out-of-scope like, "I'm moving to Japan next year, what should I prepare for?"—This query might not map to any intent (we didn't define one for general moving advice, it's too broad). Instead of Lex replying with a generic "Sorry, I didn't understand that," we can catch the fallback and hand it over to an LLM. The Lambda could forward the user's question to a SageMaker endpoint running a generative model (for instance, a fine-tuned T5 or GPT model), which can produce a helpful answer. The Lambda then returns that answer to Lex, which in turn sends it to the user. The result: the user gets a freeform, informative response like "Moving to Japan is exciting! You should secure housing, familiarize yourself with local customs and basic Japanese phrases, and ensure your visa and paperwork are in order. It also helps to set up a local bank account once you arrive." This is far beyond the scripted replies of Lex alone.

This kind of Lex/LLM integration gives a best-of-both-worlds approach: Lex handles the expected, structured queries with high precision using rules and database facts, and the LLM provides a safety net and flexibility for unexpected questions or chitchat. Essentially, Lex acts as the frontend interface and traffic cop, while the LLM (deployed via SageMaker or another service) acts as an on-demand "brain" for creative or unanticipated queries. In AWS architecture, this is often achieved with Lex's fallback intent → Lambda → SageMaker endpoint even an external API like OpenAI, but using SageMaker allows you to host your own model. Some AWS solutions incorporate frameworks like LangChain within the Lambda to manage the LLM interaction and even maintain conversation memory by retrieving the past dialogue from Lex's session and prepending it to the LLM prompt. For instance, one AWS sample project uses Lex with a Lambda that keeps track of the conversation using Lex session attributes and calls a FLAN-T5 model on SageMaker for answers, enabling multi-turn context carryover. While the inner workings can get complex, the key idea is that Amazon Lex can serve as a control system, leveraging an LLM when needed—effectively creating an *LLM-augmented conversational agent*. This addresses one of the weaknesses of a purely rule-based bot (handling the unexpected) without fully surrendering control to a large model.

Practical Example: Interacting with Lex via Code: Once a Lex bot is built using the AWS Console or APIs to define intents and Lambda hooks, it can be integrated into applications. AWS offers a REST API and SDK (`boto3` for Python) to send user input to your Lex bot and get responses, which is useful for custom client applications (web chat, mobile apps, etc.). For example, here's how you might call a Lex bot (V2) using Python:

python
Copy

```python
import boto3

# Initialize Lex V2 runtime client
lex_client = boto3.client('lexv2-runtime')

BOT_ID = "YOUR_BOT_ID"                    # Amazon Lex V2 bot ID
BOT_ALIAS_ID = "YOUR_BOT_ALIAS_ID"        # Alias pointing to a specific
                                          # bot version
LOCALE_ID = "en_US"                       # Locale, e.g., English (US)
SESSION_ID = "user-123"                   # An identifier for the
                                          # conversation session
user_input = "Hi, do I need a visa for Canada?"  # User's message
response = lex_client.recognize_text(
    botId=BOT_ID,
    botAliasId=BOT_ALIAS_ID,
    localeId=LOCALE_ID,
    sessionId=SESSION_ID,
    text=user_input
)

# The response contains what Lex understood and how it responded
messages = response.get('messages', [])
if messages:
    bot_reply = messages[0]['content']
    print("Bot:", bot_reply)
```

In this snippet, we send a text utterance to the Lex bot and receive the bot's reply. Under the hood, Lex has classified the intent as GetVisaRequirements (if our user_input matches that) and either responded with a prompt (if slot info was missing) or with the fulfillment result. The messages field in the response holds the bot's answer text. If the bot requires more input (e.g., asking for a slot), the conversation continues by calling recognize_text again with the user's next response, using the same SESSION_ID to maintain state. This simple API call demonstrates that once your Lex bot's logic is defined, integrating it into an app or a backend is straightforward. It's essentially an API-driven chatbot: your code sends user strings and receives bot strings, and Lex manages the conversational state and logic on AWS.

Before moving on, let's summarize how and when to use Amazon Lex in your dialogue system:

- **When to use Lex:** If you need a reliable, controlled conversation flow—for instance, gathering specific data like names, dates, numbers in order, or providing fixed answers to known questions—Lex is a great choice. It's especially useful if you want quick development via a GUI, multi-language support, and integration with telephony (Amazon Connect) or messaging platforms—all with the scalability and security of AWS. You don't need to train large models; Lex's built-in intent classifier does that from your examples automatically.

- **Limitations:** Lex bots can feel rigid if users go off script. Maintaining a large number of intents and utterances can become complex for broad domains. Also, Lex's responses are as good as what you script or implement via Lambda—it won't suddenly answer a question it was never designed for. That's why combining Lex with LLMs (as described) is appealing for more open-ended use cases.

- **Performance:** Because it's rule-based and cloud-hosted, Lex responds very quickly typically in hundreds of milliseconds. It's optimized for real-time interactions, which is why it's used in voice IVR systems and chatbots where quick turnaround is important. Even when a Lambda function is involved, response times can be kept low with proper design. Lambda and SageMaker endpoints can introduce some latency, but for many scenarios this remains within acceptable range, say less than one second.

Table 9-3 contrasts Amazon Lex with a purely LLM-based approach (such as hosting a generative model on SageMaker) in the context of building a chatbot on AWS.

Table 9-3. Amazon Lex vs. a Purely LLM-Based Approach

Aspect	Amazon Lex (Rule-Based)	LLM on SageMaker (Generative)
Development paradigm	Define intents, utterances, and slot-handling logic (often via a visual builder or simple JSON configs). The conversation flow is explicitly designed.	Fine-tune or deploy a pretrained language model; provide it prompts and let it generate responses. No manual enumeration of utterances or flows—the model learns from data.
Conversation control	Deterministic and predictable. Bot says exactly what it's programmed to. Easy to enforce business rules and compliance (since you script the answers or data retrieval).	Stochastic and flexible. The model can generate novel responses. Harder to strictly control output—may say unanticipated things or require filters to ensure compliance.
Handling scope	Narrow, predefined scope. Out-of-scope queries result in fallback or failure unless specifically accounted for. Good for task-specific dialogues (booking, FAQs).	Broad knowledge (if using a large model). Can handle open-domain questions or unforeseen inputs with reasonable answers, leveraging training data. Suitable for informative or general chat beyond narrow scripts.
Latency and performance	Very fast and lightweight. Each request is a quick classification and optional lookup. Ideal for real-time and high-throughput (voice calls, etc.).	Typically slower and heavier. Large models may take seconds per response and require powerful instances. However, can use smaller models or optimize for better performance (discussed in the SLM section).
Integration	Native integration with AWS services (Connect for telephony, Lambda for logic, etc.). Returns structured output (intent, slots) which can trigger backend processes directly.	Requires more custom integration. Typically you'd call the model via a SageMaker endpoint or API and then parse its natural language output. Additional tooling (like LangChain, or custom parsing) may be needed for complex workflows.

(*continued*)

Table 9-3. (*continued*)

Aspect	Amazon Lex (Rule-Based)	LLM on SageMaker (Generative)
Cost	Pay-as-you-go per text/voice request. Generally low cost for moderate usage because you're not running dedicated servers (the NLU is managed by AWS). No training cost, just design effort.	Can be costly: if fine-tuning, you pay for training compute; for deployment, you pay for a continuously running inference instance (or use serverless endpoints when possible). Using large models or high throughput can incur significant expense.
Use case examples	IVR phone menu, FAQ chatbot with fixed answers, form-filling bot (e.g., schedule a meeting by collecting date/time), simple customer service bot that handles known issues.	Virtual assistant that answers a wide range of questions (e.g., "ChatGPT"-like helpdesk), a chatbot that engages in open-ended conversation or provides detailed advice/summaries, any scenario where responses need to be generated from context or data (when integrated with retrieval).

As Table 9-3 suggests, Amazon Lex and generative LLMs address different needs. In practice, they can complement each other within the same system. A sensible approach is to use Lex for what it's good at structured dialogues and high-precision intent handling and bring in an LLM for the rest. AWS facilitates this combination, as we've seen with mechanisms like fallback intents invoking SageMaker-hosted models. The next section dives deeper into Amazon SageMaker—specifically, explaining how you can fine-tune and deploy your own language models on AWS to provide the generative brains of a chatbot.

9.2.2 Amazon SageMaker: Fine-Tuning LLMs for Dialogue

Amazon SageMaker is a fully-managed machine learning service that can train and host models at scale. In the context of dialogue systems, SageMaker allows us to take an LLM—such as *T5, BART, GPT-2/3, or newer models—and fine-tune it on domain-specific data, then deploy it behind an endpoint for real-time inference. If Amazon Lex is the rule-based "script writer," an LLM on SageMaker is the "improv actor" that can generate responses on the fly. Fine-tuning means we specialize a pretrained model on our own dataset (for example, a collection of support chat logs, or a list of question-answer pairs in our domain) so that it performs better for our application than a generic model would. SageMaker makes this process easier by providing the infrastructure (compute instances, GPUs, distributed training if needed) and integration with data storage on S3, so we don't have to set up our own ML servers from scratch.

Let's break down how you could utilize SageMaker for a dialogue system:

- **Selecting a model:** First, choose a base model to fine-tune. For dialogue generation, sequence-to-sequence models like T5 or BART are good choices because they can be trained on input/output pairs, such as a conversation history or user query → next reply. Another approach is using an auto-regressive model like GPT-2/GPT-3 style, which generates continuation given a prompt (we can format the conversation as a prompt). For instance, T5 and BART have been used in research for task-oriented dialogue generation and response generation, while GPT-2 has a variant called DialoGPT tuned for conversational response. In our travel assistant example, we might pick T5-base (220M parameters) as a starting point, since it's a relatively moderate-sized LLM with strong language understanding.

- **Preparing training data:** Fine-tuning data for dialogue could be organized in a few ways. A simple method is to create pairs of (input, output) texts. For example, the input could be a concatenation of the last few user and bot utterances, and the output is the next bot utterance. If we have transcripts of a travel advice service or FAQs, we can convert those into a conversational format. For instance: User: "What documents do I need to travel to Canada?\nBot:" → Output: "You will need a valid passport, and depending on your citizenship, possibly an eTA or visa to travel to Canada." By feeding many examples of questions and appropriate answers (especially focusing on the travel domain), we teach the model how to respond like a travel expert. Fine-tuning can also leverage multi-turn context by including several dialogue turns in the input sequence, up to a length the model can handle. The dataset might come from existing Q&A pairs, support tickets, or even scripted dialogues we create for training purposes.

- **Launching a fine-tuning job:** SageMaker can run training jobs via its Python SDK or the AWS Console. One convenient method is to use the Hugging Face Deep Learning Container in SageMaker. AWS has pre-built Docker images that come with Transformers library and can train popular models. We simply provide our training script and data. For example, using the SageMaker Python SDK within a Jupyter notebook or Python environment, you could do something like this:

python

Copy

```python
from sagemaker.huggingface import HuggingFace

# Define hyperparameters and configuration for fine-tuning
hyperparameters = {
    'model_name': 't5-base',        # which pre-trained model
                                    # to use (will download from
                                    # Hugging Face)
    'task': 'text2text-generation', # our training script can use
                                    # this to configure the pipeline
    'epochs': 3,
    'per_device_train_batch_size': 16,
    'per_device_eval_batch_size': 16,
    'learning_rate': 5e-5,
    'logging_steps': 100,
    # ... (other hyperparameters as needed)
}

# Create the HuggingFace estimator for SageMaker
huggingface_estimator = HuggingFace(
    entry_point='train.py',         # Your training script
    source_dir='scripts/',          # Directory with training script
                                    # (and possibly model code)
    instance_type='ml.p3.2xlarge',  # AWS instance type with GPU
    instance_count=1,
    role='AWS_SAGEMAKER_ROLE',      # IAM role for SageMaker with
                                    # needed permissions
```

```
            transformers_version='4.26',    # Specify versions of libraries
            pytorch_version='1.13',
            py_version='py39',
            hyperparameters=hyperparameters
)

# Launch the fine-tuning job (data is assumed to be uploaded to S3
already)
huggingface_estimator.fit({'train': 's3://my-bucket/my-dataset/'})
```

In this snippet, we set up a training job to fine-tune T5 based on our dataset. The train.py script (not shown in full here—see the GitHub for this chapter's code) would use the Transformers library to load T5-base with a sequence-to-sequence trainer. It would read our data (the train channel from S3) and train for the specified number of epochs, saving the model. SageMaker handles spinning up an EC2 instance with a GPU, running the training, and saving the fine-tuned model artifacts to S3 once done. If the model is large or the dataset is huge, we could use multiple instances or larger instances—SageMaker can distribute training across multiple machines or utilize optimized libraries like DeepSpeed for very large models, although T5-base is small enough to train on a single GPU in a reasonable time with a modest dataset.

- **Deploying the model for inference:** After fine-tuning, we want to host the model so it can generate replies to user input in real time. SageMaker allows one-click deployment of the trained model as a real-time endpoint. We could do this:

    python

    Copy

```
# Deploy the fine-tuned model to a real-time endpoint
predictor = huggingface_estimator.deploy(initial_instance_count=1,
instance_type='ml.g5.xlarge')

# Now we can use predictor to get responses from the model
user_question = "What should I prepare for a move to Japan?"
result = predictor.predict({"inputs": user_question})
print(result)
```

This would start an endpoint (on a GPU-backed instance type for speed, or even CPU if the model is small and latency isn't critical). The `predictor.predict` call serializes the input and sends it to the endpoint's REST API, which runs our model and returns the generated text. If our fine-tuning went well, `result` might contain something like: "Moving to Japan is exciting! You should research the visa process and ensure you have all necessary documents, like a Certificate of Eligibility, arrange housing, and perhaps start learning some basic Japanese. It's also important to set up health insurance once you arrive." This is a detailed answer that our model learned to give for such questions. This can now be integrated into our chatbot system. For example, our Lex bot's Lambda could call this SageMaker endpoint (via AWS SDK) whenever it needs a generative answer.

9.2.2.1 Domain-Specific Fine-Tuning Benefits

By fine-tuning an LLM on our domain (be it travel, healthcare, finance, etc.), we impart it with knowledge and style suited for our application. A general model like base T5 might know a bit about many topics (from its Internet-scale pretraining), but can be vague or not use the preferred terminology for, say, medical advice or banking. Fine-tuning on a curated dataset (and possibly instructions on how to respond) makes the model more accurate and reliable for that context. In a customer service scenario, you could fine-tune on historical chat transcripts, so the model learns the correct answers and tone for common customer queries. This often yields better results than prompting a generic model with the company info each time, and it ensures the model stays within the boundaries of what it was taught (reducing some hallucinations). SageMaker's advantage is that you can do this securely; that is, your data and model stay in your AWS environment, which is important for proprietary or sensitive data and at scale (you can use powerful hardware or even distributed clusters as needed).

9.2.2.2 Using SageMaker JumpStart

It's worth mentioning that AWS has a feature called JumpStart, which provides pretrained models and sample notebooks to deploy or fine-tune them. For instance, JumpStart offers various T5 sizes (like FLAN-T5 XL) and others like GPT-J, Llama-2, and so on. If you don't want to write a training script from scratch, JumpStart can automate some steps. In this context, you could select a conversational model from JumpStart, deploy it directly, or fine-tune it on a dataset with just a few clicks or lines of code

(JumpStart handles setting up the estimator with the right containers). This is a fast way to get an LLM up and running on SageMaker. For example, to integrate a generative model without any training, you might directly deploy a seven billion parameter model that's been pretrained for chat, if it fits your needs, and skip fine-tuning. The tradeoff is that without fine-tuning, the model might not know specifics of your domain or might not follow your desired style as closely.

Inference and Scaling: Once a model is deployed, SageMaker endpoints can scale horizontally by adding more instances to handle more traffic, and they can be integrated behind API Gateway or other services to be part of an application. You might build a simple web service where the frontend calls the SageMaker endpoint with user input and returns the model's reply. If you expect sporadic usage or want to save cost, AWS also has a *serverless inference* option or you can design a system to spin down the endpoint when not in use. There's also SageMaker Batch Transform for offline processing, which is not relevant for interactive chat, but it's useful if you need to generate responses for thousands of prompts in one go for analysis.

Two issues to manage with LLM inference are latency and cost. Large models on SageMaker might take a couple seconds to generate a response, especially in cases of extended prompts or long form answers. This is where model optimization is important (we talk about SLMs in the next subsection for speed). SageMaker does support deploying models on GPU instances for speed, or even on Inf1/Inf2 instances that use AWS Inferentia chips that are optimized for neural network inference at a lower cost. Using half-precision or model pruning/quantization techniques can also reduce response time. The bottom line is, SageMaker gives you the flexibility to choose your hardware and scale for the performance you need. For a moderate-sized model like T5-base or BART-base, an `ml.g4dn.xlarge` (with one NVIDIA T4 GPU) might be enough to serve a few requests per second. If you fine-tune a much larger model (billions of params), you might deploy on `ml.p4d.24xlarge` (with A100 GPUs) or similar, which is powerful but very expensive to run continuously. So often there's a balance: pick the smallest model that meets your quality needs to keep costs manageable.

9.2.2.3 Incorporating LLM Agents on AWS

Beyond straightforward Q&A, an LLM hosted on SageMaker can be part of more complex AI agents. For instance, you could build a system where the LLM not only chats with the user but can also call tools or query databases as needed. AWS Lambdas (or other microservices) can be those tools—like retrieving real-time data (weather, stock info) or performing an action (making a booking) upon the model's request. One way to implement this is by using frameworks like LangChain or custom logic where the LLM's output is parsed for intentions to use a tool.

For example, the model could reply with a special format like `<ACTION>lookup_flight_status("ABC123")</ACTION>`, which your system interprets and executes, then the model continues the conversation with the result. While designing such an agent is advanced, SageMaker can host the core model and AWS provides the glue for everything else—Lambda for the tools, Step Functions to orchestrate multi-step workflows, and so on. Essentially, the cloud environment lets the LLM operate with extensions rather than being an isolated brain. Many generative chatbot systems incorporate retrieval of knowledge from a document store or search engine. On AWS, you could integrate Amazon Kendra (an intelligent search service) or a simple Elasticsearch, where the LLM uses retrieved text to ground its answers in order to reduce hallucinations and improve factual accuracy. For example, a customer support bot might, for any user question, retrieve relevant product documents and feed those, along with the question, into the LLM prompt, so that the answer is based on actual support manuals. SageMaker doesn't do that retrieval part out-of-the-box, but you can build it around the model. The key point is: Amazon SageMaker provides a robust platform to train and host the generative components of a dialogue system, which you can then integrate with other AWS services (like Lex, Lambda, Kendra, etc.) to build an end-to-end conversational agent that is both intelligent and practical.

Before moving on to the next section, let's consider when it is most appropriate to use SageMaker and a fine-tuned LLM:

- If your chatbot needs to handle a lot of freeform queries or hold natural conversations that go beyond predefined flows, an LLM is likely needed, and SageMaker is how you bring that into your solution. This is especially true if you want to avoid external AI APIs to preserve data prices and reduce cost.

- If your use case is domain-specific like in the case of medical or legal documents, fine-tuning an existing model on domain data via SageMaker can dramatically improve the relevance and accuracy of its responses compared to using a generic model.

- For maintenance, you'll need to update the model if new data comes to keep it current and monitor the model's outputs. SageMaker provides ML Ops features, including model versioning, monitoring, and A/B testing endpoints that can be useful for managing a production chatbot model over time.

Finally, to complement our earlier Lex vs. LLM comparison, here's a quick summary of how they can work together on AWS: Lex provides the interface, guiding conversation and handling straightforward tasks, while SageMaker-hosted LLMs provide the brains for complex questions or generative answers. In many modern chatbot architectures, integration is quietly orchestrated by a Lambda function acting as a bridge between components. From the user's point of view, this complexity remains hidden and is ideally irrelevant. When implemented well, the transition between systems is seamless. For instance, a user might be following a structured dialogue to book a flight—handled by Amazon Lex—when they casually ask, "By the way, what's a red-eye flight?" Rather than breaking the flow, Lex can delegate the open-ended question to an LLM, which replies: "A red-eye flight is an overnight flight that arrives in the morning, named because passengers might have red eyes from lack of sleep." The conversation flows on uninterrupted, providing the user with both accuracy and flexibility. This hybrid approach—combining rule-based reliability with the generative power of LLMs is becoming increasingly common in enterprise chatbot deployments, offering a balanced blend of structure and intelligence.

With a fine-tuned LLM running on SageMaker, we have our "large brain" in place. However, not every deployment can afford or needs a large model. Often, we want something more lightweight, especially for mobile or edge scenarios, or simply to reduce costs. This is where SLMs come in, which we discuss next.

9.2.3 SLM Implementation: Using DistilGPT-2 and DistilT5 for Lightweight Deployment

SLMs are optimized, compact versions of larger models that aim to deliver decent performance at a fraction of the size and compute requirements. In dialogue systems, SLMs like DistilGPT-2 and DistilT5 allow us to deploy conversational AI in resource-constrained environments—for example, directly on a mobile app, on a single CPU server, or at the edge (IoT devices, embedded systems)—where using a giant LLM is impractical. SLMs are typically produced by techniques such as *knowledge distillation* (hence the "Distil" prefix), where a smaller model is trained to imitate a larger model's outputs, thereby inheriting some of its capabilities. They might not match the full linguistic prowess of their larger counterparts, but they can be surprisingly capable within narrower or simpler dialogue needs, and they offer low latency and low memory usage. This subsection explores how to leverage such models in practice, continuing our focus on AWS deployment as well as on-device scenarios, and provides examples to illustrate their speed and efficiency.

9.2.3.1 What Are DistilGPT-2 and DistilT5?

- **DistilGPT-2:** A distilled version of the GPT-2 model. GPT-2, originally with up to 1.5 billion parameters (for the largest version), was known for generating coherent paragraphs of text. DistilGPT-2 was created by compressing GPT-2 (usually the 124M or 355M parameter versions) into a model with about 82 million parameters, roughly 40% of the original model's size.[1] This reduction leads to a model that is faster (often ~2x speedup) and uses far less memory, while retaining a good portion of GPT-2's conversational ability. DistilGPT-2 is well-suited for generating short to medium responses and can be fine-tuned on dialogue data to improve its conversational style.

- **DistilT5:** A similarly compressed version of T5. There are various sizes of T5 (small: 60M params, base: 220M, large: 770M, etc.). A DistilT5 model might be, for example, a T5-base distilled down to ~100 million parameters or a T5-small distilled to ~30 million. One available variant is a DistilT5 for Q&A that's about half the size of T5-base.[2] Distilled T5 models maintain the ability to perform text-to-text tasks (like given a context, generate an answer) but with faster inference. For dialogue, DistilT5 can be fine-tuned to take conversation context as input and output the next response, similar to how we discussed fine-tuning T5. The model is now smaller and more efficient.

---

[1] https://aws.amazon.com/blogs/machine-learning/achieve-four-times-higher-ml-inference-throughput-at-three-times-lower-cost-per-inference-with-amazon-ec2-g5-instances-for-nlp-and-cv-pytorch-models/#:~:text=16%20Full%20651%20158%2025,5X

[2] https://aws.amazon.com/blogs/machine-learning/achieve-four-times-higher-ml-inference-throughput-at-three-times-lower-cost-per-inference-with-amazon-ec2-g5-instances-for-nlp-and-cv-pytorch-models/#:~:text=16%20Full%20651%20158%2025,5X

9.2.3.2 Why Use SLMs?

The main motivations are *performance, cost,* and *deployability*. For real-time systems such as a chatbot that needs to respond in under, say, 300 milliseconds, a giant model might be too slow, whereas a smaller one can easily meet that. If we want a chatbot inside a mobile app that works offline (imagine a travel guide app that can answer questions without Internet access), a model like DistilGPT-2 might be the largest feasible model to ship with the app (82M parameters can be a ~300MB file, which is borderline but possible; anything in the billions of params would be multiple gigabytes and impossible for mobile). On the cost side, running a large SageMaker instance 24/7 for an LLM can cost thousands of dollars a month, whereas a small model might run on a CPU instance or a much cheaper machine. Also, if we need to spawn many copies of a bot (say, for many parallel users or across many edge locations), using SLMs ensures we don't need GPU acceleration everywhere—they might run on standard servers or devices.

9.2.3.3 Deploying SLMs on AWS and Edge

There are a few ways to deploy an SLM:

- **On SageMaker (or AWS Lambda) for cost-efficient serving:**
  Because SLMs use less memory, you can host them on smaller, cheaper instances. For example, DistilGPT-2 can even be served on a CPU-only instance with acceptable speed for short responses. AWS Lambda, which has limits on memory and execution time, can sometimes host a small model for infrequent requests. Imagine a serverless architecture where a user's message triggers a Lambda function that loads a DistilGPT-2 model (possibly from Amazon ECR as a container image) and returns a response. This could auto-scale with demand and you only pay per invocation. While Lambda cold-start and loading the model might introduce a delay, you can mitigate that by keeping the model in memory across invocations (with sufficient RAM and using global scope) or using provisioned concurrency. The benefit is you don't pay for idle time, and you avoid maintaining a server. For a step-by-step real-time chat, Lambda might not be ideal due to stateful context (although you could pass conversation history each time or store it in DynamoDB between calls),

CHAPTER 9　DIALOGUE SYSTEMS

so another approach is deploying the small model to a dedicated micro-server. For instance, an EC2 t3.small instance (two vCPU, 2GB RAM) could likely run DistilGPT-2 with a lightweight web service. Or use Amazon ECS/Fargate to spin up a container running the model behind an API. The cost here is minimal compared to a large model on a p3 or p4 instance.

- **On the edge/mobile:** To run on a mobile phone or an IoT device, you would export the model to an efficient format. Tools like ONNX Runtime, TensorFlow Lite, and PyTorch Mobile allow you to optimize and run models outside of a Python environment. For example, you could convert DistilGPT-2 to ONNX and use ONNX Runtime to execute it on a device (with support for hardware acceleration like neural processing units if available on the device). There have been demos of GPT-2 variants running on smartphones—typically they generate a few tokens per second, which is slow but if the responses are short, it might be okay. DistilGPT-2 would be faster. DistilT5 (especially if starting from a small base) can be really compact and fast; something like a 30M parameter model can sometimes run in real time (tens of milliseconds per inference) on a modern phone CPU. The key limitation is memory (the model file size) and the complexity of installing it in the app. But frameworks like Core ML (for iOS) or TFLite (Android) have made it feasible to include small transformers in mobile apps for on-device NLP. Our travel assistant bot could, for instance, include a mini-dialogue model so it can answer some basic travel FAQs even without the Internet. It might not have the depth of the cloud LLM, but it can cover the most common queries offline. When online, the app could switch to using the full cloud model for more elaborate answers.

- **In combination with Lex:** Another interesting pattern is to use an SLM as a middle ground in the cloud stack. For example, Lex could call a small model first for certain tasks and only escalate to a big LLM if needed. Or you might run an SLM to re-rank or refine answers. However, if you want to avoid the complexity of fine-tuning a large model, you could directly fine-tune a smaller model on SageMaker and deploy that. For instance, instead of T5-base, fine-tune DistilT5 on your dialogues. The fine-tuning process is the same as described, just faster and cheaper because the model is smaller. The deployed endpoint will then be cheaper to run. The tradeoff is some quality loss, but it might be sufficient. Many businesses choose smaller models when the difference in answer quality is negligible for their domain or when the absolute correctness is ensured by retrieving knowledge anyway.

9.2.3.4 Example: Using DistilGPT-2 for a Quick Chat Response

To illustrate the practical difference, let's consider generating a response with a distilled model in code. Suppose we want our bot to answer a user locally, without calling the cloud. Using Hugging Face Transformers offline with DistilGPT-2 is straightforward and blazingly fast even on a CPU for short texts:

python
Copy

```
from transformers import AutoTokenizer, AutoModelForCausalLM

# Load DistilGPT-2 model and tokenizer (small and fast to load)
tokenizer = AutoTokenizer.from_pretrained("distilgpt2")
model = AutoModelForCausalLM.from_pretrained("distilgpt2")

# Construct a simple conversational prompt
prompt = "User: What should I pack for a trip to Canada?\nBot:"
input_ids = tokenizer.encode(prompt, return_tensors='pt')

# Generate a continuation for the bot's response
output_ids = model.generate(input_ids, max_length=50, pad_token_id=tokenizer.eos_token_id)
response_text = tokenizer.decode(output_ids[0], skip_special_tokens=True)

print("Bot:", response_text.split("Bot:")[-1].strip())  # extracting the part after "Bot:"
```

After execution, the model will return a continuation of the prompt. Sample output is shown here:

Bot: "You should pack warm clothing, especially if you're going in winter, as well as a good travel adapter for your electronics. Don't forget your passport and any visa documents you might need. Canada can get cold, so maybe pack a coat, gloves, and a hat!"

This response was generated in fractions of a second on a typical laptop CPU. While DistilGPT-2 might not always produce factually perfect or deeply insightful answers (it might not know the latest visa rules, for instance), it demonstrates a reasonable ability to answer a general question fluently. In a controlled environment or for less critical queries, such small models can do a fine job. If more accuracy is needed, you could incorporate a knowledge check—such as verify any facts using a database or just use the SLM for chit-chat and have rules for the factual parts.

Performance comparison: To put things in perspective, consider these model sizes and their inference speed:

- A large model like GPT-3 (which you'd access via an API rather than self-host) has 175B parameters and typically requires a multi-GPU server to run. Each query might take a few seconds and the cost per query is relatively high.

- A medium-sized model like T5-base (220M params) can be hosted on a single GPU and might respond in under a second or a few hundred milliseconds on good hardware for short prompts.

- DistilGPT-2 (82M params) can run on CPU and often generate a token in ~10-20 milliseconds on modern CPUs—meaning a short sentence of 20 tokens could be generated in under 0.5 seconds on a laptop. On a server-grade CPU, it would be even faster. And if you're using a GPU, it's essentially instant for most practical lengths.

- Even smaller SLMs (DistilBERT variants, or specially optimized models like MobileBERT, TinyGPT, etc.) might be on the order of 20-50M parameters and can run in real time on mobile devices. (For classification tasks they can be <100ms. For generation, they still have to generate word by word but can be manageable for short outputs.)

There is a clear tradeoff. As we shrink models, we lose some sophistication. For example, DistilGPT-2 may produce more generic or slightly less coherent responses than the full GPT-2 or a larger model. It might also have a narrower knowledge cutoff (since it's based on GPT-2's training, which only included data up to 2019). However, if your use case is well-defined, you can often fine-tune an SLM to compensate. For instance, fine-tuning DistilGPT-2 on a specific dialogue dataset can teach it domain-specific answers that it wouldn't know out-of-the-box.

Fine-tuning and optimizing SLMs: Because SLMs are small, fine-tuning them is fast and feasible on limited hardware. You could even fine-tune DistilGPT-2 on a decent laptop with a GPU. On SageMaker, this means you can iterate quickly. If you have, say, a few thousand example Q&As or dialogue pairs, you can fine-tune DistilGPT-2 in minutes to better suit your needs. This could be used to create a lightweight model for your chatbot that speaks in your company's tone or knows your product names, without the overhead of a large model. Another angle is quantization: Tools exist to convert the model weights to 8-bit integers (or even 4-bit), which can further reduce size and improve speed at a slight cost to precision. An 8-bit quantized DistilGPT-2 might be nearly half the size (~150MB instead of 300MB on disk) and run faster, and the drop in output quality is often minor.

9.2.3.5 Use Cases for SLM-based Dialogue Agents

- **Mobile assistants:** As an example, this could be a translation app with a built-in chatbot that works offline for basic conversation practice. An SLM can handle simple dialogues without needing cloud connectivity.

- **IoT and vehicles:** As an example, this could be a car's onboard voice assistant that can control features or answer questions without always pinging a server—a small model could be embedded for instant responses (and for privacy, keeping voice data local).

- **Mass-deployed customer service bots:** If a company wants to deploy an instance of a chatbot for each of many websites or branches, running a huge model per instance would be infeasible. Instead, a single modest server could host dozens of parallel sessions of a small model, or each branch could run its own small model on existing hardware.

- **Real-time analytics or filtering:** Sometimes SLMs are used not to respond to the user directly, but to run alongside the conversation for tasks like sentiment analysis and intent detection, or to decide when to route to a human agent. Their speed makes them suitable for such real-time NLP tasks on the side.

CHAPTER 9   DIALOGUE SYSTEMS

To illustrate with the travel assistant example, suppose we want a lightweight version of the bot that a user can use in the airport without Internet access, just to ask last-minute questions like "Do I need to declare snacks at customs?" or "How do I say 'thank you' in Japanese?". We could fine-tune DistilGPT-2 on a small dataset of travel Q&A pairs and common phrases. The resulting model might not have the full knowledge of a big cloud model, but it could cover frequently asked questions. The app on the phone could use this model to answer instantly. For anything it can't handle, it might queue it and let the user know to check when they're online or connect to the cloud if available. This way, the user has a helper anytime, anywhere.

SLMs within AWS IoT/Edge Infrastructure: AWS offers services like *AWS IoT Greengrass* that let you deploy and manage machine learning models at edge devices. For instance, you can use SageMaker to train or optimize a model, then use Greengrass to push that model to devices (say, a kiosk or a robot), which then run inference locally. SLMs are perfect candidates for such deployment because they fit on the device. Greengrass's ML Inference feature can even use SageMaker Neo to compile models for specific hardware for maximum speed. In short, AWS does provide pathways to operationalize SLMs beyond the central cloud—you train in the cloud and deploy to the edge.

To summarize, SLMs extend the reach of dialogue systems to environments where large models can't go, and they do so cost-effectively. They allow real-time interactions in low-power settings and can drastically cut down the cloud infrastructure needed for a chatbot. The penalty is a potential drop in the sophistication of responses, but for many applications, especially those with fairly routine dialogues, SLMs are up to the task. They also serve as a great starting point in development—you might prototype your chatbot with a DistilGPT-2 to test viability and user experience and only move to a larger model if necessary.

Before we conclude this section, Table 9-4 provides a quick comparison of model sizes and their typical deployment scenarios, highlighting where SLMs shine.

Table 9-4. Comparison of Large vs. Small Models for Dialogue—Showing How Model Size Influences Deployment Options

Model	Size (Parameters)	Deployment Scenario	Pros	Cons
GPT-3 (via API)	~175 billion	Cloud-only (OpenAI API or similar; not deployable on one's own easily)	Unparalleled knowledge and fluency; handles open-domain conversation impressively.	High latency (seconds per reply), very high cost per use, no offline use, and outputs need careful filtering (can be too unpredictable).
T5-base fine-tuned (SageMaker)	220 million	AWS cloud (1 GPU instance or similar)	Strong performance on domain-specific dialogues after fine-tuning; relatively fast on GPU; good balance of quality and speed for many cases.	Still requires a dedicated server/endpoint; may be slow on CPU only; not suitable for on-device; some cost to run continuously.
DistilGPT-2 (SLM)	~82 million	Edge or low-cost cloud (CPU instances, AWS Lambda, mobile devices)	Very fast inference, can run without GPUs; much lower memory and compute needs; good for real-time and high concurrency.	Limited depth of knowledge (inherits training cutoff of GPT-2, smaller internal representation); may produce simpler or more repetitive responses; might need fine-tuning to be effective for specific tasks.

(*continued*)

Table 9-4. (*continued*)

Model	Size (Parameters)	Deployment Scenario	Pros	Cons
DistilT5 (SLM)	~50-100 million (varies)	Edge or cloud (after fine-tuning, can be deployed on CPU or smaller instances)	Efficient seq2seq generation; good for tasks like Q&A or brief explanatory answers; can be fine-tuned quickly.	Shorter context window (depending on variant), meaning it might not handle very long conversations well; may struggle with very complex queries compared to larger T5 variants.
Tiny/Ultra-compact models (e.g., MiniLM, ALBERT)	10-30 million	Mobile and microcontrollers (specialized use cases)	Extremely small footprint; can even run on embedded devices or in-browser (with WebGPU, etc.); great for simple classification or keyword-spotting tasks in dialogues.	Too limited for coherent sentence generation beyond a line or two; mainly used for understanding tasks (intent detection) rather than full response generation.

This concludes the exploration of implementing dialogue systems on AWS. We saw how Amazon Lex can be used to build rule-based chatbots with ease, how Amazon SageMaker empowers us to fine-tune and serve powerful language models for generating responses, and how distilled small models provide a path to efficient, real-time deployments. In a real-world project, you might combine all these: Lex for intent handling, a SageMaker LLM for robust answers, and perhaps a distilled model as a fallback or for offline mode. AWS's ecosystem supports this modular design, letting each component handle what it's best at.

The next section covers open-source solutions for dialogue systems. We discuss frameworks like Rasa and Hugging Face Transformers, which allow building conversational agents without relying on cloud-specific services. This open-source route

CHAPTER 9 DIALOGUE SYSTEMS

offers more customization and freedom to host anywhere—a great complement to the managed AWS approach covered here. Let's now dive into how you can implement dialogue systems using these open-source tools and explore the role of SLMs in those contexts.

9.3 Open-Source Implementation

The open-source community provides a rich ecosystem of tools and models for building advanced dialogue systems. These platforms offer flexibility, transparency, and a high degree of customization, making them ideal for a wide range of applications from enterprise-grade virtual assistants to research prototypes. Unlike proprietary systems, open-source solutions allow for complete control over the data, models, and deployment environment, which is crucial for sensitive or domain-specific applications.

9.3.1 Rasa and Hugging Face Transformers

Rasa is a powerful, open-source framework for building contextual AI assistants. It provides a structured approach to building dialogue systems by separating the NLU and dialogue management layers. The NLU component, which you can customize, handles intent recognition and entity extraction, effectively translating the user's freeform text into structured data. The dialogue management component then uses a combination of stories (examples of conversational paths) and rules to decide the next action based on the conversation history and the user's intent. This architecture makes Rasa robust, transparent, and easy to debug. For fine-tuning and integrating with cutting-edge LLMs and SLMs, Rasa can be configured to use models from Hugging Face, allowing developers to leverage state-of-the-art neural networks within a well-defined dialogue framework. For example, using a pretrained model like DistilBERT for intent classification can significantly improve performance over traditional models while keeping the model footprint small.

Here's what the typical output of your custom Hugging Face intent classifier will look like when used in a Rasa pipeline. Example:

User input: "I want to book a flight to New York."

Here is the output from the classifier. The classifier modifies the `Message` object with two keys: `intent` and `intent_ranking`.

```json
{
  "intent": {
    "name": "book_flight",
    "confidence": 0.92},
  "intent_ranking": [
    {"name": "book_flight",
      "confidence": 0.92},
    {"name": "inform_location",
      "confidence": 0.05},
    {"name": "greet",
      "confidence": 0.02},
    {"name": "cancel",
      "confidence": 0.01} ]}
```

Hugging Face's Transformers library is the de facto standard for working with state-of-the-art pretrained language models. It provides a vast collection of models, including GPT-2 and T5, which are excellent for generative dialogue systems. You can use these models to create sophisticated, human-like conversations by fine-tuning them on specific dialogue datasets. This approach offers a higher degree of conversational flexibility compared to rule-based systems, as the model can generate dynamic and creative responses. The integration of Hugging Face models into Rasa allows for a powerful hybrid approach where the system can fall back to a generative model for complex or out-of-scope queries, providing a graceful handoff and a superior user experience.

The following code demonstrates how to create a custom Rasa NLU component that leverages a Hugging Face model for intent classification, a core component of a Rasa-based dialogue system.

Refer to this code on GitHub for a full illustration (hf_nlu.py and inference): https://github.com/anvcse562/ModernNLP-LLMSLM/blob/main/Ch09/ch9.3/9.3.1_Rasa-based%20dialogue%20system.ipynb.

To use the custom Hugging Face intent classifier for inference in Rasa, follow these steps after implementing the previous component code in the referenced Git code (e.g., in hf_nlu.py):

Step 1: Add the component to your project- Py file.

Step 2: Register the component in config.yml.

CHAPTER 9 DIALOGUE SYSTEMS

Make sure the import path (e.g., custom_components.hf_nlu. HuggingFaceIntentClassifier) matches your project structure.

Step 3: Run training so the classifier can train using your NLU data:

rasa train

Step 4: Run the Rasa shell for inference:

rasa shell

Step 5: Programmatic inference.

If you want to use the model outside of Rasa (e.g., for batch inference or testing), you can do something like this. Refer to GitHub for a full code illustration (Cell 6, Step 5):

https://github.com/anvcse562/ModernNLP-LLMSLM/blob/main/Ch09/ch9.3/9.3.1_Rasa-based%20dialogue%20system.ipynb.

```
# Inference function- partial code
def classify_intent(text):
    inputs = tokenizer([text], return_tensors="tf", truncation=True,
    padding=True)
    logits = model(inputs).logits
    probs = tf.nn.softmax(logits, axis=1).numpy()[0]
    predicted_label = tf.argmax(logits, axis=1).numpy()[0]
    return {
        "intent": inv_label_mapping[predicted_label],
        "confidence": float(probs[predicted_label]),
        "ranking": [
            {"intent": inv_label_mapping[i], "confidence": float(conf)}
            for i, conf in sorted(enumerate(probs), key=lambda x: x[1],
            reverse=True)
        ]
    }

# Example
print(classify_intent("I want to book a flight"))
```

9.3.2 Fine-Tuning a Generative SLM for Open-Source

Building on the foundation of open-source frameworks, we can fine-tune SLMs to create lightweight, yet powerful generative dialogue systems. Unlike rule-based systems, these models can generate dynamic responses, making them suitable for a wider range of conversational tasks. One excellent candidate for this is DistilBART, a distilled version of the BART model, which is optimized for sequence-to-sequence tasks like text summarization and generation. The core advantage of this approach lies in its ability to provide a natural, human-like feel to conversations without the prohibitive computational cost of larger models. This makes SLMs ideal for resource-constrained environments like mobile applications or edge devices.

The Hugging Face Transformers and Datasets libraries make fine-tuning a model like DistilBART straightforward. The following code snippet demonstrates how to load a pretrained DistilBART model and fine-tune it on a custom dialogue dataset.

Refer to the code on GitHub at: `https://github.com/anvcse562/ModernNLP-LLMSLM/blob/main/Ch09/ch9.3/9.3.2_fine-tune_DistilBART_dialogue_task.ipynb`.

9.4 Industrial Use Cases

This section delves into real-world applications of dialogue systems, illustrating how the concepts and technologies discussed in this chapter are implemented in practice. By examining these use cases, you can understand how to translate theoretical knowledge into practical, high-impact solutions.

9.4.1 Virtual Customer Assistants

9.4.1.1 Business Motivation

Virtual Customer Assistants (VCAs) serve as the frontline interface between users (customers or internal employees) and an organization's internal knowledge. By combining retrieval over structured and unstructured internal documents with both lightweight and generative language models, VCAs can give immediate, accurate, and context-aware answers to common queries around legal compliance, HR policies, IT support, and more. This reduces the load on human agents, shortens resolution time, and improves user satisfaction across domains like customer support, employee onboarding, and self-service troubleshooting.

9.4.1.2 Implementation Overview

This use case implements a hybrid architecture that meshes SLMs for efficient extractive grounding with LLMs for natural, explanatory answer generation. Here are the core components:

1. **Knowledge ingestion and vector indexing:**

   Internal knowledge (e.g., legal clauses, HR guidelines, IT procedures) is broken into chunks and embedded using a semantic embedding model such as `sentence-transformers/all-MiniLM-L6-v2`. These embeddings are stored in a vector store (Chroma) to enable fast similarity-based retrieval.

2. **Extractive QA with SLM (DistilBERT):**

   For low-latency, precise answers grounded in source text, a DistilBERT model fine-tuned on extractive QA tasks (e.g., SQuAD) is used to pull answer spans directly from retrieved context. This gives users confident "fact-based" snippets for straightforward queries like policy details or procedural steps.

3. **Generative QA with LLMs (Gemma and BitNet):**

   When queries need synthesis, clarification, or more conversational explanations, LLMs such as Google's Gemma (e.g., `google/gemma-7b-it`) and Microsoft's BitNet (`microsoft/bitnet-b1.58-2B-4T`) are used. These models take the retrieved context and generate fluent answers, enabling the assistant to rephrase, summarize, or extend the extractive response. Gemma provides a strong instruction-tuned generative backbone, while BitNet offers efficiency with its native 1-bit architecture.

4. **Retrieval-augmented generation (RAG) pipeline:**

   The system wraps the retrieval layer and the LLMs into Retrieval QA-style chains, so that each generative answer is contextually grounded, reducing hallucination and improving relevance.

5. **Query interface/simulated dialogue:**

   A simple loop simulates a virtual assistant: It first provides the extractive answer, then augments it with generative outputs from Gemma and BitNet. This layering mirrors production VCAs that present a concise fact and optionally expand with natural language elaboration.

9.4.1.3 Typical Business Uses

- **Instant policy lookup** (e.g., "What is our GDPR data access audit frequency?").

- **Automated responses to standard support tickets** (e.g., password reset procedures, leave policies).

- **Employee onboarding Q&A** (e.g., "What are the first steps in onboarding?").

- **Escalation triage** by combining confidence from the SLM with nuanced advice from the LLM.

9.4.1.4 Technical Notes and Best Practices

- **Hybrid SLM/LLM strategy:** Use DistilBERT for high-confidence, high-speed fact retrieval; defer to Gemma/BitNet when user queries require contextualized interpretation or multi-turn clarification.

- **Fallbacks:** If a high-capacity generative model fails to initialize (e.g., due to resource constraints or API issues), lightweight alternatives like `google/flan-t5-small` can act as substitutes to maintain service continuity.

- **Grounding:** Always prepend retrieved documents as context to reduce hallucinations and provide traceability of answers.

9.4.1.5 Deployment and Scalability

- **Local/internal deployment:** Host the vector store and QA logic on an internal service; connect via APIs to chat interfaces or internal portals.

- **Cloud deployment:** Use managed endpoints (e.g., AWS SageMaker for fine-tuned variants, or HuggingFace inference endpoints) and front them with intent recognition layers (e.g., Amazon Lex or Rasa) to process natural-language user input.

- **Containerization:** Dockerize the pipeline for reproducible environments across dev/stage/prod.

- **Extension to multi-turn dialogue:** Maintain a short-term conversation history and inject it into retrieval or prompt context to preserve coherence in back-and-forth exchanges.

- **Monitoring and feedback:** Log question/answer pairs and user feedback to continuously refine retrieval relevance and potentially fine-tune domain-specific models.

Refer to the code on GitHub for a full illustration: https://github.com/anvcse562/ModernNLP-LLMSLM/blob/main/Ch09/ch9.4/9.4.1_virtual_customer_assistant_qa.ipynb.

9.4.2 Interactive Voice Response Systems (IVR) Use Case

An Interactive Voice Response (IVR) system is a common application of dialogue systems, allowing users to interact with a computer-operated phone system via voice or touch-tone inputs. A modern IVR system can be enhanced with advanced dialogue capabilities to provide a more natural and efficient user experience.

9.4.2.1 Use Case: Automated Banking IVR with AWS and Open-Source Integration

A major bank wants to upgrade its traditional IVR system to a modern dialogue system to handle common customer inquiries more effectively and reduce the load on its call centers. The goal is to build a hybrid system that leverages the reliability of a rule-based

CHAPTER 9 DIALOGUE SYSTEMS

system for common tasks and the flexibility of a generative model for more complex, nuanced queries. The system must be able to recognize high-priority, domain-specific intents, and for more general, out-of-scope questions, it should provide a sophisticated, human-like response.

The goal is to create a web application that mimics an IVR system's logic:

1. **Intent recognition:** Recognize a user's intent from their text input.

2. **Rule-based response:** For common intents like "check balance" or "report lost card," provide a direct, rule-based response. This is handled by a system similar to Amazon Lex, which excels at predefined conversational flows.

3. **Generative response:** For more general or complex questions that don't fit a specific intent, use a generative model like those from Hugging Face to provide a flexible and natural response.

4. **Seamless handoff:** The application will demonstrate the seamless handoff between these two approaches, providing a consistent user experience.

The following code provides an HTML-based chat interface that simulates the logic of this hybrid IVR system. The core of the application resides in a JavaScript function, handleQuery, which intelligently routes user input. It first checks for specific keywords to simulate a quick, rule-based intent recognition, much like a configured Amazon Lex model would. If a high-priority intent is detected (e.g., "check balance"), it provides an immediate and precise response. If no specific intent is found, it falls back to a generative approach, leveraging a model like DistilGPT-2 or a fine-tuned T5 model. For this demonstration, we use the Gemini API to provide a dynamic and contextual response, simulating a generative model hosted on a service like SageMaker. This hybrid architecture ensures that common queries are handled with high efficiency and accuracy, while complex or out-of-scope questions are met with a sophisticated, generative response, providing a superior user experience and reducing the need for human agent intervention.

Refer to GitHub for the full code illustration: https://github.com/anvcse562/ModernNLP-LLMSLM/blob/main/Ch09/ch9.4/9.4.2_Interactive%20Voice%20Response%20dialog%20Systems.ipynb.

Figure 9-1 shows an output view of the chat interface.

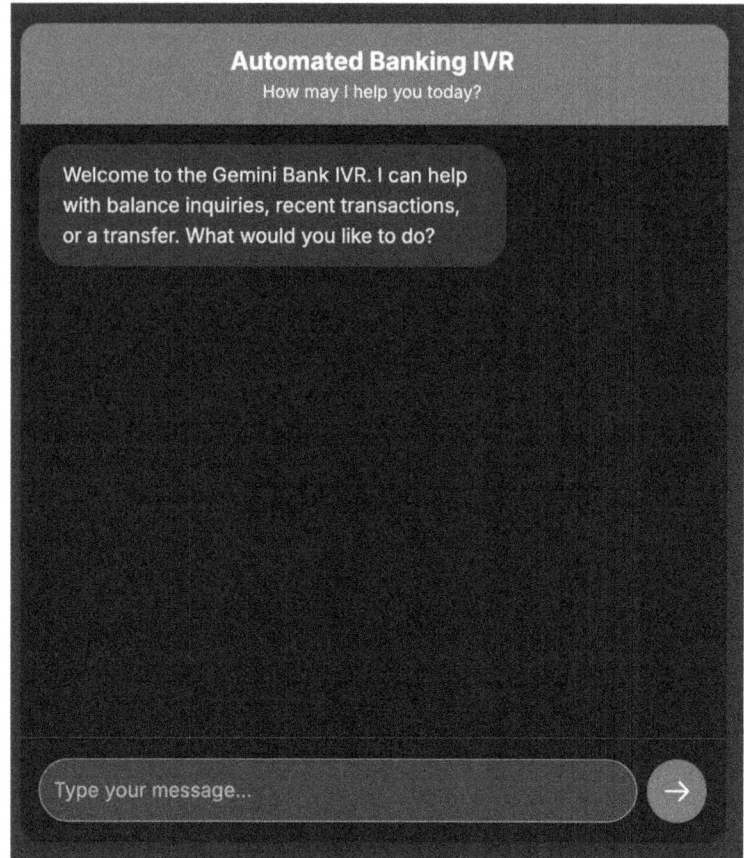

Figure 9-1. *An output view of the chat interface*

9.4.3 Healthcare Virtual Assistants

9.4.3.1 Business Motivation

Healthcare is a domain where accessibility, accuracy, and empathy are critical. As patient interactions increasingly move online, virtual healthcare assistants (HCAs) have emerged as scalable tools to answer medical FAQs, triage symptoms, guide appointment booking, and support chronic care management. These assistants must handle sensitive user inputs with safety, offer reliable information, and work seamlessly across web and mobile environments.

CHAPTER 9 DIALOGUE SYSTEMS

While traditional bots often rely on static FAQ templates, modern healthcare assistants are powered by a mix of SLMs and LLMs that offer better context understanding and conversational fluency. When deployed correctly, these systems reduce call center load, assist doctors with intake, and improve patient engagement without compromising privacy or compliance.

9.4.3.2 Implementation Overview

This use case demonstrates a modular healthcare assistant that integrates:

- Lightweight SLMs for real-time, low-latency responses
- LLMs (optionally) for fallback generative explanations
- Embedding-based intent classification (using MiniLM)
- Prompt-based response generation (using FLAN-T5-small)
- Optional integration with medical guidelines or APIs for grounding

Intent Classification with Sentence Embeddings

To detect user intents (e.g., "book appointment," "ask about symptoms," and "request medication info"), we use semantic similarity between the user's input and a curated list of labeled example sentences per intent. A transformer-based embedding model (e.g., `sentence-transformers/all-MiniLM-L6-v2`) computes sentence vectors, and the closest match determines the intent.

Generative Dialogue with Distilled LLMs

Responses are generated using FLAN-T5-small—a distilled version of T5 fine-tuned for instruction-following tasks. This small model is ideal for mobile or API-based healthcare chatbots where full-sized LLMs are too large or expensive to serve.

Here is an example pipeline (simplified):

1. User enters: "Can I take paracetamol with ibuprofen?"
2. Intent classifier matches it to `medication_advice`.

3. FLAN-T5 is prompted with: "Patient asked: Can I take paracetamol with ibuprofen? Provide a helpful and safe response related to `medication_advice`."

4. Generated response: "It's generally safe to take both together, but consult your doctor, especially if you have liver or kidney issues."

9.4.3.3 Technical Architecture

[Insert Figure: Modular architecture for Healthcare Virtual Assistant]

- **Input layer:** Accepts typed or voice-based patient queries.

- **Intent classifier:** Sentence-BERT model (MiniLM) computes similarity to predefined intent templates.

- **SLM generator:** FLAN-T5-small generates the assistant's reply using the classified intent as prompt context.

- **LLM fallback (optional):** For ambiguous cases or broad informational questions, an LLM endpoint (e.g., T5-Large or open-source Mistral) is queried.

- **Knowledge layer (optional):** Can connect to medical knowledge bases (e.g., WHO guidelines, WebMD API) to ground responses or validate facts.

- **Interface layer:** Chat UI (web or mobile) interacts with users. Can also be deployed via WhatsApp, IVR, and kiosk interfaces.

9.4.3.4 Typical Use Cases

- **Symptom checker:** Provide generic advice for common symptoms (e.g., fever, cough) and when to seek medical help.

- **Medication compatibility:** Respond to queries about drug interactions using safe, templated prompts.

- **Triage and appointment routing:** Classify the nature of the problem and route patient to the appropriate department (e.g., "You should consult a cardiologist").

- **Post-diagnosis support:** Share lifestyle tips, dietary advice, and reminders based on chronic conditions (e.g., diabetes, hypertension).

- **Vaccine FAQ:** Address concerns about eligibility, side effects, and scheduling for vaccines.

9.4.3.5 Technical Notes and Best Practices

- **Model size optimization:** Use DistilGPT-2 or FLAN-T5-small for low-latency inference. Convert to ONNX or quantize (e.g., INT8) for deployment on edge/mobile.

- **Safety first:** Responses should include disclaimers and avoid providing diagnoses. Add templates like: "This is not a substitute for medical advice. Please consult a physician."

- **Prompt engineering:** Prompt should always include intent and task scope. Example: "Respond to the user question using health guidelines. Do not offer a diagnosis."

- **Privacy and HIPAA compliance:** Avoid logging personal identifiers. Ensure data encryption, local inference (if required), or use of anonymized prompts.

- **Fallback and escalation:** If confidence score is low or intent is unclear, escalate to a human agent or display: "Let me connect you to a health expert."

9.4.3.6 Deployment and Scalability

- **Serverless (Lambda and SLM):** Deploy FLAN-T5-small behind AWS Lambda using container packaging (~400MB zipped image). Ideal for infrequent queries.

- **SageMaker endpoint:** Fine-tune FLAN-T5 or DistilGPT-2 and host as a real-time endpoint. Enables cost control via auto-scaling.

- **Mobile deployment:** Use PyTorch Mobile or TensorFlow Lite for running SLMs on-device in offline mode. Useful for remote clinics or emergency use without the Internet.

- **API-driven architecture:** The entire stack (intent classifier, generator, and fallback) can be exposed as REST APIs using FastAPI or AWS API Gateway.

- **Monitoring:** Log intent accuracy and response length and add feedback prompts ("Was this helpful?") for iterative improvements.

9.4.3.7 Code Demonstration

A full code implementation, including classification, generation, and visualization, is provided here: https://github.com/anvcse562/ModernNLP-LLMSLM/blob/main/Ch09/ch9.4/9.4.3_Healthcare_Virtual_Assistant.ipynb

It showcases:

- Classifying intents like book_appointment, symptom_check, and vaccine_info
- Using FLAN-T5-small to respond to each classified query
- Visualizing intent distribution across test inputs
- Simulated real-time console chatbot interface

9.5 Summary

This use case highlights how dialogue systems in healthcare can be lightweight, secure, and helpful using distilled models like DistilGPT-2 and FLAN-T5-small. With proper intent classification and templated prompts, SLMs can handle 80–90% of patient interactions, reducing workload on human staff while improving patient experience.

The assistant remains safe by deferring medical decisions, using disclaimers, and offering guidance rather than treatment. When needed, it escalates to LLMs or live professionals. Combined with scalable infrastructure and mobile deployment, such systems represent the future of AI-assisted preventive care and triage.

The next chapter explores another core application area of NLP—text correction and language modeling, where models are trained to detect and fix grammatical or semantic errors in freeform text.

CHAPTER 10

Text Correction and Language Modeling

This chapter explores the development of text correction systems using LLMs such as GPT-2, T5, and BART, and SLMs like DistilGPT-2, DistilT5, and TinyBERT. We focus on grammar and spelling correction, which are essential for clear communication in professional, academic, and casual writing contexts.

We cover scalable implementations using AWS services like Amazon SageMaker for training and deploying grammar-correction models. Open-source tools such as Hugging Face Transformers are discussed in depth, including fine-tuning on custom datasets and evaluating performance using standard metrics like BLEU, ROUGE, and edit distance. SLMs are highlighted for their ability to deliver fast, lightweight corrections and real-time applications such as mobile writing assistants and browser extensions.

By the end of this chapter, you'll be equipped to build and deploy intelligent, efficient text correction systems using both cloud infrastructure and open-source tools.

This chapter covers:

- **Overview of text correction:**
  - Importance of grammar and spelling correction in written communication.
  - Applications in email clients, writing platforms, and browsers.
  - Role of LLMs and SLMs in delivering real-time, high-quality corrections.

- **Implementation on AWS:**
  - **Amazon SageMaker:** Fine-tuning LLMs like T5 and BART for grammar correction tasks.

- **SLM Deployment:** Training and deploying lightweight models using SageMaker for faster, cost-efficient solutions.
- **Open-source implementation:**
  - **Hugging Face Transformers:** Fine-tuning models like GPT-2, T5, and BART on grammar correction datasets.
  - **Evaluation and feedback:** Using BLEU, ROUGE, and edit distance to assess performance; integrating user feedback loops for continual improvement.
  - **SLMs in Open-Source:** Using models like DistilGPT-2 and TinyT5 for fast, real-time corrections on local devices.
- **Industry use cases:**
  - Writing assistance tools
  - Proofreading and editing platforms
  - Educational tools for grammar and language learning

10.1 Introduction to Text Correction

This section covers a fundamental application of NLP—text correction. Just as LLMs can generate coherent dialogue, they can also help polish our writing by detecting and fixing mistakes. This section discusses the importance of grammar and spelling correction, surveys common applications, and compares traditional approaches with modern AI-driven methods. This includes but isn't limited to the role of SLMs and LLMs. This sets the stage for building some grammar and spelling correction models using both AWS and open-source tools in the upcoming sections.

10.1.1 The Importance of Grammar and Spelling Correction

Clear and correct written communication is essential for conveying ideas with precision and professionalism. Grammatical errors or spelling mistakes can distract readers and even change the intended meaning of a sentence. For example, a missing comma in "Let's eat Grandma" versus "Let's eat, Grandma" humorously illustrates how punctuation can be a matter of life and death! In professional settings, errors in an email or report can make the content look unpolished or cause misunderstandings. In fact, spelling or grammatical errors can make a proposal look unprofessional—something we all want to avoid. Good grammar ensures clarity, while correct spelling avoids confusion between similar words (e.g., "their" vs. "there"). Consistently applying basic but important punctuation is critical for credibility in business, academia, and everyday communication.

Beyond professionalism, proper grammar and spelling have broader impacts. They significantly affect readability and comprehension. For students and non-native English speakers, feedback on writing mistakes is an important learning tool. In academic papers and publications, rigorous grammar and spell checking is mandatory to meet standards. Even in casual writing like social media posts or text messages, clear writing prevents miscommunication. In short, robust text correction is critical for clarity in both casual and formal writing, ensuring that the writer's message is understood as intended.

Given this importance, it's no surprise that grammar and spell checking functionality is now ubiquitous. We find these features everywhere from word processors to web browsers, quietly improving our writing in the background. The next section examine some common applications where text correction plays a vital role.

10.1.2 Common Applications of Text Correction

Because nearly everyone writes digital text in some form, grammar and spell checkers have become standard in many applications. Here are some of the most common places we encounter text correction tools in daily life:

- **Word processors (e.g., MS Word, Google Docs):** Modern word processors have built-in spell check and grammar suggestions that highlight unknown or misspelled words. Microsoft Word, for instance, marks spelling errors and grammatical issues in the file. Google Docs similarly provides real-time suggestions; notably, Google reported that over 100 million grammar suggestions are flagged each week by its machine-learning based checker, illustrating how frequently users benefit from these features.

- **Email clients (e.g., Gmail, Outlook):** Email platforms integrate text correction to help users draft error-free messages. Gmail's smart compose and grammar suggestions, which are powered by a neural model, can propose fixes as you type. Outlook and other Microsoft 365 apps use Microsoft Editor, an AI-powered proofing tool, to catch mistakes in emails and documents. These corrections are crucial in professional correspondence where a typo or grammar slip could alter the tone or meaning of a message.

- **Web browsers:** Web browsers often include basic spell-checkers for any text input field. For example, Chrome and Firefox will underline misspelled words using a built-in dictionary. On top of this, many users install browser extensions like Grammarly and LanguageTool. These extensions provide advanced grammar and style checking on web forms, social media, blogs, and more. Grammarly, one of the most popular writing assistants, serves a massive user base (helping to improve communication for "the world's two billion English writers" according to its CEO).

- **Messaging apps and smartphones:** While not as elaborate as word processors, many messaging platforms and mobile keyboards include autocorrect and some grammar correction. Smartphone keyboard apps will fix common typos automatically and may suggest grammar improvements. For example, capitalizing "i" to "I" or adding an apostrophe in "dont" to "don't". These features improve the clarity of text and chat messages on the fly, making casual communication smoother.

- **Writing assistance tools:** Dedicated proofreading and writing enhancement software also rely on text correction. We've mentioned Grammarly, which not only checks spelling and grammar spelling/grammar but also offers style and tone suggestions. Other tools, like Ginger, ProWritingAid, and the open-source LanguageTool, provide similar services. They are used by writers, students, and professionals to self-edit documents. Such tools often combine multiple techniques, including rule-based and AI approaches, to catch as many errors as possible and even explain the corrections.

In all these applications, the goal is the same: to help the user produce clear, error-free text. The prevalence of these tools underscores how common writing mistakes are and how much we rely on automated assistance. The next section explores how these text correction systems actually work, from the earlier rule-based methods to the latest AI approaches.

10.1.3 Traditional Approaches to Text Correction (Rules and Heuristics)

Before the advent of powerful AI models, text correction was tackled with more manual or rule-based methods. Early spelling and grammar checkers followed predefined rules, lists, and heuristics to identify errors:

- **Rule-based grammar checking:** Grammar checkers dating back to the 1970s used simple pattern matching and hand-crafted grammatical rules. Linguists and developers encoded rules such as: "If a singular subject is followed by a non-singular verb form, flag an error." For example, a rule might detect "She *go* to school" and suggest "She *goes* to school," or flag a sentence starting with a lowercase letter as a potential mistake. Microsoft Word's grammar checker in the 1990s was a major breakthrough that employed a full NLP pipeline to parse sentences and apply a large set of grammar rules. Such systems can handle basic syntax errors and certain style issues. They are fast and work offline, since they're essentially running a checklist of conditions on the text.

- **Dictionary-based spell checking:** Traditional spell checkers maintain a dictionary of correctly spelled words. If a word is not found in the dictionary, it's flagged as a potential misspelling. To suggest corrections, these tools generate likely candidate words, for instance, by considering words with similar prefixes or a small edit distance from the misspelled word and rank them. A common technique is to use the *Levenshtein edit distance* algorithm to find the closest dictionary words by a few character changes. For example, if a user types "enviroment," the spell checker finds that changing one letter (inserting "n") yields "environment," which is in the dictionary and is likely what the user intended. This approach works well for obvious typos or slight misspellings (such as "recieve" to "receive"). It was and still is the backbone of most spelling correction in editors and browsers.

- **Heuristics for common errors:** In addition to strict rules, older systems include special-case heuristics for common mistakes. For instance, many spell checkers have a list of easily confused words (there/their/they're) or common typos (like "teh" for "the"). Grammar checkers might look for specific constructions such as double negatives or a missing question mark at the end of a question. These heuristics are essentially mini-rules added after observing frequent user errors.

While rule-based and dictionary-based methods laid the foundation, they have notable limitations. They lack a deep understanding of context or meaning. A rule can tell if a verb doesn't match the subject in number, but it might miss subtler issues. For example, consider the sentence: "The soup is too hot to eat my friend." A human (or an advanced AI) can see a humorous ambiguity (it sounds like the soup is going to eat your friend), but a basic grammar checker might deem the grammar technically correct and not flag it. Similarly, a spell checker will not flag real words used incorrectly—if you write "I will defiantly attend the meeting," "defiantly" is a valid word when you probably meant "definitely." Traditional systems often miss such context-dependent errors. In contrast, they might sometimes over-flag perfectly fine sentences if they don't conform to a rigid rule, leading to false alarms. Maintaining a huge list of rules for every nuance of English (let alone other languages) is very labor-intensive and error-prone.

Despite these drawbacks, traditional methods are lightweight and efficient. They don't require heavy computation—just string comparisons, regex checks, and simple algorithms—making them easy to run on personal computers even in the early days of word processors. In fact, for catching straightforward issues like typos or very basic grammar errors, these methods are quite reliable. Many modern systems still include a rule-based component (for known frequent errors) as a first pass. However, to handle the complex, nuanced errors and varied writing styles of real-world text, we need more intelligent, data-driven approaches. This is where machine learning and modern language models come into play.

10.1.4 AI-Powered Approaches (Statistical, Neural, and LLM-based)

The introduction of statistical and neural network methods revolutionized text correction by allowing models to *learn* error patterns from data rather than relying solely on handwritten rules. Broadly, AI-powered text correction can be seen as a *sequence-to-sequence problem:* transform a "bad" sentence into a "good" sentence. Early research treated grammar correction like a translation task—essentially translating from "incorrect English" to "correct English." This framing makes it natural to apply methods from machine translation to grammar/spell correction:

- **Statistical machine translation (SMT) approaches:** In the 2010s, some grammar correction systems used SMT algorithms. They required parallel corpora of sentences: one version with errors and one corrected version. From this data, the system learned probabilities of transforming phrases. However, obtaining large parallel datasets of "bad" and "good" sentences was difficult. Researchers created them by extracting sentences and their edits, for example, from Wikipedia revision histories or learner essays. SMT-based grammar correction had some success but was soon outpaced by neural models.

- **Neural sequence-to-sequence models:** The real leap came with neural networks, especially encoder-decoder (Seq2Seq) models like those used in translation. A model like T5 or BART can take an input sentence with errors (the source sequence) and generate a corrected sentence (the target sequence). For instance, you could feed in: He are moving here." and train the model to output "He *is* moving here.". Over a large training set of varied errors, the model internalizes grammar rules and spelling conventions without them being explicitly programmed. Google's grammar suggestion system for Google Docs uses such a neural machine translation approach under the hood—each suggestion is treated as translating an incorrect sentence to a correct sentence. One challenge is that, unlike translation between languages, we don't naturally have millions of "incorrect to correct" sentence pairs. To compensate, techniques like *data augmentation* are used: Google researchers generated synthetic errors by taking correct sentences and introducing errors automatically. They also mined millions of sentence corrections from Wikipedia edits. With enough training data, neural grammatical error correction (GEC) models began to achieve impressive results, often measured on benchmarks like the CoNLL-2014 shared task or the JFLEG corpus. In fact, by 2020 Microsoft's neural grammar checker achieved state-of-the-art results on CoNLL-2014 and JFLEG with a novel model, marking a major breakthrough in accuracy.

- **LLMs for correction:** With transformer-based LLMs (like GPT-3, GPT-4, and large T5 variants) demonstrating extraordinary language understanding, they can now be applied to text correction in two ways: fine-tuning or prompting. Fine-tuning means taking a pretrained LLM and training it further on a grammar/spell correction dataset, much like the seq2seq approach, but starting from a powerful base. This usually yields the most accurate results, as the model adapts specifically to the task. On the other hand, some LLMs are so capable that they can perform grammar correction *without additional training*, by means of prompting. For example, an instruction-tuned model like Flan-T5 or GPT-4 can often correct a sentence if prompted appropriately. A researcher noted that simply by prefixing "Correct grammar in the following sentence:" to an input, Flan-T5-large produced a corrected sentence on the fly. Likewise, with ChatGPT or GPT-3.5, you can literally ask, "Please correct the following text..." and it will do so, leveraging its vast training on English text and ingrained grammar knowledge. Studies have found these LLM-based corrections to be very strong in maintaining context and meaning—GPT-3.5 was shown to outperform traditional tools in fixing contextual errors and was well-received by users in comparative evaluations. The ability of LLMs to understand the *intent* of a sentence helps them avoid the pitfalls of simpler systems, such as misinterpreting what the writer meant.

- **SLMs and distilled models:** A downside of large models is their size and computational requirements. Running GPT-3 or T5-11B for real-time correction may be impractical in many cases due to cost or latency. This is where SLMs come into play. SLMs—either trained from scratch to be lightweight, or distilled versions of larger models—offer a tradeoff between performance and efficiency. For instance, DistilGPT-2 and DistilT5 are examples of models that have far fewer parameters than their originals but can still be fine-tuned for text correction tasks. These smaller models can be deployed on consumer devices or low-cost servers, enabling features like mobile keyboard autocorrect or offline grammar checking. A recent innovation by Microsoft Research introduced EdgeFormer, an on-device seq2seq model optimized to run efficiently on user devices. EdgeFormer powers a client-side grammar checker in Microsoft Editor, allowing users to get corrections even without access to the Internet, with the benefit of increased privacy and zero server cost. While SLMs may not catch every subtle nuance that an LLM like GPT-4 would, they are ideal for real-time, resource-constrained environments such as smartphone apps. Developers can fine-tune such models on grammar/spelling data and deploy them in embedded systems or as part of edge applications, as discussed for SLM use cases in the previous chapter.

To illustrate how you might use a pretrained model for grammar correction, the GitHub code shows an example using a fine-tuned T5 model (a sequence-to-sequence transformer) in Python. You can find the code on GitHub: https://github.com/anvcse562/ModernNLP-LLMSLM/blob/main/Ch10/ch10.1/10_1_4_grammar_correction.py.

In this snippet, we load a fine-tuned T5-base model for grammar correction. The correct_grammar function passes an input sentence to the model and decodes the model's suggested correction. For the sentence "He are moving here," the model would output "He is moving here."—fixing the grammatical number agreement (*are* to *is*). Behind the scenes, the model learned to make such corrections by training on a large corpus of wrong/right sentence pairs. You can imagine extending this approach: the same model would correct spelling errors ("recieve" to "receive") or more complex grammatical errors, including adding missing articles, fixing tense consistency, and more.

Finally, it's worth noting that AI-driven correction isn't limited to just fixing surface errors. Advanced systems can also explain the corrections (e.g., "Replace 'affect' with 'effect'—*effect* is the noun form you need here"), which is immensely helpful for users to learn from mistakes. Some research even involves using LLMs to *generate explanations* for grammar fixes. This moves these tools from mere fixers to educational assistants, bridging into the territory of intelligent tutoring systems.

In summary, modern text correction leverages the best of both worlds: the *knowledge and flexibility* of large-scale language models, and the *speed and efficiency* of optimized smaller models or rule-based checks for simpler tasks. By combining these, we can create robust systems that correct a wide range of errors with high accuracy and in real time. The next section explains how to implement such text correction models on AWS—fine-tuning LLMs or SLMs using SageMaker to deploy our own grammar and spelling correction service.

10.2 Implementation on AWS: Building Text Correction Models with SageMaker

After exploring the fundamentals of text correction and how LLMs and SLMs can detect and fix errors, this section turns to a practical implementation. This section leverages Amazon Web Services (AWS)—specifically the managed machine learning platform Amazon SageMaker—to build, fine-tune, and deploy our grammar and spelling correction models. Cloud platforms like AWS provide the heavy lifting for training large models on powerful hardware and serve them at scale to end users. We cover how to prepare a grammar correction dataset, fine-tune LLMs as well as SLMs using SageMaker, and deploy the models for real-time use. Throughout, we discuss best practices like using the right instances or distributed training for LLMs and even touch on how LLM-based agents can be integrated for intelligent decision-making in the pipeline. By the end of this section, you'll see how AWS can take a text correction system from an idea to a production-ready service.

10.2.1 Why Use AWS SageMaker for Text Correction?

Amazon SageMaker is a fully-managed service that covers the entire machine learning workflow—from data preparation to model training to deployment. For text correction task, SageMaker offers several clear advantages:

CHAPTER 10　TEXT CORRECTION AND LANGUAGE MODELING

- **Scalable infrastructure:** SageMaker makes it easy to train on powerful GPU instances like NVIDIA V100 or A100 GPUs or even clusters of instances if needed, without having to manage the underlying servers. If the grammar correction model (say a T5 or GPT variant) is large, we can simply choose a bigger instance or multiple instances. SageMaker supports distributed training, including data parallelism and model parallelism, to handle models that don't fit on one GPU.[1] This means even billion-parameter LLMs can be fine-tuned by partitioning the work across devices, a difficult feat to manage on-premise.

- **Integrated Hugging Face support:** AWS has pre-integrated support for Hugging Face Transformers, which means we can bring in state-of-the-art text models (like BART, T5, GPT-2, etc.) and fine-tune them with minimal setup. SageMaker provides ready-made Docker containers with Transformers, so our code can focus on modeling, not on the environment. This drastically simplifies fine-tuning LLMs on custom data—companies can fine-tune models without building everything from scratch.

- **Secure and managed training environment:** With SageMaker, the data (for example, a corpus of incorrect and corrected sentences) can reside in Amazon S3 and be loaded in a secure environment for training. SageMaker handles the IAM roles, encryption, and networking needed to keep the data and model secure. For enterprises concerned about data privacy like sensitive documents for grammar correction, SageMaker can train within a private VPC with no Internet access, ensuring compliance.

- **MLOps and deployment pipelines:** Beyond training, SageMaker has tools to monitor experiments, perform hyperparameter tuning, and easily deploy models as scalable APIs. Once we train a grammar correction model, SageMaker can host it on an HTTPS endpoint with auto-scaling, so users anywhere can send text and receive

---

[1] https://docs.aws.amazon.com/sagemaker/latest/dg/model-parallel-intro.html#:~:text=Introduction%20to%20Model%20Parallelism%20,devices%2C%20within%20or%20across%20instances

corrections. There's also support for monitoring model performance in production through logs and metrics and updating the model with zero downtime—critical for a continuously improving application.

In short, SageMaker provides a one-stop solution to transition from a notebook prototype of a grammar fixer to a robust cloud service. Many organizations choose SageMaker to fine-tune and deploy their domain-specific LLMs because it handles the "heavy lifting" of infrastructure, offers optimized hardware, and ties into AWS' ecosystem for reliability. The next sections walk through using SageMaker step-by-step: preparing the dataset, fine-tuning a large and small model, and deploying the result.

10.2.2 Data Preparation: Grammar Correction Dataset on S3

The first step in building a SageMaker training job is preparing the dataset of "incorrect" and "correct" sentence pairs. A quality dataset is the foundation for a good grammar correction model. In practice, assembling this data can be one of the hardest parts, but AWS doesn't prescribe how you get it—you can use any text data you have, then upload it to the cloud.

As far as the sources of the training data, if you're lucky enough to have a proprietary corpus (for example, a company might have a collection of sentences with edits from human reviewers), that data can be used directly. Otherwise, there are a few common approaches to get a large parallel corpus of mistakes and corrections:

- **Publicly available GEC corpora:** The NLP community has compiled datasets from language learners and other sources. For instance, the Cambridge Learner Corpus (FCE) and the NUS Corpus of Learner English (NUCLE) contain essays by English learners annotated with corrections. The BEA-2019 shared task released a large, combined dataset of corrected texts. Although these are not hosted by AWS, we can download them and then place them on S3 for training. Another source is Wikipedia edit histories—researchers have mined millions of sentence revisions from Wikipedia where an editor fixed a grammatical or spelling error. One such resource, the WikEd Error Corpus, consists of tens of millions of sentences of revision histories from Wikipedia articles,[2] effectively a massive collection of human-made corrections. This can serve as a rich pretraining dataset for grammar correction.

- **Synthetic data generation:** An effective strategy is to generate your own training pairs by starting with correct text and introducing errors. For example, you could take a well-formed sentence like *"I received the package yesterday."* and randomly distort it to *"I recieved the package yesterday."* or *"I have receive the package yesterday."* By applying a repertoire of such mistakes to a large corpus of correct sentences, you create a synthetic "incorrect → correct" dataset. This can be done with a simple data processing script. The benefit is that you can generate virtually unlimited training data. In fact, modern systems often use a blend of real errors and synthetic errors—for instance, Microsoft researchers used GPT-3.5 to generate high-quality training sentences with errors, which significantly boosted their model's performance.[3] On AWS, you could use a SageMaker processing job or an AWS Lambda function to automate this dataset generation at scale. For example, take the case of a Lambda that takes a batch of correct sentences and returns them with random errors.

- **Task-specific data augmentation:** Depending on your application, you might include domain-specific text. If you're building a grammar checker for formal writing, you'd include a lot of formal text in your corpus—such as news articles and academy text—and ensure that common formal writing errors are covered. If it's for casual chat or social media, you'd include more conversational snippets and maybe focus more on spelling and casual grammar mistakes. In any case, ensure your "correct" sentences truly represent the style you want the output to be in. You might also leverage user feedback data—for example, initial versions of your model might be deployed in a

---

[2] https://www.researchgate.net/figure/most-frequent-edits-in-the-WikEd-09-corpus_tbl1_266079736

[3] https://www.microsoft.com/en-us/research/blog/achieving-zero-cogs-with-microsoft-editor-neural-grammar-checker/#:~:text=which%20allows%20us%20to%20ship,to%20more%20customers

limited setting and any manual corrections done by users can be fed back into the training data. This idea is later in a section on user feedback.

Once you have your dataset ready (let's say it's a `train.csv` or `train.jsonl` file containing sentence pairs), the next step is to upload it to Amazon S3, which is AWS' object storage service. SageMaker will read the training data from S3 during the job. For example, using the AWS CLI or `boto3`, you can run:

bash

```
aws s3 cp train.csv s3://your-bucket-name/text-correction-data/train.csv
```

It's often useful to also upload a validation dataset, say `val.csv`, containing pairs of incorrect/correct sentences that the model didn't see during training. We use this to evaluate how well the model is generalizing. The CSV might be structured with two columns ("bad sentence" and "good sentence"), or you might choose JSON Lines format with entries like `{"input": "He are a doctor.", "output": "He is a doctor."}`. The exact format can be based on how your training script is written (we discuss that next).

Since our task is essentially a translation from "incorrect English" to "correct English", many implementations simply treat this like a translation task. A common format if using Hugging Face Transformers is to concatenate input and output with a special marker in text. For example, you could preprocess each pair into a text like so:

swift

```
"Incorrect: He are moving here. \n Corrected: He is moving here."
```

and train the model to generate everything after the `Corrected:` prompt. However, a more standard approach is to feed the model the input sentence as a sequence and expect it to output the corrected sentence as a sequence (this is the Text2Text framework that T5 and BART use naturally). We don't need to add special tokens beyond perhaps a task prefix. In some implementations, adding a prompt like `"fix grammar: <sentence>"` helps the model know what to do. This idea is explored in the section on user feedback. For example, if you're using an instruction-tuned model, you might actually provide input as: "Correct the following: He are moving here." and train it to output "He is moving here." This aligns with how you'd prompt an LLM agent. We keep things simple and assume our model knows it's making corrections based on either a prefix or the training context.

With data on S3 and formatted appropriately, we are ready to configure a SageMaker training job. Next, we discuss fine-tuning the model itself—starting with an LLM to maximize accuracy, and then considering a smaller model for efficiency.

10.2.3 Fine-Tuning an LLM on SageMaker

First, you need to choose a base model. For grammar and spelling correction, popular choices of base model include encoder-decoder transformers that have been pretrained on text. Models like T5 and BART are suitable since they were designed for text generation tasks and have been used in previous research for GEC. For instance, T5-base with 220 million parameters or BART-large with 400 million parameters are reasonable starting points that can capture a lot of language nuances. If you want to go really big, T5-3B or even 11B might push accuracy further—but those demand significantly more compute hours. To keep this example grounded, we pick `t5-base` as our LLM to fine-tune the grammar.

10.2.3.1 Setting Up the Training Script

In SageMaker, we typically need a training script (let's call it `train.py`) that contains the code to load data, load the model, and execute training (using an optimizer, loss, etc.). If we are using the Hugging Face Transformers library, we can rely on the `Trainer` API to do a lot for us. Our script might do the following:

- Parse input arguments (SageMaker will allow passing hyperparameters like `epochs`, `batch_size`, etc. to the script).

- Use the `Datasets` library to load our CSV from the `SM_CHANNEL_TRAIN` directory (an environment variable that SageMaker sets to point to the S3 data we specified for training).[4]

- Load the pretrained T5 tokenizer and model (from `t5-base` checkpoint on the Hugging Face Hub, which the script will download).

---

[4] https://huggingface.co/docs/sagemaker/en/train#:~:text=

- Set up training arguments (epochs, learning rate, output directory, etc.) and instantiate a `Trainer` with our model, dataset, and possibly evaluation metrics.

- Call `trainer.train()` to fine-tune the model on our data, and then `trainer.save_model()` to save the model to the `SM_MODEL_DIR` (which SageMaker will upload back to S3 when training is done).

Conceptually, this is similar to how we fine-tuned models using Hugging Face, except now SageMaker is orchestrating it. The big advantage is that SageMaker will spin up, say, a GPU instance (e.g., `ml.p3.2xlarge` which has one NVIDIA V100 GPU) for the job, handle all the setup, run our script there, and then terminate the instance. We only pay for the training time we use, and we don't have to manually manage any servers.

10.2.3.2 Launching the SageMaker Training Job

We can use the SageMaker Python SDK from our notebook or local environment to kick off the job. Here's a simplified snippet using the Hugging Face estimator to fine-tune our model (for illustration purposes):

python

```python
from sagemaker.huggingface import HuggingFace

# Define hyperparameters for training
hyperparameters = {
    'model_name_or_path': 't5-base',  # which pre-trained model to
                                      fine-tune
    'epochs': 3,
    'per_device_train_batch_size': 32,
    'learning_rate': 5e-5
}

# Create the SageMaker estimator
huggingface_estimator = HuggingFace(
    entry_point='train.py',           # our training script
    source_dir='src',                 # directory of the script (and possibly
                                      data processing code)
    instance_type='ml.p3.2xlarge',    # GPU instance for training (V100, 16GB
                                      GPU memory)
```

```
    instance_count=1,                    # single-machine training
    role=aws_role,                       # IAM role with SageMaker permissions
    transformers_version='4.26',         # using a recent Transformers version
    pytorch_version='1.13',
    py_version='py39',
    hyperparameters=hyperparameters
)

# Launch the training job
huggingface_estimator.fit({'train': 's3://your-bucket/text-correction-data/train.csv', 'validation': 's3://your-bucket/text-correction-data/val.csv'})
```

This code can be run in a SageMaker Notebook or any environment with AWS credentials. It configures an environment with our `train.py` script, points it to the training and validation data on S3, and specifies the resources (one `ml.p3.2xlarge` instance). When `.fit()` is called, SageMaker will start the job. Under the hood, it pulls an appropriate Docker image (one that has Hugging Face Transformers 4.26 and PyTorch 1.13 in this case), downloads our script and the data into that environment, then executes `train.py`. The hyperparameters we provided are passed to the script (e.g., as `--model_name_or_path t5-base --epochs 3 ...`).

During training, we would see logs of the loss and, if we set up evaluation, maybe metrics like BLEU or accuracy on the validation set. After the three epochs (or however long we set it), SageMaker will automatically save the model artifacts. By default, anything written to the `/opt/ml/model` directory in the container is zipped up to `model.tar.gz` and stored in S3.[5] This `model.tar.gz` contains our fine-tuned T5 model weights, ready for deployment.

10.2.3.3 Handling Larger Models and Distributed Training

The previous example uses a single GPU instance, which is sufficient for a model the size of T5-base and a reasonably sized dataset. If we wanted to fine-tune a larger LLM, we might want to use multiple GPUs or even multiple machines. SageMaker makes this feasible by simply changing `instance_count` to more than 1 and specifying a `distribution` strategy. For example, we could choose data parallelism or model

---

[5] https://huggingface.co/docs/sagemaker/en/train#:~:text=

parallelism. AWS provides the SageMaker Model Parallel Library, which integrates with Transformers; it can automatically partition a large model across GPUs in a cluster.[6]

For instance, to train a 3B parameter model, we might use `instance_type='ml.p4d.24xlarge'` and enable model parallelism so each GPU holds a chunk of the network. This way, even huge models can be fine-tuned. The tradeoff is complexity and cost: multi-GPU training needs coordination (SageMaker handles a lot of it, but debugging can be trickier) and obviously more compute hours. In many cases for grammar correction, a single GPU with a moderately large model is sufficient to achieve good performance. Fine-tuning is usually not as lengthy as training from scratch—we might fine-tune for a few epochs, which could be on the order of an hour or two for a couple million sentences on one GPU.

While the job runs, SageMaker streams logs so you can watch training progress. It's good practice to early-stop if you see divergence or adjust learning rate if needed. You can also use SageMaker's hyperparameter tuning jobs to try different hyperparameters automatically and pick the best model. This is optional but SageMaker makes it relatively easy to perform such experiments in parallel on the cloud.

As an example, suppose after training, we evaluate the model on a sentence from our validation set: "He aremoving here." This has grammatical errors and should rather read "He is moving here". Our fine-tuned T5 model should output "He is moving here." on that input. We would consider that a successful correction—an exact match with the reference. For a slightly harder test, consider "We enjoyed the music, it was a great concert"—this is a comma splice, that is, two independent classes joined by just a comma. A strong model might output "We enjoyed the music. It was a great concert." (fixing it by splitting into two sentences or using a semicolon). If it does this, it's not just about fixing obvious grammar; it's actually making a stylistic improvement that requires an understanding of sentence boundaries. During training, if our dataset had such errors and corrections, the model could learn to handle them. We could measure quality by using BLEU or ROUGE between model output and reference corrections across many sentences. Typically, a BLEU score in the 50-60+ range on a dataset like CoNLL-2014 indicates a very good GEC model.

---

[6] https://huggingface.co/docs/sagemaker/en/train#:~:text=

In summary, fine-tuning an LLM on SageMaker is straightforward and powerful: we get the benefits of a large model's accuracy, and SageMaker handles the grunt work of using high-end hardware to do it. Of course, using a large model comes with the cost of inference speed and usage cost. This is where considering a smaller model like an SLM can be useful, especially if we need real-time corrections on a tight budget or in resource-constrained environments. Next, we explore fine-tuning an SLM and compare the two approaches.

10.2.4 Fine-Tuning a SLM and the Tradeoffs

Not every application needs a huge model, and often there are latency or cost constraints that make an SLM attractive for grammar correction. SLMs are models with far fewer parameters—typically in the range of 50 million to a few hundred million. Examples include DistilGPT-2 or T5-small, which are "distilled" or smaller versions of larger models. These models can be trained and run much more efficiently, albeit usually with some drop in absolute performance on complex tasks.

The process is the same. The good news is that from SageMaker's perspective, training an SLM is no different than training an LLM. We use the same dataset and the same `train.py` logic. We simply change the model checkpoint. For instance, if we set `model_name_or_path` to `"t5-small"` and perhaps tweak hyperparameters, we can run the same Hugging Face estimator call as before. This is because a smaller model might benefit from a slightly lower learning rate or more epochs since it has less capacity. Ultimately, the job will likely finish faster because each epoch processes data more quickly because the model itself does less computation per token. Also, an SLM uses much less memory, so we could even run it on a smaller, cheaper instance—possibly an `ml.g4dn.xlarge`, which has a modest NVIDIA T4 GPU or even on CPU if we enable CPU training. Although GPU is still recommended for speed.

As an anecdotal example, fine-tuning T5-small on a million sentence pairs might only take around 30-40 minutes on a single GPU, whereas T5-base might take a couple of hours, and T5-large might take four to six hours on the same hardware (these numbers can vary, but this should give you a sense). The faster training means lower cost and quicker iteration. If you realize you need to adjust hyperperameters, it's less painful to retry with an SLM. This agility is great during development.

The key question is how well does the smaller model perform? Generally, larger models have an edge in capturing subtle grammatical nuances or less common error patterns. A small model might change "She are a doctor." to "She is a doctor." easily because that's a very common pattern it can learn, but it could struggle with something like "Hardly he had arrived when it started to rain." which is a more subtle error (should be "Hardly had he arrived…"—an inversion rule). An LLM with its vast capacity might catch that, whereas an SLM might miss it or even produce an ungrammatical correction. However, with sufficient training data, SLMs can still achieve strong results on frequent error types. They tend to make more mistakes on long or complicated sentences and might not generalize to as many linguistic edge cases.

To illustrate the differences and tradeoffs, consider Table 10-1, which compares LLMs and SLMs for this grammar correction scenario.

Table 10-1. *Tradeoffs in Capability and Efficiency When Using LLMs and SLMs for Grammar Correction*

Aspect	LLMs	SLMs
Model size	Very large (billions of parameters). Examples: T5-3B, GPT-2 XL.	Compact (tens or hundreds of millions). Examples: T5-small, DistilGPT-2.
Infrastructure	Requires powerful hardware. Fine-tuning often needs multi-GPU or high-memory instances (e.g., 16+ GB GPU RAM).	Runs on modest hardware. Fine-tuning can be done on a single GPU or even on a CPU for very small models.
Training speed	Slower. Each epoch takes longer due to more computations. (May still converge in a few epochs, but overall wall-clock time is higher.)	Faster. Fewer computations mean quicker epochs. Can iterate and experiment more quickly.
Inference latency	Higher per sentence. May need GPU at inference to get acceptable response times (especially for real-time use).	Low latency. Can often run in real-time on CPU. Suitable for on-device or high-throughput scenarios.
Accuracy and coverage	Typically higher. Captures subtle grammar issues and long-range context better. More likely to fix rare or complex errors.	Good on common errors, but might miss complex or rare issues. Could occasionally produce incorrect fixes due to limited capacity.

(*continued*)

Table 10-1. (*continued*)

Aspect	LLMs	SLMs
Cost (compute)	More expensive to train and deploy. Uses more AWS compute resources (thus higher AWS bill). Inference scaling is costlier (each instance can handle fewer requests per second).	Cost-effective. Cheaper training and cheaper to host (can use smaller or fewer instances). Easier to scale to many users with limited budget.
Use cases	Ideal for mission-critical proofreading, where maximum accuracy is needed (e.g., an editing service for professional writers). Also useful when the text may be complex or specialized.	Ideal for real-time assistance in user interfaces (e.g., mobile keyboard suggestions, browser extensions) where speed is crucial and errors are mostly straightforward. Also good for scenarios with limited compute (embedded systems, on-device agents).

As shown in Table 10-1, there is no one-size-fits-all approach; the choice depends on requirements. It's worth noting that research and engineering can narrow the gap between LLM and SLM performance. Techniques like knowledge distillation can train an SLM to mimic an LLM's behavior, yielding a small model that performs surprisingly well. For example, Microsoft's EdgeFormer, a lightweight seq2seq model for grammar checking, was optimized to run on-device while still maintaining competitive performance to the larger server model.[7] This was achieved through specialized training and model optimization. In practice, you might fine-tune a large model first to serve as a teacher, then use its corrections to train a smaller model. SageMaker could facilitate this multi-step process: first run a job on the LLM, then a job on the SLM with the LLM's outputs as additional training data.

10.2.4.1 An Example of Fine-Tuning an SLM

Let's say we fine-tuned DistilGPT-2 on the same data. DistilGPT-2 is a smaller decoder-only model. It generates text continuation, so how do we use it for correction? One way is to formulate the input as a prompt, such as, "Incorrect: I has a apple. Corrected:", and

---

[7] https://huggingface.co/docs/sagemaker/en/train#:~:text=

train it to continue with "I have an apple." Essentially, the model learns to take the part after "Incorrect:" as input and produce the continuation after "Corrected:". This is similar to how we'd do with T5, except T5 actually encodes the input separately. With careful training, DistilGPT-2 can learn to output corrected sentences. It may not *understand* grammar rules explicitly, but by example it learns that "I has" should become "I have," and so on. After training, it will likely do well on simple sentences. On complex sentences, it might occasionally fail. In an interactive application, you could implement a confidence check or fall back to a larger model if the small model's output seems questionable—effectively creating a hybrid system. We talk more about such deployment strategies soon.

To sum up this section, SLMs fine-tuned on SageMaker are a great option for deploying grammar correction in constrained settings. They're fast and cheap, and with the right data they can handle the majority of everyday writing errors. LLMs, on the other hand, give the best performance especially for tricky sentences, at the cost of more resources. Depending on our product needs, we might choose one or even use both in a tiered fashion. AWS provides the flexibility to train and deploy either type (or both in concert).

10.2.5 Deploying the Model on AWS SageMaker for Scalable Inference

Training a model is only half the battle—we need to make it available to end users (writers, students, customers, etc.) in a reliable and scalable way. AWS SageMaker shines here as well, by providing multiple options to deploy our fine-tuned model as a web service. We focus on SageMaker's real-time inference endpoints, which are well suited for a grammar correction API that applications can call on-demand.

10.2.5.1 Real-Time Endpoint Deployment

After our training job is finished, we have a `model.tar.gz` in S3 containing the model weights and configuration. We can deploy this with just a few lines of code using the SageMaker SDK or via the AWS console. Using code, it looks like this, a continuation from the previous Hugging Face estimator example:

Python

```
# Deploy the fine-tuned model to an endpoint
predictor = huggingface_estimator.deploy(
    initial_instance_count=1,
    instance_type='ml.g5.xlarge'  # for example, 1 GPU instance for
                                  inference
)

# Invoke the endpoint with a test payload
input_text = "I has finish my homework"
response = predictor.predict({"inputs": input_text})
print(response)
```

SageMaker will take the model artifacts, spin up the specified instance—which is an `ml.g5.xlarge` with an NVIDIA 10C CPU and ample memory with the Hugging Face inference container—and deploy our model behind a HTTPS endpoint. The `predictor.predict` call serializes our input as a JSON with `"inputs"`: `"I has finish my homework"` and sends it to the endpoint. It then receives the model's output. The response in this case should contain the corrected sentence—"I have finished my homework.". The exact format of the response depends on the inference container; typically for text generation models, it might return a list of generated sequences. The Hugging Face container, in particular, will return something like `{'generated_text': 'I have finish my homework.'}` or the fully corrected text. We'd parse that accordingly.

Ensuring the model output is fully correct, the model might output "I have finish my homework." missing the -ed on finished. If we see such issues in testing, it means our model might need more training data on verb tenses, or perhaps we should incorporate a grammar rule-based post-processing to add the -ed. There's always room to combine rule-based checks with the AI model for final polish, especially in critical applications.

10.2.5.2 Scaling and Instance Choice

We deployed on one GPU instance for simplicity. But SageMaker endpoints can be scaled to many instances, such as if our grammar checker service has to handle thousands of requests per minute. To use an example, imagine a large enterprise integrating this system into their email client for all employees. They could increase `initial_instance_count` to 3 or 5 and enable auto-scaling. Auto-scaling can add or remove instances based on metrics like CPU usage or number of requests, ensuring low latency without over-provisioning during quiet periods.

We could also choose different instance types. If we deployed an SLM that runs fast on a CPU, we might use cheaper CPU-only instances, such as `ml.c5.2xlarge`. On the other hand, a big LLM might even need a multi-GPU server for a single model if we didn't distill it—though for inference, usually one GPU is enough for models up to a few billion parameters, possibly with some optimizations. AWS also introduced Inf1/Inf2 instances powered by AWS Inferentia chips, which are specialized for deep learning inference. These instances can significantly reduce cost per inference for Transformer models. For example, Amazon claims that the Inf2 instances provide "up to 4x lower cost for LLM inference compared to GPU-based instances."[8] If our grammar model can be compiled to run on Inferentia (AWS Neuron SDK supports many Hugging Face models), we could cut down on operating costs. This is especially relevant if we are running a high-volume service—saving cost on each request multiplies out. It's exactly these kinds of optimizations (along with algorithmic ones) that allowed Microsoft to serve trillions of grammar checking requests per year at reasonable cost.[9] On AWS, we can throw hardware at the problem, where we can scale out with instances using Inferentia for efficiency and optimize the model.

10.2.5.3 Latency Considerations

Real-time grammar checking should ideally be very fast (sub-second) to feel instantaneous to users. Our deployed endpoint's latency will depend on the model size and instance type. A smaller model on a decent CPU might give ~50 ms response for a single sentence. A larger model on a GPU might take ~200-500 ms. These are rough figures; we should test and measure. If latency is high, we have options—we could enable model batch inference (processing multiple requests together on the GPU for efficiency), or as mentioned, go for optimized hardware. SageMaker also offers serverless inference endpoints that scale to zero when not in use, which could be handy if usage is sporadic. Serverless endpoints currently have memory limits, so they might only fit smaller models. For a steady load, persistent endpoints are fine.

---

[8] https://huggingface.co/docs/sagemaker/en/train#:~:text=
[9] https://huggingface.co/docs/sagemaker/en/train#:~:text=

10.2.5.4 Integration into Applications

Once the endpoint is up, any application can integrate by making a network request. For example, a web app could use API Gateway or an AWS Lambda function as a façade to invoke the SageMaker endpoint (the Lambda would call the SageMaker runtime API). Or a desktop app could directly call the AWS SDK to invoke the endpoint. The response—corrected text and perhaps additional info like the confidence or the list of corrections made—would then be shown to the user. Because it's an AWS endpoint, it's highly available and can be placed in multiple regions globally to serve users with low latency. We could deploy one model per region (SageMaker is region-specific) if we have a worldwide user base, or use a CDN-like approach to route to the nearest region.

10.2.5.5 A/B Testing and Continuous Improvement

SageMaker allows us to deploy multiple models as separate endpoints or even behind a single endpoint variant. This means we could, for instance, deploy our LLM-based corrector and an SLM-based corrector and compare them. Perhaps we direct 10% of traffic to the LLM model and 90% to the SLM to monitor differences in user feedback. If the LLM yields significantly better user satisfaction, we might justify switching more users to it despite the cost. If the SLM is almost as good, we save resources. This kind of experimentation is valuable in a production scenario and SageMaker's MLOps tools or even simple CloudWatch metrics help track it.

10.2.5.6 LLM Agents in the Loop

It's worth noting how an LLM agent concept can be applied here. Instead of a single model doing everything, we can have a system where multiple components work together intelligently. For example, we might create a small classifier model that acts as a gate (like a mini-BERT trained to judge if a sentence is grammatically correct or not). This is analogous to the acceptability classifier discussed earlier. When a piece of text comes in, the classifier (an SLM that's extremely fast) can decide whether it "looks good" or "needs correction." If it's good, we don't need to waste time or cost on the heavier correction model—we can immediately return "No issues found." If it's bad, then we pass the text to the big correction model (LLM) to fix it.

This two-stage pipeline is essentially an LLM-driven agent: the first stage is a lightweight agent deciding *whether* to act, the second stage is the heavy-duty agent deciding *how* to act (by producing a correction). We could further enrich the agent with rules. For example, if the model corrects "There is many reasons" to "There are many reasons", the agent could attach an explanation: "Subject-verb agreement corrected (should use plural verb with plural noun)." The explanation could come from a rule system or an additional model. AWS doesn't provide this logic out of the box, but we can build it by orchestrating multiple SageMaker endpoints or Lambda functions. SageMaker endpoints can be called from one another as well, or all called from a central Lambda that acts as the agent controller. This modular design is powerful—imagine extending the agent to not just correct grammar, but to also check for style or toxicity by routing text through different models. In an enterprise setting, such an agent might even have access to a knowledge base—for example, verifying company-specific terminology. While this is beyond basic grammar correction, the point is that AWS's ecosystem allows us to seamlessly integrate our model into larger intelligent systems.

10.2.5.7 Testing the Deployed Model

After deployment, we should test our SageMaker endpoint with a variety of inputs to ensure it works as expected. We send sentences with errors and see if the response is correct. We also test edge cases: empty input, very long sentence input, or sentences with no errors (the model should ideally return the original sentence if no change is needed, or perhaps a tagged response indicating "no correction"). These tests help verify that the model is ready for real users. We might find minor issues—for example, the model might sometimes overly correct (changing things that were actually fine, a false positive). If those are frequent, we might address it by training on more "clean" sentences (so it learns not to change correct sentences) or by adding a confidence threshold (the model can output a probability or score for how sure it is, and the agent only applies corrections above a certain confidence). Such tweaks often make the difference between a useful tool and an annoying one (nobody likes an overzealous grammar checker that "corrects" stylistic choices incorrectly).

Finally, once deployment is successful and tested, we have a live grammar checking service! This could be used internally (maybe a company adds it to their documents workflow) or externally (offered as a feature in a consumer app). The heavy lifting—managing the servers and scaling—is handled by AWS, so our focus can shift to monitoring quality and improving the model. We can set up logging to record instances where the model's suggestion was rejected by users, feed that back into a new training dataset, and periodically retrain (using SageMaker again) to make the model better—a continuous improvement cycle.

This section showed how to implement a text correction system on AWS SageMaker, from data prep through training LLMs/SLMs to deploying an endpoint for real-time use. SageMaker provides an end-to-end pipeline to do this efficiently and at scale, which is why it's a popular choice for companies fine-tuning language models.[10] We also discussed how LLM agents and multi-model pipelines can be incorporated for smarter and more efficient processing, which foreshadows design patterns in real-world AI applications.

The next section explores how to achieve similar outcomes using open-source tools on local and non-cloud environments. We look at training with Hugging Face Transformers outside of SageMaker and how to evaluate and improve models with user feedback. This open-source implementation path is useful for those who want to prototype locally or who prefer not to rely on a cloud platform. Let's now dive into the open-source side of building grammar correction models, and later, we compare the approaches and highlight use cases in the industry.

10.3 Open-Source Implementation: Training with Hugging Face

This section provides a hands-on guide to building a text correction model using open-source tools. We leverage the Hugging Face ecosystem, which has become the industry standard for working with pretrained transformer models. By fine-tuning a pretrained sequence-to-sequence model like T5, we can quickly adapt it to our specific task of correcting grammatical errors and typos, all while running on standard hardware with powerful libraries.

The process involves:

1. **Dataset preparation:** Creating a parallel corpus of "incorrect" and "correct" sentences.

2. **Model selection:** Choosing a suitable pretrained model from the Hugging Face Hub (e.g., `t5-small`).

---

[10] https://huggingface.co/docs/sagemaker/en/train#:~:text=

3. **Fine-tuning:** Training the model on our custom dataset to learn the transformation rules.

4. **Evaluation and inference:** Testing the model's performance and demonstrating how to use it for new text.

In the context of text correction, the distinction between an LLM and an SLM is primarily about scale and computational requirements.

- **LLM:** A model with billions of parameters (e.g., T5-large, T5-3B). Fine-tuning an LLM requires significant computational resources (multiple GPUs with large VRAM), but often results in a highly accurate and general-purpose correction model capable of handling a wide variety of errors and stylistic nuances.

- **SLM:** A model with millions of parameters (e.g., T5-small, T5-base). Fine-tuning an SLM is computationally much cheaper and faster, making it ideal for resource-constrained environments or for deploying on edge devices. While slightly less powerful than its larger counterparts, a well-tuned SLM can still achieve excellent performance on specific, well-defined tasks like correcting common grammatical errors.

10.3.1 Fine-Tuning an LLM or SLM for Grammar Correction

Consider these key considerations for fine-tuning a grammar correction model:

1. **Data quality and diversity:**

   The heart of any successful fine-tuning project is the data. A diverse and high-quality dataset is crucial.

   - **Breadth over depth:** Your dataset should contain a wide variety of grammatical errors, not just one or two types. This includes subject-verb agreement, tense errors, punctuation mistakes, run-on sentences, spelling corrections, and more.

- **Domain specificity:** The grammar of a technical manual is different from a young adult novel. If your goal is to correct manuscripts in a specific genre (e.g., science fiction, academic papers), your training data should reflect that style and vocabulary.

- **Realistic errors:** Synthetic data can be useful, but for best results, include real-world examples of text written by humans. This helps the model learn to correct the kinds of mistakes people make.

2. **Defining "correctness":**

   Grammar is not always black and white. A human editor makes judgment calls. You must decide what constitutes a "correct" output for your model.

   - **Style guides:** Adhering to a specific style guide (e.g., *The Chicago Manual of Style*, the AP Stylebook) can provide a consistent standard for your "correct" examples.

   - **Minor vs. major edits:** Will your model only fix blatant errors, or will it also suggest stylistic improvements and rephrasing? The scope of its corrections should be clearly defined in your training data.

3. **Model selection:**

   While the T5 family of models is an excellent choice for this task, the size of the model matters.

   - **Size vs. performance:** A smaller model like `t5-small` is faster and requires fewer computational resources, but a larger model like `t5-base` or `t5-large` will likely achieve higher accuracy and handle more complex corrections. The choice depends on your budget and performance requirements.

   - **Pretraining objective:** T5's text-to-text pretraining objective makes it highly suitable for tasks like grammar correction, where the input and output are both text sequences. This is why it's a popular choice.

4. **Evaluation and metrics:**

   Training a model is only half the battle; knowing if it's working is the other.

   - **Standard metrics:** Metrics like ROUGE and BLEU are standard for text generation tasks and are good starting points. They measure the overlap of n-grams between the corrected text and the reference text.

   - **Human-in-the-loop:** The most reliable form of evaluation is human review. No automated metric can fully capture the nuance of a grammatically correct and stylistically pleasing sentence. After training, a human should always review the model's output on a held-out test set to ensure quality.

   - **Error analysis:** Instead of just looking at the overall score, perform an error analysis. What types of errors does the model still make? Does it over-correct, or miss subtle mistakes? This will guide your next steps in data collection and training.

By meticulously considering these points, you can transform a basic fine-tuning exercise into a professional-grade grammar correction tool that is accurate and reliable.

The script in this example demonstrates how to fine-tune a T5 model. You can switch between an SLM (t5-small) and an LLM (t5-base or t5-large) to see the differences in training time and model size. The script first prepares a small dataset of incorrect and correct sentences. It then uses a T5 tokenizer to preprocess this data into a format the model can understand, specifically by adding a grammar correction: prefix. Next, it uses the Hugging Face Trainer class to fine-tune the t5-small model over a specified number of epochs. During this training, it evaluates the model's progress using ROUGE and BLEU metrics. Finally, after the training is complete, the script saves the fine-tuned model and its tokenizer to a local directory so it can be used later to correct new sentences. A brief example at the end demonstrates how to load this saved model and perform inference on a new sentence.

Refer to GitHub for the full code at: https://github.com/anvcse562/ModernNLP-LLMSLM/blob/main/Ch10/Ch10.3/10.3.1_Fine-tune_SLM_LLM_Grammar_Correction.ipynb.

10.3.2 Evaluation and User Feedback

Evaluating the performance of a text correction model is crucial for understanding its effectiveness, while incorporating user feedback ensures its continuous improvement. This section details the key metrics and the process for building a self-improving system.

1. **Evaluation metrics:**

   For generative tasks like text correction, standard classification metrics (like accuracy) aren't sufficient because there can be multiple "correct" outputs. Instead, we use metrics that compare the model's generated text to a reference text.

   - **BLEU:** Originally for machine translation, BLEU measures the precision of the model's output by counting the number of n-grams (contiguous sequences of words) in the generated text that also appear in the reference text. A higher BLEU score indicates a closer match to the reference.

   - **ROUGE:** This metric focuses on recall, measuring the overlap of n-grams between the generated text and the reference. ROUGE is particularly useful for tasks where you want to ensure the generated text captures all the key information from the reference.

   - **Edit distance:** Also known as Levenshtein distance, this metric calculates the minimum number of single-character edits (insertions, deletions, or substitutions) required to change one text into another. A lower edit distance indicates that the model's output is very similar to the reference, meaning that fewer changes are needed.

2. **Incorporating user feedback:**

   While automated metrics are useful, they don't capture subjective quality or context-specific nuances. A *user feedback loop* is the most effective way to address this, creating a cycle of continual learning.

3. **Feedback collection:**

   The application presents the model's correction to the user and asks for a response. Common methods include:

   - **Binary feedback:** A simple 👍 or 👎 (accept/reject) option.
   - **Corrected input:** The user manually edits the text, and their final version is saved as the new "ground truth."

4. **Data curation:**

   The collected feedback is logged and stored. This raw data is then cleaned and formatted into a new dataset of "incorrect sentence" to "corrected sentence" pairs. This new dataset is highly valuable because it reflects real-world use cases and user preferences.

5. **Model retraining:**

   The new, curated dataset is periodically combined with the original training data to retrain the model. This process allows the model to learn from its mistakes, correct its biases, and adapt to evolving language patterns and user needs.

This code demonstrates how to set up a simple system to collect and log user feedback. This feedback is critical for building a continual learning pipeline where the model is periodically retrained on a new, refined dataset.

The provided code describes a system for collecting user feedback to improve a proofreading tool. The `main_app_loop` function simulates the user interface, capturing the original text, the model's correction, and the user's final preferred version. This data is then logged by the `record_feedback` function into a `feedback_data.jsonl` file. The `create_retraining_dataset` function then processes this log file to extract pairs of incorrect and correct sentences, creating a new, valuable dataset for continually retraining and improving the model based on real-world user interactions.

Refer to GitHub for the full code at: https://github.com/anvcse562/ModernNLP-LLMSLM/blob/main/Ch10/Ch10.3/10.3.2_UserFeedback_Proofreading_Continual%20training.ipynb.

10.4 Industrial Use Cases
10.4.1 Writing Assistance Tools
10.4.1.1 Use Case Overview

Writing assistance tools help users produce clearer, more correct, and more effective text by automatically detecting and correcting grammar, spelling, and stylistic issues. In professional and academic contexts—emails, reports, documentation, onboarding material, and code comments—such corrections reduce misunderstandings and improve perceived quality. These systems blend SLMs for fast, focused error detection (e.g., grammatical acceptability) with generative models (LLMs or sequence-to-sequence models) that propose corrected rewrites. This hybrid setup balances efficiency, interpretability, and fluency.

10.4.1.2 Challenges Addressed

- **Error detection vs. correction:** Identifying whether a sentence is acceptable (SLM) and then generating a corrected version (LLM) are distinct tasks; combining both yields a more reliable assistant.[11]

- **Limited context and precision:** Simple rule-based checkers miss nuanced errors; learned models can generalize while still handling common patterns.

- **Evaluation of corrections:** Automatic metrics like BLEU/ROUGE quantify fidelity to reference corrections, while edit-distance (Levenshtein) captures the magnitude of change; these help monitor drift and quality over time.

10.4.1.3 Implementation Overview

The minimal demonstration uses a small synthetic parallel corpus of erroneous sentences paired with their corrected forms; in production this would be replaced or augmented with established grammar correction corpora such as JFLEG, BEA, or GECToR-style datasets.

---

[11] https://model.aibase.com/models/details/1915694112227090433?utm_source=chatgpt.com

SLM: Grammatical Acceptability Detection

An SLM (e.g., a DistilBERT variant fine-tuned on the Corpus of Linguistic Acceptability—CoLA) is used to judge whether a sentence is grammatically acceptable. This model provides a quick signal about whether a user's input likely contains an error, enabling triaging or highlighting before full correction.

Generative Model (LLM-style): Grammar Correction

A sequence-to-sequence model (T5-small) is fine-tuned to perform grammar correction: inputting a possibly flawed sentence and generating a corrected rewrite. T5's flexible text-to-text framework makes it well-suited, and prior work has shown its effectiveness on grammar correction tasks when fine-tuned appropriately.

Here is the data pipeline:

1. **Input:** User submits a sentence.

2. **Acceptability check:** The SLM evaluates whether the sentence is grammatically acceptable. If the sentence is flagged as problematic, the pipeline proceeds to correction.

3. **Correction generation:** The fine-tuned T5 model (LLM-style) generates a candidate corrected sentence.

4. **Post-hoc metrics:** Differences between original, corrected, and reference (if available) are quantified with BLEU/ROUGE for overlap similarity and Levenshtein distance for edit magnitude.

5. **Suggestion:** The system surfaces an acceptability judgment and corrected rewrite, with edit distance indicating severity.

The evaluation metrics include the following:

- **BLEU/ROUGE**: Measure n-gram overlap between model output and reference correction, offering a proxy for fidelity. BLEU emphasizes precision; ROUGE emphasizes recall.

- **Levenshtein (edit) distance:** Quantifies the number of atomic insertions/deletions/substitutions needed to transform the model output to the reference or vice versa—useful for gauging extent of change.

- **Acceptability signal:** The SLM provides binary/soft judgments about grammatical well-formedness based on CoLA-style training.

Code Integration Summary

The provided notebook implements this pipeline using Hugging Face Transformers and Datasets. Consider these key components:

- **Data preparation:** Synthetic sentence pairs are packaged into a dataset and formatted for T5.

- **SLM setup:** Loading a DistilBERT-based CoLA classifier to obtain acceptability scores for input sentences.

- **LLM Fine-tuning:** T5-small is fine-tuned on the synthetic grammar correction examples with Hugging Face's `Trainer` API.

- **Correction pipeline:** Functions produce corrections, combine signals, and compute evaluation metrics including BLEU, word overlap, and edit distance.

- **Interactive assistant:** A wrapper function shows how a sentence is evaluated and corrected, enabling easy integration into UI or APIs.

Deployment Options and Scalability

- **Local/internal deployment:** Serve the models inside a corporate network as part of a writing assistant (e.g., integrated into email clients, documentation portals).

- **Cloud deployment:** Host fine-tuning and inference on managed GPU/CPU instances (e.g., AWS SageMaker, given the broader chapter's orientation) for scalability and shared access.

- **Containerization:** Package the pipeline into Docker to ensure consistent environments across development, staging, and production.

- **Hybrid edge/cloud:** Use the lightweight SLM on-device for immediate acceptability feedback while offloading heavier generative correction to cloud-hosted T5 variants.

You can find the code on GitHub at: https://github.com/anvcse562/ModernNLP-LLMSLM/blob/main/Ch10/ch10.4/10_4_1_writing_assistance_tools.ipynb.

10.4.2 Proofreading Applications

This section delves into a practical, industry-grade use case for a writing assistance tool, demonstrating how the concepts of text correction and language modeling can be applied to build a robust commercial product. This use case outlines the architecture, workflow, and benefits of a comprehensive proofreading application designed for enterprise environments.

10.4.2.1 The ProseGuard Enterprise Writing Assistant

A fictious multinational corporation, called GlobalTech, struggles with inconsistent and error-prone written communication across its various departments. Marketing materials, legal documents, and internal memos often contain grammatical errors, misspellings, and inconsistent terminology. The manual proofreading process is slow, costly, and fails to enforce a uniform brand voice. GlobalTech needs a scalable, intelligent solution to ensure high-quality, consistent, and error-free content.

We propose developing *ProseGuard*, an enterprise-level writing assistant. This platform will go beyond basic spell-checking by leveraging a fine-tuned language model (similar to the T5 model described earlier in this chapter) to provide real-time, context-aware suggestions for grammar, spelling, style, and terminology specific to GlobalTech's brand and legal standards.

10.4.2.2 Technical Architecture

- **Frontend integration:** ProseGuard will be deployed as a browser extension and as a plugin for popular enterprise software (e.g., Microsoft Word, Google Docs, a custom CMS). This allows for seamless integration into existing workflows.

- **Backend microservice:** A dedicated backend service, likely running on a cloud platform like AWS, will host the core text correction model. This service will expose a REST API endpoint that receives text and returns a JSON payload containing corrections and suggestions.

- **Core text correction model:** This is the heart of the application. It is a fine-tuned T5 model trained not only on general grammar datasets but also on a custom corpus of GlobalTech's internal documents, style guides, and legal precedents. This ensures the model understands and enforces company-specific terminology and writing conventions.

- **User feedback and data pipeline:** An integral part of the system is a feedback loop. User actions (e.g., accepting a suggestion, rejecting a suggestion, or manually editing text) are captured and sent to a data storage layer (e.g., a S3 bucket or a database).

- **Model retraining pipeline:** Periodically, the collected user feedback data is used to create an updated training dataset. An automated pipeline retrains the core text correction model on this new data, ensuring the model continually learns and adapts to the corporation's evolving needs and writing styles.

10.4.2.3 Workflow and Features

1. **Real-time analysis:** As the employee types in their chosen application, the ProseGuard frontend sends snippets of the text to the backend API.

2. **Intelligent suggestions:** The API, powered by the fine-tuned T5 model, analyzes the text for errors and provides a range of suggestions:

    - **Grammar and spelling:** Corrects common mistakes like "I has gone" to "I have gone."

    - **Brand voice:** Flags sentences that don't match the company's formal or professional tone.

    - **Terminology consistency:** Highlights instances where a product name is misspelled or an outdated term is used, suggesting the correct, approved version from a company-specific dictionary.

3. **User interaction:** The suggestions are presented in a user-friendly interface. The user can accept or reject each suggestion with a single click.

4. **Feedback capture:** When the user interacts with a suggestion (e.g., accepts a change), this action is logged and sent to the feedback pipeline. For example, if the model suggests "The team is working," and the user accepts it, this positive reinforcement is recorded. If the model makes an incorrect suggestion that the user rejects, this negative feedback is also logged.

5. **Continuous improvement:** The collected feedback data is used to retrain and update the model quarterly. This iterative process ensures the model's accuracy and relevance grow over time, making ProseGuard a self-improving asset.

10.4.2.4 Outcomes and Benefits

- **Improved quality:** Ensures all corporate communications are free of errors and maintain a professional, consistent tone.

- **Increased efficiency:** Drastically reduces the time and effort spent on manual proofreading, allowing employees to focus on content creation.

- **Enhanced brand consistency:** Enforces a single brand voice and terminology across all departments and external communications.

- **Cost reduction:** Minimizes the need for a dedicated proofreading staff and reduces the risk of costly errors in legal or marketing materials.

- **Scalability:** The cloud-based microservice architecture allows the tool to scale effortlessly to support thousands of employees and millions of documents without performance degradation.

You can find the code on GitHub at: https://github.com/anvcse562/ModernNLP-LLMSLM/blob/main/Ch10/Ch10.4/10.4.2_Proofreading_Enterprise%20Writing%20Assistant.ipynb.

10.5 Summary

This chapter explored text correction using LLMs and SLMs. You learned how to build grammar and spelling correction systems that enhance clarity and accuracy in written communication. LLMs like T5 and BART offer powerful capabilities for complex correction tasks, while SLMs such as DistilGPT-2 and TinyT5 provide efficient solutions for real-time, lightweight applications.

You also gained hands-on experience deploying these models using AWS services like Amazon SageMaker, and working with open-source tools such as Hugging Face Transformers. Industry use cases included writing assistance tools, proofreading applications, and educational platforms focused on grammar improvement.

You should now have a solid understanding of text correction techniques and how to implement them at scale. The next chapter dives into another key modern NLP task: coreference resolution and text entailment.

CHAPTER 11

Coreference Resolution and Text Entailment

This chapter examines the development of systems for coreference resolution and text entailment, utilizing LLMs such as BART, RoBERTa, and DeBERTa, as well as SLMs, like DistilRoBERTa, TinyBERT, and MiniLM. These tasks are critical for deep language understanding—*coreference resolution* identifies relationships between entities across a text, while *text entailment* determines whether one sentence logically follows from another.

We cover scalable implementations using AWS services, such as Amazon SageMaker, to train and deploy models for coreference and entailment tasks. Open-source tools, such as Hugging Face Transformers, are emphasized, along with strategies for improving model performance through techniques like mention detection, entity linking, and data augmentation. SLMs are highlighted for offering efficient alternatives to real-time or resource-constrained environments.

By the end of this chapter, you'll be able to build and deploy high-performing coreference and entailment models using both cloud-based and open-source solutions.

This chapter covers:

- **An overview of coreference resolution and text entailment:**

  - **Coreference resolution:** Identifying and linking references to the same entity across the text.

  - **Text entailment:** Evaluating if one sentence logically follows (is *entailed* by) another.

  - **Importance:** Enables semantic understanding in complex documents, which is essential for downstream NLP tasks.

- **Implementation on AWS:**
  - **Amazon SageMaker:** Build and fine-tune LLMs like BART or DeBERTa for coreference and entailment.
  - **SLM deployment:** Use lightweight models like DistilRoBERTa for faster, low-latency applications in production.
- **Open-source implementation:**
  - **Hugging Face Transformers:** Leverage pretrained models for both coreference and entailment tasks.
  - **Model optimization:** Use coreference and entailment.
  - **SLMs in open-source:** Deploy efficient models like TinyBERT or MiniLM in real-time environments.
- **Industry use cases:**
  - Enhance AI-driven customer support with improved context tracking
  - Perform legal document analysis and summarization
  - Create context-aware chatbots and virtual agents
  - Create academic research tools for semantic analysis

After enhancing text clarity through correction and language modeling in the previous chapter, we now turn to deeper comprehension tasks. The next logical step is to understand the meaning and context of text. This brings us to coreference resolution and textual entailment—tasks that allow AI systems to interpret who or what the words refer to, and how statements relate logically to each other. Both are crucial for advanced language understanding and are the focus of this chapter.

11.1 Overview of Coreference Resolution

Coreference resolution and textual entailment are two fundamental natural language understanding (NLU) tasks that enable AI systems to grasp context and semantics beyond the surface level of text. Coreference resolution involves identifying when two or more expressions in the text refer to the same entity (e.g., linking "she" to Alice). At the same time, textual entailment (also known as *Natural Language Inference—NLI*) involves determining if one statement logically follows from another (entailment), contradicts it, or is unrelated. Mastering these tasks is essential for applications ranging from dialogue agents to question-answering systems, because it enables the AI to maintain coherent context and reason about information. We first discuss why these tasks are so important for context and semantics, then delve into what each task entails, with simple examples and code snippets to illustrate how they work in practice.

11.1.1 Importance of Understanding Context and Semantics

Understanding language requires more than just correct grammar—it requires tracking *who* or *what* the sentences are about and how different statements relate in meaning. Coreference resolution and textual entailment play a pivotal role in this deeper understanding:

- **Maintaining context through coreference:** In any document or conversation, pronouns and references abound. Humans effortlessly know that "he" in "John went home because he was tired" refers to John. For machines, resolving such coreferences is vital for maintaining coherence. Without it, an AI might interpret sentences in isolation and lose track of *which entities* are being discussed. Coreference resolution enables continuity, ensuring that a sequence of sentences is understood as a connected whole rather than disjoint pieces. This is especially critical in conversational AI, where an assistant needs to remember earlier mentions. For example, if a user says "I met Sara yesterday. She gave me her email," the system must link "She" and "her" to *Sara* to respond meaningfully. In fact, coreference resolution enables machines to comprehend and track references throughout a conversation, ensuring continuity and context preservation. Many downstream tasks, such as summarization, information extraction, and translation, achieve improved accuracy with effective coreference handling.

- **Semantic consistency through entailment:** Textual entailment addresses the logic and semantic relationship between statements. It asks: given a *premise* sentence and a *hypothesis* sentence, does the premise imply the hypothesis, contradict it, or neither? This evaluation of the *truth relationship* is the key to reasoning. For example, if the premise is "Alice owns a car," a hypothesis "Alice has a vehicle" is an *entailment* (it is logically true given the premise), whereas "Alice has no vehicle" would be a *contradiction*. Recognizing these relations is crucial for tasks like question answering (the answer's sentence should entail the question's premise when stated as a sentence), fact-checking (does a source statement contradict a claim?), and maintaining consistency in dialogues. NLI is an important aspect of natural language understanding because it requires the model to grasp the meaning of two texts and how they relate. By handling entailment, AI agents can avoid providing inconsistent answers (for instance, not asserting something that contradicts earlier information) and can perform reasoning, such as deducing conclusions from given facts.

In summary, coreference resolution ensures an AI *knows what each pronoun or noun phrase refers to*, anchoring words to real entities in context. Textual entailment ensures that an AI *understands the logical implications* between pieces of text. Together, these capabilities allow language models and agents to move beyond surface corrections into genuine comprehension of context and semantics.

Note LLMs are often implicitly trained on these aspects. For example, transformer-based models like BERT and GPT learn to handle pronoun references and logical relationships through their vast training data. In fact, BERT's pretraining included a *Next Sentence Prediction* task where the model learned to predict whether one sentence logically followed another—a concept closely related to entailment and context continuity. Today's LLM-based agents leverage such built-in knowledge, sometimes supplemented by explicit coreference modules or fine-tuned entailment classifiers for greater accuracy.

11.1.2 Coreference Resolution

Coreference resolution is the task of finding all expressions in the text that refer to the same entity and linking them. In other words, if a text mentions "Alice" and later says "she", a coreference system determines that "she" refers to Alice. This is essential for machines to interpret text in the same way humans do, since we naturally associate pronouns and descriptions with their proper referents in our minds.

At its core, coreference resolution involves two subtasks: *mention detection* (identifying the spans of text that potentially refer to entities, such as proper names, pronouns, and noun phrases) and *mention linking* (determining which mentions refer to the same entity). Mentions that refer to the same real-world entity are grouped into a *coreference chain* (or cluster). For example, consider the following:

> "**Alice** noticed that the weather was getting cold. **She** pulled her jacket tighter."

Here, "Alice" in the first sentence and "She" in the second are part of the same coreference chain (they refer to the same person, Alice). A coreference resolution model would output something like: *{Alice, She}* as one cluster, indicating that these two mentions co-refer.

Coreference can occur with pronouns (he, she, it, they, etc.), proper nouns (Alice, Bob), noun phrases (*the tall man, the company, her jacket* referring to *Alice's jacket*), and more. Pronoun resolution (a subset often called *anaphora resolution* when the pronoun refers backward) is the most common case. There are also cases like *cataphora* where a pronoun appears first and the noun later (e.g., "When *he* arrived, *John* was happy"—*he* refers to John but came before John was mentioned).

11.1.2.1 Challenges

Coreference resolution is challenging due to the ambiguities in language. A pronoun like "it" or "they" might have multiple possible antecedents, and the system must choose the correct one using context and common sense. For instance, in "The dog chased the cat because it was scared," does "it" refer to the dog or the cat? Humans use semantics (dogs chasing cats likely scare the cat, so "it" probably refers to the cat) and sometimes world knowledge to make decisions. Models need to mimic this reasoning. Other challenges include varying expressions (the same entity might be referred to as "Dr. Smith" in one sentence and "the doctor" in another) and contextual nuances (sometimes the closest noun is not the correct referent; understanding the situation is necessary).

11.1.2.2 Why LLMs Help

Modern deep learning models, particularly transformer-based LLMs like BERT, have significantly enhanced coreference resolution. They create rich, context-aware representations of each word, which helps in determining whether, for example, "he" and "John" are related. Fine-tuning such models on coreference datasets (e.g., OntoNotes, the dataset used in the CoNLL shared task) yields state-of-the-art results. Large models can encode subtle clues, such as gender/number agreement (e.g., "she" should not link to a male name) and semantic compatibility. There are also hybrid approaches that combine rules (for explicit constraints, such as gender agreement) with machine learning for added flexibility. In practice, advanced coreference systems often use *span-prediction models* (where the model directly predicts if two spans are coreferent) or a clustering approach that groups mentions.

Let's walk through a basic example to see coreference resolution in action.

In this illustrative example, the coreference resolution system identifies that the pronoun "He" and the possessive "his" in the second sentence both refer to "John" in the first sentence. This linking of pronouns to the noun "John" ensures the text is understood coherently. By resolving these references correctly, an NLP system knows that it is John who loves exploring new trails on his bike and who went on the challenging ride, rather than introducing a new person. This example highlights how coreference resolution preserves context: all these sentences are about the same person (John), and the system can treat information about "he" as information about John.

The result of a coreference resolution process might be output as clusters or even by replacing pronouns with their referents. For instance, the system could transform the text into "*John* is an avid cyclist. *John* loves exploring new trails on *John's* mountain bike. Last weekend, *John* went on a challenging ride...", which makes the references explicit. (Of course, a human wouldn't usually repeat names like that; this is just to illustrate clarity.)

Consider this code example (coreference with spaCy). Some libraries facilitate coreference resolution. For example, spaCy with the `coreferee` or `neuralcoref` plugin can identify coreference chains in text.

CHAPTER 11   CORE REFERENCE RESOLUTION AND TEXT ENTAILMENT

Refer to the code on GitHub for a simple illustration using spaCy with a coreference resolver: https://github.com/anvcse562/ModernNLP-LLMSLM/blob/main/Ch11/ch11.2/11_1_2_Coreference_with_spaCy.py.

When run, this code would output something like this:

```
0: Alice(0), her(7)
1: Bob(3), He(5)
```

This indicates two coreference chains identified in the text. *Chain 0* links Alice (token 0) and her (token 7), showing that "her" refers to Alice. *Chain 1* links Bob (token 3) and He (token 5), showing that "He" refers to Bob. With this information, an application could, for example, replace pronouns with names to avoid ambiguity, or simply use it internally to ensure that any actions or attributes mentioned (such as thanking) are correctly attributed to the right person.

From an LLM agent perspective, coreference resolution can be handled implicitly by the model's large context window and training (since models like GPT-4 often correctly infer pronoun referents in coherent text). However, for very long documents or more reliable handling, you may need to integrate an explicit coreference module. For instance, an AutoGPT-style agent reading a multi-document report might utilize a coreference resolver to track entities across the documents. Ensuring that the agent knows who or what each pronoun refers to prevents errors, such as mixing up characters in a story or attributing an action to the wrong entity.

To summarize coreference resolution, here's a quick reference table:

Task	Typical Input	Output/Goal	Example
Coreference resolution	One text (sentence or document) with pronouns or other references.	Identify and link all mentions that refer to the same entity; they often produce clusters of references.	"Bob took his dog out. *He* walked for an hour." The system links "He" to Bob, resolving that Bob is the one who walked. All mentions of {Bob, He} form one coreference cluster.

This next section examines the other key task—textual entailment—and explains how it complements understanding by capturing the logical relationships between sentences.

11.1.3 Text Entailment

Textual entailment, or text entailment, refers to the task of determining whether a given *hypothesis* sentence can be inferred (entailed) from a given *premise* sentence. This task is often framed as NLI: a model must decide if the relationship between two pieces of text is *entailment* (the hypothesis is true given the premise), a *contradiction* (the hypothesis is false given the premise), or *neutral* (the hypothesis is neither definitely true nor false given the premise).

In simple terms, suppose we have two statements: A (the premise) and B (the hypothesis). If every time A is true, B must also be true (in the understanding of an average person), then A entails B. If A being true would make B definitely false, there's a contradiction. If B could be true or false independently of A, the relationship is neutral. Unlike strict logical entailment, textual entailment is a softer, language-based notion: "t entails h if, typically, a human reading t would infer that h is most likely true".

For example, consider the premise: "A group of people are playing football in the park." Now, examine some hypotheses:

- "Some people are playing outdoors." This is likely an entailment. Given the premise, a reader can infer this hypothesis is true (football is a sport played outdoors by people, so indeed some people are outdoors).

- "Nobody is outside." This is a contradiction. It directly conflicts with the premise (the premise says a group is outside playing).

- "It is raining heavily." This is neutral. The premise doesn't tell us anything about the weather, so we cannot infer this either way; it's unrelated information in this context.

These examples show how entailment captures semantic understanding. The model must recognize, for instance, that "playing football" implies "playing outdoors" (entailment requires background knowledge that football is played outside), or that "a group of people" means at least one person is outside (so "nobody is outside" is incompatible). At the same time, it knows that "raining" has no apparent relation to playing football in the park (it's possible to play in rain or shine, so the premise doesn't guarantee or rule out rain).

Textual entailment is widely used as a benchmark for a model's understanding of language. It has practical use in question answering (checking if a candidate's answer sentence entails the question), summarization (ensuring the summary's statements are entailed by the source text, not hallucinated), information retrieval (finding text that entail an answer the user is looking for), and debate/chatbots (to maintain consistency and detect contradictions in a conversation). For example, in a customer support chatbot, if the bot previously stated, "Your order was delivered on Monday," and later its knowledge base indicates, "The package is still in transit," the system should recognize the contradiction and resolve it, rather than providing the user with conflicting information. Entailment-checking can flag such inconsistencies.

Approaches to textual entailment have evolved from symbolic to neural:

- Earlier systems attempted to utilize logical inference by transforming sentences into formal logic and employing theorem provers. They worked but required heavy linguistic and knowledge engineering (and struggled with the ambiguity and variability of real language).

- Modern approaches predominantly use machine learning, especially deep learning. A standard method is to use a sequence-pair model: feed the premise and hypothesis into a neural network (often formatted with a special separator token) and have the network output one of the three labels (entailment, contradiction, neutral). Models like BERT, RoBERTa, and T5 fine-tuned on large NLI datasets (such as SNLI or MNLI) achieve very high accuracy on benchmark tasks. These models learn subtle cues—for example, negation ("no", "not"), which often leads to contradiction, or hyponym-hypernym relations (knowing that "dog" entails "animal", etc.).

- Because NLI models learn from numerous examples, they also acquire knowledge of the world. For instance, they might recognize that "acquiring a car" entails "owning a vehicle" due to the frequency of such patterns in training data, even if they are not explicitly taught as logical rules.

Consider this code example (entailment with transformers). Using Hugging Face Transformers, we can quickly test entailment with a pretrained model. For instance, the `roberta-large-mnli` model is a RoBERTa model fine-tuned on the Multi-Genre NLI corpus. We can use the pipeline API to get entailment predictions.

See the code on GitHub at: https://github.com/anvcse562/ModernNLP-LLMSLM/blob/main/Ch11/ch11.2/11_1_3_entitlement_with_coherence.py.

This will output a label with a confidence score, for example:

[{'label': 'ENTAILMENT', 'score': 0.99}]

Here, the model correctly predicts ENTAILMENT, since if a cat is on the mat, an animal is indeed on the mat (a cat *is* an animal). We concatenated the premise and hypothesis as a single string for the pipeline; under the hood, the model knows how to separate the two and evaluate the relationship between them. In practice, you can also format the input explicitly with a separator token or use the pipeline's tuple input format; however, the approach here works due to how this particular model expects input.

To test a contradiction, you could change the hypothesis. For example, hypothesis = "The mat is completely empty." The model should then output a CONTRADICTION label (since a cat being on the mat contradicts the mat being empty).

Entailment models typically produce one of three labels. If needed, you can map them to more user-friendly terms or scores. In many use cases, just knowing "yes, entails" or "no, contradicts" is enough. Some systems also utilize the neutral category for cases that are irrelevant or unknown, which can serve as a signal that more information is needed or that the hypothesis cannot be concluded from the premise.

The following table recaps textual entailment:

Task	Input	Output	Example
Textual entailment (NLI)	Two texts: a premise and a hypothesis.	A classification of the relationship—typically entailment, contradiction, or neutral.	Premise: "Bob's dog is a Golden Retriever." Hypothesis: "Bob has a dog." Result: Entailment (the hypothesis is true given the premise).

From this example, knowing Bob's dog breed implies Bob indeed has a dog, so the entailment holds. If the hypothesis were "Bob has no pets," the relation would be contradiction (premise says he has a dog, contradicting the claim of no pets). If the hypothesis were "Bob's favorite music is classical," that would be unrelated to the premise, hence neutral.

LLMs are quite adept at NLI when fine-tuned, and even zero-shot, many LLMs can guess entailment relations by virtue of their training on huge text corpora. An LLM-based agent might internally use an NLI step to, for example, verify that a potential answer answers a question (entails the question) or to ensure consistency in a generated narrative. Some advanced agents use *chain-of-thought* prompting, where the model is asked to reason: "Does sentence B logically follow from sentence A?" to leverage the model's reasoning capabilities in checking entailment.

11.2 Implementation on AWS

Now that you have an idea what coreference resolution and text entailment are—including their importance and examples of how pronouns link to entities and how one statement can imply or contradict another—this section explains how to implement these capabilities using Amazon's cloud platform. In this section, we leverage Amazon SageMaker to develop, train, and deploy models that perform coreference resolution and textual entailment at scale. We explore setting up custom models on SageMaker, utilizing LLMs such as BART or RoBERTa for high accuracy, and deploying SLMs for efficiency. Real-world examples (continuing the Alice, Bob, and John narratives) illustrate how these models work in an AWS environment, and code snippets demonstrate key steps. By the end of this section, you'll see how AWS enables you to go from idea to a production-ready endpoint for coreference resolution and entailment. Let's dive in.

11.2.1 Custom Model Development on SageMaker

This subsection explains how to build and fine-tune your own coreference resolution and entailment models using Amazon SageMaker. We start with the basics of SageMaker—Amazon's fully managed machine learning service—and how it streamlines training and deployment. We then walk through the typical workflow: preparing data on S3, using pretrained Transformer models (instead of training from scratch) to save time, and launching a SageMaker training job. Finally, we illustrate the deployment of the fine-tuned model as an endpoint and test it with an example. The goal is to show a clear, step-by-step path from raw data to a hosted model that can be integrated into applications.

11.2.1.1 Why SageMaker for NLP

Amazon SageMaker is a cloud service that handles the heavy lifting of machine learning so you can focus on your model and data. For NLP tasks such as coreference resolution and text entailment, which often require large models and extensive data, SageMaker offers a scalable environment. Instead of worrying about setting up servers or GPUs, you define your training job and SageMaker provisions the infrastructure, downloads data, trains the model, and can even automatically deploy the result. SageMaker integrates with other AWS services (S3 for data storage, CloudWatch for logging, IAM for security) to provide a seamless pipeline from development to production. This means you can train a model on a big dataset and, once it's ready, deploy it with a few lines of code as a live API endpoint. In short, SageMaker enables you to transition from a Jupyter notebook experiment to a globally accessible model without leaving the AWS ecosystem.

11.2.1.2 The SageMaker Workflow

The typical workflow for developing a custom model on SageMaker is as follows (also illustrated in the following figure):

Figure caption: High-level SageMaker workflow. Training data is stored in S3, a SageMaker training job runs on managed compute instances to produce a model artifact (saved back to S3), and then SageMaker hosts the model on an endpoint for inference. CloudWatch captures logs and metrics during training and serving.

- **Data preparation:** First, gather and prepare your dataset. For coreference resolution, the OntoNotes corpus (used in the CoNLL-2012 shared task) may be a suitable option, as it contains documents annotated with coreference links. For entailment, you could use a dataset like SNLI or MultiNLI, where each pair of sentences is labeled as *entailment, contradiction,* or *neutral.* Typically, you'll upload your data (e.g., in CSV, JSON, or TFRecord format) to Amazon S3, which is SageMaker's go-to storage for input data. For example, you might have `s3://my-bucket/coref/train.jsonl` and `s3://my-bucket/coref/val.jsonl` for your training and validation sets. SageMaker will pull this data into the training instance when the job starts.

- **Choosing a pretrained model:** Training complex NLP models from scratch is usually unnecessary and expensive. Instead, we start from a pretrained transformer (recall from Chapter 5 that transfer learning is key in NLP). You might choose RoBERTa-base or BART-large as a base for entailment, or a model like SpanBERT (a variant of BERT specialized for spans of text) for coreference. These models have already learned language patterns from vast amounts of text. Your job is to fine-tune them on the specific task. SageMaker makes this easy through its integration with Hugging Face Transformers. In fact, AWS provides Deep Learning Containers (DLCs) with popular libraries pre-installed. For instance, there's an Hugging Face DLC that includes Transformers and the Hugging Face training toolkit. By using these, you don't need to build a custom Docker image—you simply specify the pretrained model name and your training script.

- **Launching a SageMaker training Job:** Once the data is ready and you have decided on a model, you write a training script (e.g., train.py) that will fine-tune the model on your data. This script might use the Transformers Trainer API or PyTorch directly. You then configure a SageMaker *estimator* to run this script. For example, using the SageMaker Python SDK, you can define a HuggingFace estimator with the desired instance_type (such as a GPU instance like ml.p3.2xlarge for heavy models) and hyperparameters (learning rate, epochs, batch size, etc.). Here's a simplified code snippet to illustrate how you would set up a training job for a Hugging Face model on SageMaker:

python

CopyEdit

```
from sagemaker.huggingface import HuggingFace

# Define hyperparameters for fine-tuning
hyperparameters = {
    'model_name': 'roberta-base',    # pre-trained model to
                                     fine-tune
    'task': 'entailment',            # just an example, you can
                                     define your task in script
    'epochs': 3,
    'train_batch_size': 16
}

# Create the Hugging Face estimator
```

```
huggingface_estimator = HuggingFace(
    entry_point='train.py',          # your training script
    source_dir='src/',               # directory of your script
                                     and any helper code
    instance_type='ml.p3.2xlarge',   # type of instance for
                                     training (GPU instance)
    instance_count=1,                # number of machines (use >1
                                     for distributed training)
    role=aws_role,                   # IAM role with SageMaker
                                     permissions
    transformers_version='4.20',     # version of HF
                                     Transformers to use
    pytorch_version='1.12',          # version of PyTorch
    py_version='py38',               # Python version in the
                                     container
    hyperparameters=hyperparameters
)

# Launch the training job (data is on S3)
huggingface_estimator.fit({
    'train': 's3://my-bucket/datasets/coref/train/',
    'validation': 's3://my-bucket/datasets/coref/val/'
})
```

In this snippet, SageMaker will spin up the specified instance(s) with the Hugging Face container, download your train.py and the dataset from S3, and execute the script. Under the hood, this might run a command such as:

```
python train.py --model_name roberta-base --task entailment --epochs 3 ....
```

You can pass any arguments this way. SageMaker handles provisioning the machine, installing the correct software, and even tearing it down after training is complete. This means you only pay for the training time you use, and you don't need to manage persistent servers.

During training, you can monitor progress. SageMaker streams logs from the training job to Amazon CloudWatch so that you can watch metrics or any printouts (like validation accuracy per epoch) in real time. Once training is complete, the model artifacts (typically the model's weights, e.g., a .bin or .pt file, plus any other assets, such as tokenizers) are saved to an S3 location designated by SageMaker. The huggingface_estimator.model_data property in the code will give the S3 path to the model file.

11.2.1.3 Deployment as an Endpoint

After fine-tuning, the next step is to deploy the model for inference. SageMaker can create a *managed endpoint*—an HTTPS API that handles incoming requests and runs the model to generate predictions. To deploy, we define a HuggingFaceModel with the artifact from our training job and then call deploy() to launch it. For example:

python
CopyEdit

```
from sagemaker.huggingface.model import HuggingFaceModel

# Create a SageMaker model object, pointing to the trained model artifacts
huggingface_model = HuggingFaceModel(
    model_data = huggingface_estimator.model_data,   # S3 path to the
                                                      trained model
    role = aws_role,
    transformers_version = '4.20',
    pytorch_version = '1.12',
    py_version = 'py38',
    env = {
        'HF_TASK': 'text-classification'  # for entailment, we could use
                                            sequence classification
        # (for coref, we might need a custom inference script or handler,
        explained below)
    }
)
# Deploy the model to an instance for real-time inference
predictor = huggingface_model.deploy(
    instance_type = 'ml.g4dn.xlarge',
    initial_instance_count = 1
)
```

A few things to note in this code snippet:

- We used `HF_TASK='text-classification'` as an environment variable, which is a hint to the container about what kind of pipeline to use for inference. Hugging Face's Inference Toolkit can automatically set up a pipeline for specific tasks (text classification, question-answering, etc.) so that you don't have to write an inference server from scratch. For a textual entailment model (which is essentially a text pair classification task), using a `text-classification` or `sentiment-analysis` pipeline (with three labels) works—it will feed input through the model and return the label (entail/contradict/neutral). For coreference resolution, which is not a standard pipeline, you would likely provide a custom inference script or specify a custom handler. In a custom script, you would override how input is parsed and output formatted (for example, outputting the clusters of coreferent mentions). SageMaker allows this by letting you subclass the default handler or specify `entry_point` for inference separately, but that's an advanced detail. For simplicity, let's assume we wrap our coreference model so that it returns a list of coreference links given a text.

11.2.1.4 Testing the Deployed Model

Once the endpoint is up (which usually takes a few minutes to provision), we can invoke it. The `predictor` object in the code provides a convenient `.predict()` method for testing. For example:

python
CopyEdit

```
data = {
  "inputs": "Alice thanked Bob for all his help. He gave her a gift in return."
}
result = predictor.predict(data)
print(result)
```

If this were an entailment model, data["inputs"] might need to be a pair of sentences (depending on how the model expects input). For instance, Hugging Face's zero-shot classification uses a separator, but in our fine-tuned entailment, we might format it as "premise: ... hypothesis: ..." inside the model. However, since we set HF_TASK, the model server will likely expect either a single string (for single-sentence tasks) or a dictionary with inputs. The output for an entailment request might look like {'label': 'ENTAILMENT', 'score': 0.98} or a list of labels for each input. For a coreference resolution model, if we were to write a custom handler, the output might be something like a list of clusters. For example, given the text about Alice and Bob, a coreference model could return a result identifying that "Alice" and "her" refer to the same entity, and "Bob" and "He" refer to the same entity. It might look like this:

json
CopyEdit

{ "coref_clusters": [["Alice", "her"], ["Bob", "He"]] }

This indicates that the model links Alice to her and Bob to him, which aligns with our expectations. In an application, you could use this to highlight all mentions of an entity when one is clicked, or to help an LLM-based agent keep track of who is who in a conversation.

11.2.1.5 Cost and Scaling Considerations

SageMaker endpoints hosting LLMs can be resource-intensive (a BART-large or DeBERTa-large model with hundreds of millions of parameters might require several GB of memory). One advantage of SageMaker is that you can choose an appropriate instance type for inference. For example, you could use a GPU instance for low latency or a cheaper CPU instance for cost savings if real-time speed isn't critical. You can also enable autoscaling on endpoints to handle variable loads or use batch transform jobs for large offline predictions. If you only need the model occasionally, SageMaker provides a serverless inference option, or you could deploy on-demand and shut it down when done (to avoid paying for idle time). The key takeaway is that AWS provides flexibility: you can train once and then deploy the model in a way that best suits your production needs and budget.

11.2.1.6 Example Recap

Suppose we want to implement a support ticket analyzer that reads incoming text and decides if the new message is related to an existing issue. This might involve coreference (to link pronouns to the correct earlier entity in the conversation) and entailment (to see if the customer's latest message entails that their problem is resolved or if it contradicts previous info). Using SageMaker, we could fine-tune an entailment model (e.g., RoBERTa) on a dataset of support dialogues labeled for resolution status. We could also fine-tune or use a pretrained coreference model (e.g., one based on BERT) to help maintain the context of who the customer and agent are referring to. Both models can be deployed as endpoints. The application's backend could first call the coreference endpoint to preprocess the conversation text (ensuring all references are explicit), and then call the entailment model to infer if the latest message *"entails"* that the issue was fixed or *"contradicts"* a previous solution attempt. In this way, SageMaker-hosted models become building blocks in a larger intelligent system—even an LLM-driven agent can use these endpoints as specialized tools (for instance, an LLM agent might not have been explicitly trained on the company's jargon, but a fine-tuned entailment model could reliably detect satisfaction or dissatisfaction from the conversation).

11.2.2 LLMs for Coreference and Entailment

In this work, we focus on utilizing LLMs for our tasks on AWS. "LLM" in this context refers to models with a large number of parameters or complex architectures that achieve state-of-the-art results, such as BART, RoBERTa-large, or DeBERTa. We discuss why these models are chosen, how they can be fine-tuned for coreference resolution or text entailment, and what special considerations apply when using them on SageMaker. We use examples to illustrate the power of LLMs (like higher accuracy in understanding nuance) and mention how LLM-based agents might leverage these models. Additionally, we cover any AWS-specific tips for handling big models, such as using optimized hardware or distributed training.

LLMs have shown remarkable capabilities in understanding context and semantics—exactly what coreference resolution and entailment require. For instance, RoBERTa-large (with 355 million parameters) and DeBERTa (a relative of BERT that uses disentangled attention and enhanced mask decoding) have topped leaderboards for tasks like the GLUE benchmark and the SuperGLUE (which includes an entailment task). These models come pretrained on massive corpora, providing them with a general understanding of language. Fine-tuning them on a specific task typically yields very high accuracy because they can capture subtle patterns. For example, that "X went to the store. He bought milk." implies X is male. He refers to X, or that "not arriving on time" contradicts "arrived exactly at the scheduled time."

11.2.2.1 Using LLMs for Coreference Resolution

Coreference resolution can be formulated in different ways for model training:

- **One approach is a span classification model:** The LLM's encoder (such as BERT/RoBERTa) produces contextual embeddings for all possible mention spans in a text, and then a classifier (often a feed-forward network on top of span pairs or a clustering algorithm) determines which spans refer to the same entity. For example, a RoBERTa model can encode a whole document, and then we extract embedding vectors for each detected mention (e.g., proper nouns, pronouns, etc.). A pairwise score can be computed for every pair of mentions to predict if they co-refer. This was the basis of one of the pioneering neural coref models by Lee et al. (2017), and when updated with BERT or SpanBERT encodings, it became robust.

- **Another approach is fine-tuning an LLM in a question-answering style:** You can transform coreference into a question-answering task. For example, ask the model, "Who does *he* refer to in the sentence 'Alice met Bob and he gave her a gift'?" and train it to answer "Bob." However, this isn't the most common method, as it requires generating text answers. More straightforward is classification or clustering.

- **You can also fine-tune a model to output a structured prediction directly:** For instance, with a sequence-to-sequence LLM like BART, you could train it to rewrite text by resolving pronouns (turn "Alice met Bob. He gave her a gift." into "Alice met Bob. Bob gave Alice a gift."). This generative approach uses the LLM's capacity to produce coherent text and can implicitly solve coreferences. The advantage is that it yields a human-readable result (resolved text), though it might struggle with very complex cases without additional constraints.

On SageMaker, training a large coreference model requires a powerful instance (e.g., ml.p3.16xlarge with multiple GPUs), especially if the text input (documents) is lengthy. AWS also offers distributed training libraries—if your dataset is huge or you want to fine-tune a really large model (like if someone attempted a 1.5 billion parameter DeBERTa-XLarge), SageMaker can leverage data parallelism or model parallelism to split the work. In many cases, for coreference and entailment, a single GPU with sufficient memory (such as 16GB) can handle a batch of examples, making single-instance training feasible.

As an example of training an LLM for coreference, imagine that we take SpanBERT-large (an LLM variant of BERT known to be suitable for coref) and fine-tune it on the OntoNotes coreference dataset. On SageMaker, we load the OntoNotes data (after some preprocessing to JSON lines) and use a training script that implements the span classification approach. After fine-tuning, we deploy it. When we send a sentence like "John went home because he was tired," the model might return an output indicating that "John" and "he" are the same entity. If the model is designed to output clusters, it could respond with something like {"clusters": [["John", "he"]]}. Because it's a large model, we expect it to handle even tricky cases by leveraging its deep understanding. For example, given, "The dog chased the cat because it was frightened," a strong LLM-based coref model might correctly link "it" to "the cat" (understanding that likely the cat was frightened, not the dog). This decision requires common-sense reasoning. Smaller or older models might get that wrong, but an LLM fine-tuned properly has a better shot at getting it right.

11.2.2.2 Using LLMs for Text Entailment

Textual entailment has been a showcase task for transformer models. BERT itself was introduced with an NLI-style next sentence prediction in pretraining, and models like RoBERTa-large and DeBERTa achieved very high accuracy on the MNLI (Multi-Genre Natural Language Inference) benchmark. Fine-tuning an LLM for entailment is typically done by adding a simple classification head on top of the model's [CLS] token (or equivalent) representation:

- **The input to the model is formatted to include the premise and hypothesis:** With RoBERTa or BERT, a standard format is: [CLS] premise [SEP] hypothesis [SEP]. The model then outputs a three-way classification: entailment, contradiction, or neutral. LLMs excel here because they capture nuanced language. For example, they learn that "not" or negation words often indicate contradiction, or that certain word relationships (car and vehicle) indicate entailment by synonymy/hypernymy.

- **Another type of LLM that can be utilized is a sequence-to-sequence model, such as T5 or BART, where the task is posed as a generation task:** For example, feeding in the premise and hypothesis in a text prompt allows the model to generate one of the options: entailment, contradiction or neutral. Indeed, the T5 model was fine-tuned on NLI, among other tasks, by simply asking it to output the label as a word. This generative classification can be convenient if you want a single unified model that does multiple tasks (as T5 can) or if you prefer the flexibility of prompts.

On AWS SageMaker, fine-tuning a large entailment model is straightforward because it's essentially a classification problem. You can use the built-in text-classification training example from Hugging Face. For instance, to fine-tune DeBERTa-large on MNLI, you'd specify the model and dataset and likely utilize a data parallel strategy if you're using multiple GPUs (since MNLI has ~400,000 examples, training on a single GPU might be slow). AWS's new generation instances, such as ml.p4d.24xlarge (with eight NVIDIA A100 GPUs) can significantly accelerate the training of such large models.

As an example of deploying an LLM for entailment, say we fine-tuned RoBERTa-large on a mix of general NLI data and domain-specific entailment data (such as legal entailment, if we want to use it for contract analysis). We deploy it on SageMaker. Now we have a powerful endpoint that, given any two sentences, returns their logical relation. If we send:

json
CopyEdit

```
{
  "inputs": "premise: Alice owns a car. Hypothesis: Alice has a vehicle."
}
```

The model would likely return a prediction of entailment with high confidence, because in its training, it has seen patterns where "X owns a car" implies "X has a vehicle." Similarly, if we flip the hypothesis to "Alice has no vehicle," it should return contradiction. These subtle differences in wording are where LLMs shine—they have trained on countless examples of factual relationships and linguistic cues. A model like

DeBERTa, for instance, encodes relative position and word importance in a clever way that helps it rank first on SuperGLUE (a benchmark that includes entailment). By using such an LLM, the SageMaker endpoint becomes a very accurate logical reasoning tool for text.

As far as AWS considerations for LLMs, remember that large models do require more resources:

- **Memory and compute:** A 300M+ parameter model typically requires a GPU with more than 8GB of memory just to load, and even more to handle batch inference. AWS offers GPU instances in various sizes—for real-time inference, even an `ml.g4dn.xlarge` (which has one NVIDIA T4 GPU, 16GB) can host a RoBERTa-large for moderate loads. If you need to handle a high volume of requests per second, you may choose a larger instance or multiple instances behind an endpoint autoscaler.

- **Inference optimization:** SageMaker's inference toolkit (using a multi-model server, as mentioned earlier) can be configured with the number of workers per model, which essentially determines how many processes handle requests in parallel. For LLMs, you might set `SAGEMAKER_MODEL_SERVER_WORKERS` to 1 or 2 per GPU (since each process will use the GPU for the model). You can also enable features such as Amazon Elastic Inference (if you are using TensorFlow models) or NVIDIA's TensorRT for optimized serving, although these are advanced options. Another strategy is model *quantization* (reducing precision from 32-bit to 16- or 8-bit) to save memory and speed up inference—you could prepare a quantized version of the model offline and deploy that if needed.

- **Distributed inference for very large LLMs:** While entailment and coreference don't usually require billion-parameter models, it's worth noting that SageMaker can serve huge models through features like model partitioning across multiple GPUs (SageMaker Large Model Inference). This is used to host models like GPT-J or GPT-NeoX. If you were to use a generative LLM (such as GPT-3-sized) to perform coreference by prompting (zero-shot), it might require such an infrastructure. However, our focus is on fine-tuned, task-specific models, which are manageable in terms of size.

11.2.2.3 LLM Agents on AWS

When building an LLM-driven agent, these high-accuracy models for coreference and entailment can be integrated as tools the agent uses. For instance, consider an agent that answers questions based on documents. The agent could call a SageMaker *entailment model* to double-check that its draft answer is supported by the source text (preventing hallucinations by ensuring an entailment relation exists between source and answer). Likewise, in a complex conversation, an agent could call a SageMaker *coreference resolver* to clarify references, ensuring that when the user says, "Where is *he* now?" the agent knows who "he" refers to from earlier context. This kind of modular tool use is made possible by SageMaker endpoints—your agent can call a REST API for a specific sub-task. Indeed, AWS's robust cloud environment means you could have a suite of endpoints (NER, coref, entailment, sentiment, etc.) and orchestrate them as needed. The agent (which might itself be an LLM like GPT-4 running via an API or AWS's Bedrock service) benefits from the specialist models: it can delegate subtasks to them. In summary, LLMs fine-tuned for coreference and entailment on AWS give you powerful building blocks to weave into larger AI solutions.

11.2.3 The SLM Approach

This subsection introduces SLMs and their applications on AWS for coreference resolution and entailment tasks. SLMs are lightweight models—often distilled or compressed versions of larger models—that offer faster inference and lower resource usage at the cost of a slight drop in accuracy. We discuss why SLMs are helpful, especially in production systems where latency and cost are significant concerns. Techniques such as knowledge distillation, which produce these smaller models, are covered conceptually. We also compare some popular SLMs (DistilBERT/DistilRoBERTa, TinyBERT, MiniLM) to their larger counterparts and guide you on deploying SLMs in SageMaker for real-time applications (e.g., a low-latency chatbot that still requires coreference or entailment).

Bigger isn't always better in practice. While LLMs achieve higher accuracy, SLMs can be "good enough" for many cases and are much more efficient. In scenarios like real-time customer support, where you have to handle dozens of requests per second, a model that is 5% less accurate but responds twice as fast could improve the overall user experience. Moreover, smaller models cost less to run (you might fit an SLM on a cheaper instance, or run more of them in parallel on the same hardware). AWS SageMaker makes it easy to deploy SLMs just like LLMs—the process is the same, but you reap the benefits in performance and cost.

CHAPTER 11 COREFERENCE RESOLUTION AND TEXT ENTAILMENT

SLMs typically originate from a process called *knowledge distillation*. In knowledge distillation, a large "teacher" model (say BERT-base or RoBERTa-large) is used to train a "student" model with fewer parameters. The student model tries to mimic the teacher's outputs (its predictions or its internal representations) on a large amount of data, often the original training set or additional unlabeled data. The idea is that the student learns the essential patterns without needing as many parameters. SLMs can also result from architecture tweaks (for example, using fewer transformer layers, a smaller hidden size, or weight quantization).

Here are some well-known SLM examples relevant to our tasks:

- **DistilBERT (and its variant DistilRoBERTa):** A distilled version of BERT-base. It has about 40% fewer parameters than BERT and runs ~60% faster, while retaining 95-97% of BERT's performance on language understanding benchmarks. DistilBERT achieves this by reducing BERT's layers from 12 to 6 and using a special distillation training where it learns from BERT's softer predictions. For instance, if BERT assigns a 90% probability to "entailment" and a 10% probability to "neutral" for a pair, DistilBERT is trained to produce a similar distribution, not just the hard label. This helps it generalize almost as well as BERT.

- **TinyBERT:** An even smaller model; one version of TinyBERT has only four layers (roughly 14.5 million parameters). Notably, TinyBERT achieves ~96-97% of its teacher BERT's accuracy on GLUE tasks, while being ~7.5 times faster (see `arxiv.org`). TinyBERT uses a two-stage distillation (one during pretraining, one during fine-tuning) and carefully matches not just outputs but also intermediate layer representations with the teacher. For tasks like entailment, a TinyBERT fine-tuned model may be slightly less precise on tricky, challenging, yet still usable cases. For coreference, if a TinyBERT is fine-tuned on OntoNotes, it will be much faster at inference time, which can be vital if you need to resolve coreferences in, say, every sentence of a large document quickly.

- **MiniLM:** A family of distilled models from Microsoft that focuses on refining the self-attention knowledge of larger models. A MiniLM with six layers and approximately 22 million parameters has been shown to outperform older distillations, such as DistilBERT, on many tasks.[1] MiniLM achieves this by not requiring layer-to-layer teacher-student alignment; instead, it only looks at the last layer's attention and value relations. In plain terms, it squeezes the "essence" of how the big model attends to words into a much smaller model. For entailment, Microsoft reported that MiniLM's accuracy was on par with BERT-base, despite being approximately four times smaller. Indeed, the widely used all-MiniLM-L6-v2 model (which is six layers) is popular for tasks like semantic search due to this outstanding balance of speed and performance.

So, how do you use SLMs on AWS? The good news is that once an SLM is trained (often by open-source contributors or researchers), you can fine-tune or directly deploy it just like any other model. For example, Hugging Face provides weights for DistilBERT, DistilRoBERTa, TinyBERT, and MiniLM, among others. You can pick one of these as your base model in SageMaker and fine-tune it on your specific data:

- If you have limited data or need fast training, SLMs train faster as well (fewer parameters to update). Fine-tuning DistilRoBERTa for entailment on SNLI might take a fraction of the time it would take to fine-tune RoBERTa-large.

- The deployment is particularly cost-effective. An SLM can often be served on a CPU-only instance with acceptable latency. For example, DistilBERT can run inference quite quickly on an `ml.c5.xlarge` (4 vCPU) instance without a GPU, especially when model optimizations are enabled. AWS even has a feature called *Inferentia* (custom AI chips) that works well with distilled models—you could compile a DistilBERT model to run on an Inferentia instance, which is a cost-effective way to deploy at scale.

Let's consider a real example. Suppose we want a coreference resolution step in a web application that highlights references when you hover over a name. The user might upload a document, and we need to process it and return the coreference clusters quickly. A large model like BART might give the best accuracy, but if it takes two seconds to process a paragraph, the user experience suffers. Instead, we might use a DistilBERT-

---

[1] sh-tsang.medium.com

based coreference model, which could process the exact text in under 0.5 seconds. The slightly lower accuracy may mean it occasionally misses a link or incorrectly links a pronoun. However, with intelligent post-processing or training on relevant data, it can still perform very well. In practice, many NLP-powered products use these smaller models behind the scenes to meet latency requirements. On SageMaker, we would deploy this distilled coref model on a small instance, possibly even leveraging serverless inference to avoid paying for an idle server when no documents are being analyzed.

To make these differences concrete, Table 11-1 provides a comparison of some LLMs vs. SLMs as they pertain to these tasks.

Table 11-1. *LLMs vs. SLMs for Language Understanding Tasks*

Model	Size (Parameters)	Speed (Relative)	Performance (NLI/GLUE or Coref)	Suitable Use Case
BERT-base (teacher)	~110M	1× (baseline)	100% (baseline performance) on GLUE/NLI	Base reference: good accuracy
RoBERTa-large (LLM)	~355M	~0.5× (slower, needs GPU)	~90% accuracy on MNLI (outperforms BERT-base)	Highest accuracy entailment; research setups
DeBERTa-large (LLM)	~400M (depends on version)	~0.5× (slower, high memory)	SOTA on SuperGLUE (entailment ~90%+)	When top accuracy is needed, if resources allow
DistilBERT (SLM)	~66M (40% of BERT)	~1.6× faster than BERT	~97% of BERT's NLI performance	Real-time inference, limited-resource env
DistilRoBERTa (SLM)	~82M (6-layer)	~2× faster than RoBERTa-base	~95% of RoBERTa-base performance (estimated)	Fast entailment checks in production
TinyBERT-4L (SLM)	~14.5M (tiny!)	~9× faster than BERT	~96-97% of BERT on GLUE	Mobile or high-volume applications
MiniLM-6L (SLM)	~22M	~4-5× faster than BERT	~99% of BERT on many tasks	When you need the best of both worlds (accuracy and speed)

LLMs generally achieve higher absolute accuracy (especially on nuanced language understanding) but are slower and more resource-intensive. SLMs are much faster and cheaper to run, while retaining a very high percentage of the teacher model's performance. The choice depends on the application's requirements.

In the context of coreference resolution, an SLM like DistilRoBERTa fine-tuned for coreference will consume less memory and respond faster when given text, making it suitable for an interactive system. For textual entailment, a TinyBERT fine-tuned on entailment could be deployed in a serverless manner, so that whenever a piece of text needs verification, the check occurs in an instant, perhaps even in parallel across multiple microservices.

11.2.3.1 Deploying SLMs on SageMaker

This follows the same steps described earlier. One additional advantage is that you can pack multiple small models on one machine. For example, SageMaker offers a multi-model endpoint option, which allows a single endpoint to serve multiple models (loading each on demand). If each model is small (say 20MB), this is feasible. This could be useful, for example, if you have one coref model for English, one for French, and one for Spanish—all are DistilBERT-sized. They could theoretically live on one endpoint (since they're small), and you select which one to use per request. With a huge model, you'd need separate endpoints or a much bigger machine.

11.2.3.2 When to Choose SLMs Over LLMs

If you're working in a resource-constrained environment (like deploying a feature to a mobile app or an edge device), SLM is the only choice—you might even go as far as TinyBERT or a six-layer MiniLM to fit in memory and run on CPU. On the other hand, if you're on the server side but need to handle scale (many requests) or low latency (e.g., every web page load triggers a model inference), using an SLM means you can use smaller instance types or serve more requests per second on the same instance. Many companies employ a two-tier approach, utilizing a fast SLM for most cases and resorting to a powerful LLM for borderline cases or periodic rechecks. With AWS, you can deploy both and route traffic accordingly (for instance, 95% of requests go to the cheaper, faster model, and 5% go to the more expensive, accurate model when high confidence is needed).

In summary, the SLM approach on AWS is about efficiency: you leverage the work done by large models (through distillation) to get almost the same capabilities at a fraction of the cost. This democratizes coreference resolution and entailment, allowing even smaller teams or applications with tight budgets to incorporate these advanced NLP features. Whether it's a chatbot requiring instant pronoun resolution or a news feed filter performing on-the-fly entailment checks to avoid fake news, SLMs on SageMaker can be the silent heroes under the hood, delivering results quickly.

You have now seen how to implement coreference resolution and text entailment using AWS infrastructure. From setting up custom models on SageMaker to leveraging large and small pretrained models, AWS provides the tools to train at scale and deploy in a production-ready manner. You can mix and match LLMs for their accuracy and SLMs for their speed, even integrating them into larger AI systems or agents. The next section shifts the focus to performing similar tasks using open-source frameworks on local or non-cloud setups, including harnessing Hugging Face Transformers libraries and discussing how to improve these models with techniques such as mention detection, entity linking, and data augmentation. This will provide a rounded view of our two tasks, both in cloud and open-source contexts.

11.3 Open-Source Implementation for Coreference Resolution and Text Entailment

This section provides a hands-on guide to implementing coreference resolution and textual entailment models using the Hugging Face ecosystem. These tasks are more challenging than simple classification or NER and require a deeper understanding of the document's semantic structure. This section builds on the foundational concepts of fine-tuning language models by addressing more complex and nuanced linguistic challenges in legal documents, specifically coreference resolution and textual entailment. Coreference resolution is essential for understanding who or what is being referred to throughout a document (e.g., distinguishing between "the Company," "Party A," and "the Defendant"). Textual entailment enables us to determine whether a specific legal obligation is implied or explicitly stated, which is crucial for compliance and risk analysis. We explore how to fine-tune open-source models for these tasks, a more sophisticated SLM approach, and techniques to improve their accuracy.

11.3.1 Coreference Resolution

Coreference resolution is vital for clarity in legal documents. For example, "Party A," "the Company," and "it" might all refer to the same legal entity.

Consider these coreference models:

- **Pipeline models:** Hugging Face offers pretrained pipelines for coreference resolution, often built on models like SpanBERT or RoBERTa. These models are trained to predict which spans of text (mentions) refer to the same entity.

- **SLM approach:** Fine-tuning a smaller, specialized model (SLM) for this task can be highly effective. A model like roberta-base can be fine-tuned on a legal corpus annotated for coreference.

- **Improving model accuracy:**

  - **Mention detection:** The first step is accurately identifying all potential mentions (e.g., "the Company," "its obligations," "the aforementioned party").

  - **Entity linking:** Linking these mentions to a canonical entity is crucial. For instance, linking "Party A" to a specific company name. This can be enhanced by using an external knowledge base of parties involved in a document or a legal dictionary.

  - **External knowledge bases:** Integrating an external knowledge base, such as WordNet or a custom legal ontology, can help the model understand that "the Defendant" and a specific person's name refer to the same entity.

This code example demonstrates how to use a pretrained coreference resolution pipeline. We utilize a powerful model, such as Coref-BERT-Large, to analyze a text with a legal-like structure.

Refer to the GitHub for the code illustration: https://github.com/anvcse562/ModernNLP-LLMSLM/blob/main/Ch11/ch11.3/11.3.1.1_Coreference_Resolution_Pipeline.ipynb.

11.3.2 Textual Entailment

Textual entailment is the task of determining whether a hypothesis is true, false, or neutral in light of a given premise. In a legal context, this could mean verifying whether a specific obligation (the hypothesis) is implied by a clause in a contract (the premise).

Consider these entailment models:

- **SLM approach:** A fine-tuned SLM, such as `t5-small` or `distilbert`, can be repurposed for this task. The model is trained on a dataset of premise-hypothesis pairs labeled as "entailment," "contradiction," or "neutral."

- **Improving model accuracy:**
  - **Data augmentation:** Since legal entailment datasets are rare, creating synthetic data by paraphrasing clauses or generating hypothetical scenarios can be a powerful technique.
  - **Domain-specific fine-tuning:** Training the model on a corpus of legal texts (e.g., court decisions, regulations, contracts) will equip it with the necessary legal vocabulary and reasoning patterns that are specific to the domain.

This code example illustrates how to set up a fine-tuning task for textual entailment using a model like `t5-small`. The goal is to classify a premise-hypothesis pair as "entailment," "contradiction," or "neutral."

Refer to the GitHub site for this code illustration: https://github.com/anvcse562/ModernNLP-LLMSLM/blob/main/Ch11/ch11.3/11.3.1.2_Textual%20Entailment_with_FinetunedSLM.ipynb.

11.3.3 Improving Model Accuracy

When dealing with NLP models, particularly in specialized domains such as law, *model accuracy* is crucial. The performance of these models can be significantly improved by focusing on specific components of the NLP pipeline. Two critical areas that often require targeted enhancements are coreference resolution and entailment. Coreference is foundational for understanding who or what is being discussed in a text. Entailment determines if one statement can be inferred from another. The strategies outlined next provide a roadmap for improving the accuracy of models tackling these complex linguistic tasks.

11.3.3.1 Coreference

Improving coreference resolution requires a multi-faceted approach.

- **Mention detection:** The first step is to accurately identify all potential mentions (e.g., "the Company," "its obligations," "the aforementioned party"). Errors in this initial step propagate through the entire pipeline.

- **Entity linking:** After identifying mentions, you must link them to a canonical entity. For example, "Party A" and "the Company" should be related to the same entity. This can be enhanced by using an external knowledge base or a simple dictionary of parties within a specific document.

- **External knowledge bases:** Integrating external data, such as a custom legal ontology or a glossary of terms, can provide the model with additional context, helping it to distinguish between similar entities or acronyms.

Improving coreference resolution accuracy involves a pipeline approach—you enhance the initial stages of the process to achieve better final results. The code here demonstrates mention detection and entity linking, which are the building blocks of coreference.

Refer to GitHub for the code example: https://github.com/anvcse562/ModernNLP-LLMSLM/blob/main/Ch11/ch11.3/11.3.2.1_coreference%20resolution_Linking_Detection_improvingaccuracy.ipynb.

11.3.3.2 Entailment

Improving entailment accuracy is challenging due to the scarcity of high-quality, labeled datasets for legal purposes.

- **Data augmentation:** Generating synthetic data by paraphrasing legal clauses or creating hypothetical legal statements can significantly expand the training dataset. This helps the model learn a broader range of linguistic variations.

- **Domain-specific fine-tuning:** Instead of using a general-purpose model, fine-tuning a model on a large corpus of legal texts (e.g., court decisions, regulations, contracts) enables it to acquire the necessary legal vocabulary and reasoning patterns, resulting in more accurate predictions.

Improving entailment accuracy focuses on utilizing better data and fine-tuning models. The code on GitHub shows how data augmentation and domain-specific fine-tuning are implemented. See https://github.com/anvcse562/ModernNLP-LLMSLM/blob/main/Ch11/ch11.3/11.3.2.2_Entailment_Model_tuning_boost_accuracy.ipynb.

11.4 Industry Use Cases

11.4.1 Enhancing AI-driven Customer Support

Modern customer support teams manage a diverse array of FAQs, policy documents, API references, and billing guides. A retrieval-augmented assistant that genuinely understands what "it," "they," or "this" refers to, and that can check whether its answers are actually supported by documentation, can slash response times and improve answer quality.

The business value here includes:

- **Faster resolution:** Agents (and customers) get precise, policy-aligned answers within seconds.

- **Lower workload:** Routine queries (refund windows, rate limits, SLA times) are automated.

- **Consistency and compliance:** Answers are grounded in approved sources; entailment checks flag weak or unsupported claims.

- **Context handling:** Coreference-aware queries reduce "hallucinated" context and cut follow-ups.

- **Onboarding:** New agents learn faster with instant, cited explanations.

CHAPTER 11   COREFERENCE RESOLUTION AND TEXT ENTAILMENT

This use case combines SLMs and LLMs in a hybrid QA pipeline:

- **Data loading and indexing:** Sample support documents are chunked using the `RecursiveCharacterTextSplitter`, embedded via `HuggingFaceEmbeddings` (`all-MiniLM-L6-v2`), and indexed in Chroma for retrieval.

- **Coreference (Chapter 11 link):** A lightweight spaCy heuristic resolves ambiguous pronouns by substituting the most recent named entity (ORG/PRODUCT/PERSON/GPE). This boosts retrieval quality before answering.

- **Extractive QA (SLM):** `distilbert-base-uncased-distilled-squad` extracts answer spans directly from the top retrieved passage—fast and cheap for well-scoped questions.

- **Generative QA (LLM):** Uses the Hugging Face Inference API via `HuggingFaceEndpoint` with `google/gemma-7b-it` (fall back to a BitNet repo if available, then to a tiny local model) wrapped in a RetrievalQA chain.

- **Text entailment (Chapter 11 link):** An NLI classifier (`cross-encoder/nli-deberta-base`) scores whether the retrieved context entails the answer. Low entailment can be logged, used to trigger safe responses ("I don't know"), or routed for review.

Here is what's in the notebook:

- **Turn-key setup:** Installs `langchain`, `langchain-chroma`, `langchain-huggingface`, `chromadb`, `transformers`, `sentence-transformers`, `spacy`, and downloads `en_core_web_sm`.

- **Indexing:** Builds a persistent Chroma store with synthetic but realistic support content (accounts/security, billing/plans, SLA, API limits, privacy/compliance, cancellations).

- **Coref and entailment modules:**
  - `resolve_coref(question, chat_history)` (heuristic)
  - `nli_entailment(premise, hypothesis)` (returns label and score)

- **QA paths:**
  - ask_question_extractive(...) (DistilBERT over top context)
  - ask_question_generative(...) (RetrievalQA and hosted Gemma/BitNet fallback/local `distilgpt2`)
- **Demos:** A handful of common support queries, with answers, sources, and entailment scores.

Consider these deployment and scalability options:

- **Local/internal:** Keep indexes and endpoints private; enforces RBAC and audit trails.
- **Cloud:** Managed vector DBs and hosted LLM endpoints for elasticity; add tracing (retrieval hits, latency, entailment distribution).
- **Docker:** Containerize the retriever and QA service, utilizing health checks and autoscaling; monitor entailment and deflection rates.

See the code on GitHub at: `https://github.com/anvcse562/ModernNLP-LLMSLM/blob/main/Ch11/ch11.4/11_4_1_ai_customer_support.ipynb`.

11.4.2 Industry Use Cases: LegalTech and Contract Review

This section outlines a practical, industry-grade use case for legal document analysis, demonstrating how these concepts can be applied to build a robust commercial product for a law firm or a corporate legal department. This use case will detail the architecture, workflow, and benefits of a comprehensive contract review application.

11.4.2.1 The ClauseFinder Automated Contract Review Platform

A corporate legal department, called LawCorp, handles thousands of contracts annually. The manual review process is time-consuming, expensive, and prone to human error, resulting in missed clauses, compliance risks, and prolonged deal closures. They need an intelligent, scalable solution to automate the initial contract review, highlighting critical clauses and potential risks.

We propose developing *ClauseFinder*, an automated contract review platform. This system will leverage fine-tuned language models for coreference resolution and textual entailment to provide in-depth, context-aware analysis of legal documents. It will significantly reduce the time spent on routine contract review tasks, allowing legal professionals to focus on high-level strategic work.

Workflow and Features

1. **Document ingestion:** A user uploads a legal document. An OCR service converts the document to plain text for processing.

2. **Coreference resolution:** A fine-tuned coreference model analyzes the text to identify all mentions of legal entities (e.g., "the Parties," "the Company," "it") and links them. This ensures that subsequent analysis understands who is being referred to at all times.

3. **Textual entailment analysis:** The platform utilizes a fine-tuned entailment model to verify specific conditions and obligations. For example, it can check if a "Confidentiality Clause" truly entails the obligation of "non-disclosure" by comparing the clause against a predefined hypothesis.

4. **Risk highlighting:** The platform uses the model's output to automatically flag clauses that deviate from standard templates or contain unusual language, alerting the legal team to potential risks.

5. **Interactive interface:** The user interface displays the document with the model's classifications and extracted entities overlaid, enabling quick navigation and review.

Outcomes and Benefits

- **Time savings:** Drastically reduces the time required for the initial contract review, from hours to minutes.

- **Risk mitigation:** Automatically flags non-standard clauses and potential risks, minimizing the chance of human error.

- **Increased consistency:** Ensures that all contracts are reviewed against a consistent set of criteria and legal standards, ensuring uniformity and accuracy.

- **Scalability:** The cloud-based architecture enables the tool to scale effortlessly, supporting thousands of documents without performance degradation.

This inference and legal document analysis code snippet demonstrates how to load a fine-tuned model and use it to perform complex analysis on a new legal document. This is the core functionality of the ClauseFinder use case. It shows how to integrate the concepts of coreference resolution and textual entailment into a cohesive system.

Model Loading

- A coreference resolution model (`coref-bert-large`) is loaded to identify and group all mentions of the same entity (e.g., "the Company," "ABC Corp.," and "Its") in the text.

- A pretrained T5 model is loaded and fine-tuned for textual entailment to determine if a specific legal hypothesis is true, false, or neutral based on a clause from the document.

Analysis Function

The `analyze_legal_document` function takes the legal text and performs the following two steps:

1. **Coreference resolution:** The coreference model processes the text to find clusters of referring expressions. This helps the system understand which words refer to the same person or company.

2. **Textual entailment:** The fine-tuned T5 model evaluates specific legal clauses (premises) against predefined statements (hypotheses) to determine if a legal obligation exists. For example, it checks if the statement "The contract is subject to the jurisdiction of New York" is entailed by a clause about Delaware law.

Execution

Finally, a sample legal document is defined and passed to the `analyze_legal_document` function to demonstrate the end-to-end process of using these models to extract insights from legal text.

See the code on GitHub at: `https://github.com/anvcse562/ModernNLP-LLMSLM/blob/main/Ch11/ch11.4/11.4.2_Inference_Legal%20Document%20Usecase_ClauseFinder.ipynb`.

11.5 Summary

This chapter explored coreference resolution and text entailment using LLMs and SLMs. You learned how to build systems that understand entity references across text and determine logical relationships between sentences—critical for tasks requiring deep semantic understanding.

LLMs, such as BART and RoBERTa, offer robust capabilities for high-accuracy context modeling, while SLMs, including DistilRoBERTa and TinyBERT, enable efficient and real-time applications. You also gained practical experience in deploying these models using AWS services, such as Amazon SageMaker, and working with open-source tools like Hugging Face Transformers. Techniques that enhance model performance—like mention detection, entity linking, and domain-specific fine-tuning—were introduced.

Industry use cases included AI-powered customer support, legal document analysis, and advanced language understanding in chatbots. You now have a strong foundation in building models that comprehend and reason over text. The next chapter is the last chapter of the book; it explores emerging trends and future directions in NLP.

CHAPTER 12

Emerging Trends and Future Directions in NLP

This final chapter explores cutting-edge developments in large and small language models that are shaping the next generation of NLP systems. We examine how LLMs have evolved beyond text-only prediction to handle multimodal inputs, perform zero-shot and few-shot learning, and engage in more advanced reasoning. We then delve into new training paradigms that emphasize responsible AI alignment—from instruction-tuning and Reinforcement Learning from Human Feedback (RLHF) to proactive red teaming for safety. Finally, we turn to innovations that make SLMs more efficient and accessible: techniques for model compression, on-device edge AI deployment, and hybrid systems that combine SLMs and LLMs. Throughout, we highlight key buzzwords and concepts—including multimodality, in-context learning, alignment, bias mitigation, human-in-the-loop, edge intelligence, privacy-preserving modeling, and more—while linking them back to the foundations established in earlier chapters. By the end of this chapter, you will understand the trajectory of modern NLP: where we started with pretrained transformers like BERT, how we arrived at today's instruction-following LLMs, and where we are headed in terms of model scalability, fairness, and real-world deployability.

12.1 LLM Evolution: Multimodality, Zero/Few-Shot, and Reasoning

The last five years have transformed NLP from a field dominated by task-specific models into one led by foundation models capable of multitask generalization. LLMs now act as universal reasoning engines, offering *in-context learning*, multimodal understanding, and dynamic alignment with human values. These advances mark a decisive departure from the paradigm of "train once, fine-tune forever" toward a more agile, prompt-driven, and reasoning-oriented future.

12.1.1 From BERT/T5 Pretraining to Prompting Paradigms

Earlier chapters (see Chapter 5 on pretraining foundations) outlined how models such as BERT (2018) and T5 (2019) demonstrated the power of large-scale self-supervised learning. BERT excelled at *masked language modeling*, while T5 reframed every NLP task as *text-to-text translation*. This era was dominated by pretraining and fine-tuning strategies, in which each downstream task required a new round of supervised training.

The arrival of GPT-3 in 2020 shifted the ground: task adaptation no longer required parameter updates. Instead, the model could be prompted with instructions or examples, achieving *zero-shot* or *few-shot* performance. This ushered in the prompting paradigm, where task specification is embedded directly into natural language input. Table 12-1 summarizes these paradigms.

Table 12-1. Comparison of NLP Task Handling Paradigms

Paradigm	Example Models	How Tasks Are Handled	Key Benefits	Limitations
Pretrain and fine-tune	BERT (2018), T5 (2019)	The model is pretrained on generic text and then fine-tuned on each specific task using labeled data.	High task performance after fine-tuning; leverages pretraining for better sample efficiency than training from scratch.	Requires a sizable, labeled dataset and a new training run for each task. Not flexible to new tasks without retraining; deployment of many task-specific models can be cumbersome.
Prompt-based learning	GPT-3 (2020)	A single large model is given a textual prompt describing the task (and optionally a few examples), and it generates the answer without parameter updates.	No additional training is needed for new tasks, enabling *zero-shot* and *few-shot* learning. Tasks can be specified in natural language, making the model very flexible.	Requires huge models to work well (GPT-3's success was driven by scale). Prompt design can be challenging (the model is sensitive to phrasing). May be less reliable on tasks with very specialized requirements or where small, fine-tuned models still outperform prompt-alone approaches.
Unified text-to-text and fine-tune	T5, mT5 (2019)	All tasks are formatted as text-to-text (e.g., "Translate English to French: …" → "…"). The model is pretrained on multiple text tasks and then fine-tuned on a mixture of tasks or a specific task.	Flexible input-output format that can cover many tasks; multi-task fine-tuning can give a single model some ability to generalize to related tasks.	Still requires fine-tuning for optimal performance; it is not truly zero-shot for tasks dissimilar from those seen during training. Essentially, it is a precursor to complete prompt-based zero-shot learning, but it requires more supervision.

12.1.2 In-Context Learning and Instruction-Tuning

In-context learning (ICL) enables a model to condition on demonstrations placed inside the prompt, effectively treating them as ephemeral training data. This reduces reliance on static fine-tuning and moves adaptation into *real-time inference*. Instruction-tuning, exemplified by FLAN-T5 and InstructGPT, extends this by explicitly training models on curated instruction-response pairs. The result is improved alignment with user intent, producing responses that are not only accurate but also helpful and safe.

Case Study   Instruction-Tuned GPT Models: Instruction-tuning transformed GPT-3 into InstructGPT, drastically improving usability for general users. Whereas GPT-3 often produced verbose but irrelevant outputs, InstructGPT learned to follow human phrasing more faithfully. This adaptation reduced the cognitive burden of prompt engineering and catalyzed widespread adoption.

12.1.3 RLHF, Multimodal Integration, and Tool Use

The next leap was Reinforcement Learning from Human Feedback (RLHF), which enabled the iterative fine-tuning of LLMs to better reflect human values. Human annotators scored outputs, and reinforcement learning adjusted the reward model accordingly. This made LLMs more "aligned" with human preferences—safer, less toxic, and more context-aware.

Simultaneously, multimodality emerged as a defining feature. Models such as GPT-4V and PaLM-E can integrate text with images, video, or even robotic sensor data. This enables holistic reasoning across modalities, which is essential for applications such as autonomous driving, radiology, and education.

Finally, tool use—the ability for models to call external APIs, calculators, or databases—expanded LLMs from passive text generators to active reasoning agents. This aligns with the broader trend of *agentic AI* discussed later in this chapter.

12.1.4 Long-Context Reasoning and Chain-of-Thought Advances

A long-standing limitation of LLMs has been their finite context window. Recent advances in models such as Claude 2 and GPT-4 (128K) have enabled reasoning over entire books, complex documents, and long conversations.

Equally transformative has been *chain-of-thought (CoT) prompting*, where models generate intermediate reasoning steps. This encourages transparency and enhances performance on tasks requiring logical depth, such as mathematics or legal reasoning.

12.2 New Training Techniques and Responsible Alignment

As LLMs grow in power, the challenge has shifted from pure performance to responsible alignment. The frontier of research is not just about *what models can do*, but *how they should behave*.

12.2.1 Instruction-Tuning, RLHF, and Red Teaming

The three dominant approaches to alignment are:

- **Instruction-tuning:** Fine-tuning on curated instruction datasets.
- **RLHF:** Reinforcement based on human preference ratings.
- **Red teaming:** Adversarial testing by dedicated groups probing for unsafe or biased behaviors.

Together, these strategies form a multi-layered defense against hallucination, bias, and misuse. Table 12-2 compares these strategies.

Table 12-2. Comparison of Alignment Strategies

Strategy	Strengths	Limitations
Instruction-tuning	Improves task following; scalable with data	Quality depends on dataset diversity
RLHF	Encodes human preferences; adaptive	Expensive, time-consuming annotation
Red teaming	Exposes vulnerabilities; improves safety	Reactive rather than proactive

12.2.2 Bias Mitigation, Fairness-Aware Learning, and Governance

LLMs risk encoding societal biases from their training data. Bias mitigation techniques include *counterfactual data augmentation, adversarial debiasing,* and *fairness-aware objectives.* Beyond technical fixes, AI governance frameworks, such as the EU AI Act and the NIST AI Risk Management Framework, are establishing accountability norms for responsible AI.

12.2.3 Empathy-Informed Reinforcement Learning and Human-in-the-Loop Protocols

A new horizon in responsible AI is empathy-informed RL, where human raters emphasize not only factual correctness but also *tone, empathy, and inclusivity.* This human-in-the-loop design emphasizes the relational aspect of AI systems, which is particularly crucial for healthcare, education, and counseling applications.

12.3 SLM Innovation: Efficiency and Accessibility

While LLMs dominate headlines, the parallel revolution in SLMs has democratized AI. SLMs bring language intelligence to edge devices, IoT platforms, and privacy-preserving settings.

12.3.1 Edge AI, Distilled/Quantized Models, and IoT Deployments

SLMs such as DistilBERT, TinyBERT, and MobileBERT are optimized for efficiency. Techniques like *quantization* (reducing numerical precision) and *distillation* (compressing knowledge into smaller models) make them deployable on mobile phones, wearables, and embedded IoT systems.

Case Study   DistilBERT in Real-World Use: DistilBERT retains 97% of BERT's performance while being 60% smaller and twice as fast. This efficiency enabled companies to deploy real-time NLP on applications like spam filtering, on-device voice assistants, and industrial monitoring.

12.3.2 Benchmarking Efficiency: Latency, Memory, and Context Tradeoffs

Unlike LLMs, SLMs must strike a balance between performance, latency, memory footprint, and context handling. Standard benchmarks now consider energy per inference and sustainability metrics, aligning with global priorities for producing green AI.

12.3.3 Transfer Learning with SLMs: Knowledge Distillation, Task Adapters, and LoRA

SLMs benefit from *lightweight adaptation methods*. Low-Rank Adaptation (LoRA) enables task fine-tuning with minimal additional parameters, while adapters plug into pretrained models without retraining the entire network. This modularity ensures scalability and accessibility even for small enterprises (see Table 12-3).

Table 12-3. SLM Efficiency Techniques

Technique	Description	Benefits	Limitations
Distillation	Transfer knowledge from LLM to SLM	High compression, fast	Some loss of accuracy
Quantization	Reduce precision (e.g., FP16 to INT8)	Smaller footprint	Risk of degraded reasoning
Pruning	Remove redundant weights	Energy efficiency	May harm robustness
LoRA/adapters	Low-rank fine-tuning modules	Fast adaptation, modularity	Adds minor complexity

12.3.4 Hybrid Collaboration: SLM–LLM Router Systems and Cascade Pipelines

The frontier is *hybrid pipelines,* where SLMs handle lightweight tasks (e.g., classification, routing), while LLMs tackle complex reasoning. This yields both efficiency and scalability. Router systems dynamically decide which model to invoke, balancing cost, latency, and accuracy.

12.4 Scaling Models: Tradeoffs Between Size, Accuracy, and Efficiency

Throughout this book, we have observed an ongoing tension in the NLP model design: larger models tend to yield higher accuracy and more capabilities, yet they incur significant costs in terms of computational resources and latency. This section steps back and examines how to scale models thoughtfully—balancing model size, data, and compute to meet various practical goals. Scaling is not just a race to build the biggest model; it involves strategic tradeoffs. We discuss an evidence-based framework for determining when a gigantic, dense LLM is warranted, when a sparse mixture-of-experts (MoE) approach might be more efficient, and when a compact SLM is actually the more intelligent choice. The key idea is that there are three axes of scaling—parameters, data, and compute—and finding the "sweet spot" depends on the task and deployment scenario. We also explore how *sparsity* (activating parts of a model on demand) and *compression techniques* allow us to scale differently, achieving strong results at a fraction of the usual cost. Throughout this section, we tie these concepts to real-world use cases, ranging from large-scale AI agents that require top-tier reasoning to mobile applications that need instant responses. The goal is to provide practical guidance on optimizing NLP systems along the size-accuracy-efficiency frontier. By reflecting on these tradeoffs—and linking them back to the themes of multimodal integration, reasoning, and efficiency from earlier in the chapter—we aim to conclude with a holistic perspective on how modern NLP is scaling up and out in a sustainable and accessible way.

12.4.1 The Three Axes of Scaling

When we talk about "scaling up" a language model, we often intuitively mean making the model larger (i.e., increasing the number of parameters). Indeed, the dramatic increase in capabilities from early GPT models to GPT-3 and beyond was primarily driven by the rise in parameter count. However, model size is only one axis of scaling. Equally important are the amount of data the model is trained on and the amount of compute (processing power and time) expended during training.

In Chapter 4, we introduced the concept of *scaling laws*, which are empirical relationships discovered by researchers that relate model performance to specific factors. Let's revisit that with a broader lens and updated insights. The pioneering work by Kaplan et al. (2020) found power-law relationships between model size, training data quantity, and the resulting loss/accuracy—in general, more of everything (parameters, data, compute) yields better performance, following a predictable curve of diminishing returns. Hoffmann et al. (2022) (the *Chinchilla* study) refined this by demonstrating that many earlier models had been off-balance—they had too many parameters for the amount of data they were trained on—and that for a given compute budget, there is an optimal way to allocate between training more data versus building a bigger network. In Chinchilla's case, a 70B model trained on 1.4 trillion tokens outperformed a 175B model (GPT-3) trained on 0.3 trillion tokens, establishing a new norm: don't just scale up parameters, scale up data in tandem, roughly linear in this instance.

However, one of the key lessons of recent research is that "compute-optimal" choices are not one-size-fits-all. The ideal balance of parameters and data depends on the task or skill of interest. A 2025 study by Roberts et al. explored *skill-dependent scaling laws*, comparing knowledge-based tasks (such as factual question answering) to reasoning-intensive tasks (with coding serving as a proxy). They found striking differences—*knowledge-heavy QA tasks tend to be capacity-hungry*. They benefit disproportionately from larger models (with more parameters to store and retrieve facts), whereas *code generation tasks are data-hungry*. Given a fixed compute budget, having more training examples of code yields a greater benefit than extra model size. In practical terms, this means if you're building a model primarily to answer trivia and world knowledge questions, you might lean toward a bigger model even if you can't use as much data. Still, if your model's main job is something like writing code or solving math, you'd be better off using the compute to feed it more examples (even if that means using a slightly smaller model).

These differences can significantly shift the optimal parameter count. In fact, Roberts et al. showed that depending on the composition of the validation set, the estimated best model size can vary by 30–50%. That's huge—imagine choosing between a 10 billion or a 15 billion parameter model purely based on whether your target task is more about stored knowledge or about pattern generalization.

The takeaway is that the scaling strategy should be informed by what "skills" you need: some skills emerge with more parameters, others with more data. This also suggests a nuanced approach in multi-skill models: a balanced training data mix is crucial, and you must carefully choose how to evaluate the model (the validation set) so it represents all desired skills; otherwise, you might over-invest in size or data for one skill at the expense of another.

Another often overlooked axis is *inference-time compute*—the amount of computation used *per query* when the model is serving requests. Thus far, we talked about training, but once the model is trained, you still have options at inference: you could use a bigger model with a single forward pass to answer a question, or a smaller model but allow it to do more work (like multiple passes, or an ensemble of models voting). There's a tradeoff between model size and inference procedures. For example, one strategy to improve a model's answer quality is to use *self-consistency*: have the model generate multiple solutions and then pick the most common answer. This increases the inference compute (since you run the model several times), but can boost accuracy without changing the model's size. Similarly, an ensemble of three medium models might rival a single large model in accuracy, at the cost of three times the computation per query. Recent analyses formalize this by extending scaling laws to account for inference demand. If you expect an NLP system to handle very high query volumes, the cost of inference becomes as important as the training cost. In such scenarios, *it may be optimal to train a smaller model for a longer period* (with more data) so that each inference is cheaper, even if training that smaller model slightly longer requires extra computation. This counterintuitive idea was exemplified by Meta's LLaMA-2 and LLaMA-3 models, which were trained on substantially more tokens than Chinchilla-scaling would dictate.

Why "over-train" a 7B or 13B model on enormous data? Because if that yields the same quality as a 20B model trained less, the smaller model will be much cheaper to deploy at scale. In other words, *for inference-heavy applications, a compute-optimal solution often involves a smaller, more efficiently trained model.* An illustrative figure from Sardana et al. (2024) shows that if you anticipate on the order of 10^9 or more queries, the total cost (training and inference) is minimized by a model significantly smaller than the classical optimum, trained on many more tokens. This logic inspired the LLaMA series and others to prioritize *runtime efficiency*, acknowledging that a model spends most of its life in deployment, not in training.

To summarize these axes and tradeoffs, consider the following guidance matrix:

- **If your goal is maximum accuracy on broad tasks and you have ample compute, a dense, large model (with many parameters, trained on as much data as possible) is the straightforward path.** This was the recipe behind GPT-3 and GPT-4—pushing model size to the limit of available hardware and filling it with vast amounts of data. Such models excel in their generality and emergent capabilities, but they are costly to train and deploy.

- **If you have a fixed big compute budget but also high demands for efficiency, consider redistributing that compute.** For instance, instead of a 20B model trained on X tokens, a 10B model trained on 2X tokens could achieve similar or better performance. The latter will answer questions faster (smaller model) and use less memory, which can be crucial for real-time applications or deploying at scale (think of a virtual assistant handling millions of users—shaving off 100 ms per response matters).

- **If your tasks are specialized (e.g., scientific QA versus code completion), adjust your scaling strategy accordingly.** A knowledge-intensive QA system may derive more benefits from increasing model depth/width (for stronger representations).

In contrast, a code model should prioritize seeing diverse code examples (even if that means reusing compute to cycle through more data with a moderate-sized network). In practice, this could mean selecting different model sizes for various domains within the same compute limit—a one-size-fits-all approach will be suboptimal.

- **At inference time, consider compute as a flexible resource.** If using a robust model is too slow or expensive, consider whether a smaller model boosted with a clever method (such as reasoning chains, ensembling, or tool use) could achieve the required accuracy. Conversely, if latency is not a concern (for example, offline batch processing for research), using multiple passes of a model or larger ensembles may yield extra performance without necessitating the design of a whole new model.

This multi-axis view of scaling also connects to earlier topics; for example, in-context learning effectively utilizes inference-time computation (longer prompts with examples, which incur additional token costs to process) to compensate for not fine-tuning the model on a task. Meanwhile, chain-of-thought prompting is another case of spending a bit more inference compute (having the model generate a rationale) to improve accuracy. The broad lesson is that scaling isn't only about making a model bigger—*it's about allocating resources across model complexity, training data, and inference strategies to get the best outcome for your particular use case.*

12.4.2 From Dense to Sparse: Mixture-of-Experts (MoE)

So far, we've implicitly discussed scaling in terms of *dense* models, where every parameter in the network is activated for every input. An alternative approach is to make models *sparse,* so that at any given time, only a subset of the parameters is used. The leading paradigm in this context is known as *Mixture-of-Experts (MoE)*. We first encountered MoE briefly in Chapter 5 when talking about efficient transformer architectures; now, with the context of scaling tradeoffs, we explore why MoEs are a powerful idea for achieving massive model capacity without a commensurate increase in computation.

The core concept of an MoE is simple: you have many "expert" sub-models (for instance, many feed-forward networks inside a transformer layer), and a gating mechanism that routes each input token to only one or a few of these experts. In effect, the model learns to specialize parts of itself—one expert might handle mathematical questions, another might handle casual dialogue, and so on, although specializations in practice are not so interpretable. The key is that if only (say) 2 out of 16 experts are active for a given token, you've just used 1/8th of the full model's computations for that token, even though the total parameter count might be enormous. *Sparse MoE architectures thus offer a way to significantly increase the model's total parameters while keeping the per-token computation (FLOPs) much lower than that of an equivalently sized dense model.* For example, Google's early MoE model (Switch Transformer) had over a trillion total parameters but only activated a tiny fraction of them per input, requiring compute comparable to a dense model with a few billion parameters. In essence, MoE lets you separate "model capacity" from "compute cost per inference."

Why is this useful? Because not every input needs the full breadth of a model's knowledge—if the gating works well, it can send each query to just the right experts. Intuitively, this is akin to having a large team of specialists (the experts), but consulting only one or two for a particular problem, rather than having the entire team weigh in on everything. The challenge, of course, is training the gating mechanism to route inputs effectively and ensuring that all experts receive training (you don't want some experts to be never used and therefore undertrained—a problem known as load imbalance).

Over the past few years, researchers have refined techniques (like load-balancing losses and better gating functions) to make MoEs trainable at scale. The result has been some impressive efficiency gains: one report noted that a 16 billion total parameter MoE model, using only 2.8B parameters per token (sparsity ~18%), matched the performance of a 7B dense model—effectively achieving the quality of a model over twice its dense size at the exact inference cost. Another way to say it: this MoE achieved a parameter efficiency gain of approximately 2.5 times. Many companies and research groups have deployed MoE variants (e.g., Google's GLaM, NVIDIA's and Microsoft's MoE in MT-NLG, Amazon's Djinn, etc.), especially when aiming to scale to massive dimensions where training a dense model is prohibitively expensive.

A large-scale empirical study conducted in 2024 aimed to systematically map out the "scaling laws" for MoE models, similar to those done for dense models. The findings give practical guidance on how to optimize MoE design:

- **The activation ratio (fraction of experts used per token) is the primary driver of efficiency.** The study introduced a metric called *efficiency leverage (EL)* to quantify the amount of compute saved by an MoE. It exhibited a stable power-law relationship, where using fewer experts per input (i.e., increasing sparsity) yielded predictable gains in efficiency. In plain terms, if you only activate one expert instead of two (halving the activation ratio), you might expect, say, a consistent doubling of EL (these numbers are illustrative). This means MoE performance scales nicely as you make the model more sparse—albeit with the assumption that the model is "well-trained" so that the most relevant expert is indeed sufficient to get the answer right.

- **Expert granularity (i.e., the size of each expert and the number of experts) has an optimal range.** Using many tiny experts or a few giant experts is suboptimal—there is a sweet spot. Empirically, they found the best results when each MoE layer had on the order of 8 to 12 experts. Fewer than that, and you don't get much sparsity benefit; more than that, and each expert becomes too small or the gating becomes too diffuse, hurting quality. This aligns with earlier observations that often top-two expert routing (using two experts per token out of maybe 16) was a robust choice. It seems that having a dozen or so experts per layer strikes a balance between specialization and capacity.

- **The advantage of MoE grows with scale.** The efficiency gains from MoE become more pronounced as the training compute budget increases. In other words, if you're only training on a small dataset or with limited steps, a sparse model might not show much benefit (and could even underperform if experts don't get enough data to specialize). But at massive training scales, MoEs shine—their relative efficiency follows a power law as compute increases. This is a crucial point: MoE is a strategy that yields the most benefits from substantial pretraining efforts, which is why it is employed in frontier models. If you only have, say, a few billion token dataset, a dense model might be more straightforward and just as effective. But if you have trillions of tokens to process and the hardware to do it, MoE can give you more bang for your buck.

- **Other architectural factors (shared experts, arrangement of MoE layers, etc.) are secondary.** The study found that once the main sparsity parameters (i.e., the number of experts and the number to use per token) are set, the rest is relatively straightforward. Some designs share experts between layers or intermix dense and MoE layers, but these details tend to have only minor effects on overall efficiency. In practice, this simplifies the use of MoE: you can, for example, alternate dense and MoE layers or place MoE in every other layer, and it usually works without the need for exhaustive tuning. The gating algorithm (traditionally based on token embeddings and a softmax to pick top experts) has also converged to a few standard choices that work.

Putting these together, MoE offers a clear tradeoff option—if you need a model with a *capacity* equivalent to a 100B-parameter dense network, you could build an MoE that is, say, 100B total but only 10B active at a time. This might run at a similar cost to a 10B dense model while (ideally) performing closer to the 100B one. The tradeoffs are that MoEs are more complex to implement and train (distributed training is almost a must, as each expert can reside on a different device). Inference becomes less predictable (sometimes two tokens may be assigned to various experts, which could be on different machines, introducing communication overhead). Additionally, specific tasks may not naturally lend themselves to being split among experts—if a single input requires knowledge spread across multiple experts, forcing only one expert to handle it might limit performance. Nonetheless, for many broad data scenarios, MoEs have proven effective. Notably, one of the world's largest open models, Google's Switch-C, used an MoE to reach trillions of parameters, and Microsoft's MT-NLG (a massive MoE model) powered some of their Azure AI services with improved efficiency.

To provide a concrete scenario, consider a multilingual model designed to handle dozens of languages. A dense model must juggle all languages within its single parameter space. An MoE could, in theory, learn experts that specialize in specific languages or groups of languages. For instance, an input in Spanish may prompt an expert focused on Romance languages, while an input in Japanese may activate an expert specializing in East Asian languages. This way, the capacity (i.e., the number of parameters) allocated to each language can be greater than if a single monolithic model handled everything. Google, in fact, explored this with their GLaM model, finding that MoE experts sometimes organized by language or topic. This specialization can be both a blessing (in terms of efficiency, as not every parameter is loaded with every language) and a challenge (if the gating misroutes an input, performance could drop).

In summary, sparse models, such as MoEs, represent a different path on the scaling roadmap. Where dense scaling says, "throw more compute to use all parameters for every task," sparse scaling says, "throw more parameters in, but only use what you need each time." The frontier of NLP is likely to see a combination of both. Some teams will prefer the simplicity of dense models (especially for moderate scales where MoE overhead isn't justified), while others pursuing extreme scale will leverage MoE to push boundaries.

Importantly, MoEs remind us that *effective parameter count* and *actual compute used* can diverge—a theme that resonates with our earlier discussion on inference-time tradeoffs. MoEs push that idea into the architecture itself. As you proceed to the following subsection, which focuses on small, efficient models, keep in mind that many of the compression techniques discussed there can also be applied to MoEs (for example, you could quantize the experts or distill a dense model into a sparse one, or vice versa). All these tools—scaling laws, MoEs, compression—are complementary pieces in the grand puzzle of how to get the most language intelligence for the least cost.

12.4.3 SLMs: Efficiency First

In contrast to colossal models and MoE behemoths, this subsection shifts the perspective to the other end of the spectrum: SLMs and how we can maximize efficiency with limited parameters. Throughout the book, we emphasized the growing importance of efficient models that can run on everyday hardware or under tight latency requirements. Here, we will highlight how a combination of *compression techniques* and clever training can make a small model punch above its weight. The motto is "efficiency first"—prioritize the practical deployability of a model, then see how far we can scale *down* while preserving accuracy. This involves methods like distillation, pruning, quantization, and the use of lightweight adaptation modules. We also mention how even advanced tricks (like the chain-of-thought reasoning we discussed earlier) are being distilled into smaller models, bridging some of the gap in complex reasoning tasks between SLMs and their LLM big siblings.

In Chapter 8, you learned about *knowledge distillation,* where a larger "teacher" model is used to train a smaller "student" model by transferring knowledge. Distillation is a cornerstone of creating SLMs that retain much of the accuracy of LLMs. For example, DistilBERT compresses BERT by ~40% while preserving around 97% of its performance on language understanding benchmarks. How? By having the student mimic the teacher's predictions (soft logits) on a large corpus, and also by hint-training the student on intermediate representations of the teacher. This way, the student doesn't just learn from the final task labels (which are often limited), but from the rich behavior of the teacher model across many inputs. In recent years, researchers have enhanced distillation with richer forms of supervision.

One exciting development is to use rationales or chain-of-thoughts from a teacher model as part of the training data for the student. For instance, a large model like GPT-4 might not only provide the correct answer to a commonsense question but also a step-by-step explanation. If we then train a smaller model on those explanations (as input-output pairs), the smaller model can learn to produce similar reasoning chains and arrive at answers more accurately than if it were trained on the answer alone.

This was demonstrated in 2023 by a technique called *Symbolic Chain-of-Thought Distillation (SCoTD)*. In that work, a 1.3B-parameter student (roughly the size of GPT-2) learned to solve tricky commonsense QA problems significantly better by training on the "thinking process" of a 175B teacher model. The distilled student even generated its own chain-of-thought that humans judged to be as valid as the teacher's, despite the massive gap in size. This approach—providing the student with more than just the bare answer—is like teaching a math student not only with homework answers, but also with complete worked solutions, so that they can understand the method. For small models, it significantly improves their reasoning ability, helping to close the gap on tasks such as logical reasoning, mathematical word problems, and multi hop questions that typically *require* a large model to solve correctly. Similarly, researchers have found that distilling multiple skills or steps (sometimes referred to as *mixed distillation*) can yield a student that combines abilities, such as reasoning and factual knowledge, in one compact network.

Another technique in the efficiency toolbox is iterative pruning and retraining. Pruning means removing weights or neurons that contribute little to the model's predictions (often those near zero). If you naively prune, say, 30% of the model's parameters and then continue using it, the accuracy will drop. However, if you prune gradually—removing a small percentage of weights and then fine-tuning the model on the training data to recover accuracy—and repeat this cycle, you can often cut away a substantial portion of the model *without significant performance loss*. It's akin to slimming down a network by carefully trimming fat while keeping the muscle. For example, you could start with a 100M parameter model and end up with 50M that performs almost the same on a task, after a few rounds of pruning and retraining. This works because deep networks exhibit a high degree of redundancy. In Chapter 5, we mentioned the "Lottery Ticket Hypothesis"—that within a large model, there exists a smaller sub-network that can be trained to achieve similar results. Iterative pruning is a method for identifying such sub-networks *post hoc*. Modern approaches even use *structured pruning* (dropping entire neurons or attention heads, rather than individual weights) to make the resulting model smaller and faster on standard hardware. By combining pruning with distillation (first distill to shrink the model somewhat, then prune to shave off even more), researchers have achieved impressive compression ratios. There have been cases of taking a monstrous model (with billions of parameters) and ending up with a model ten times smaller that retains, say, 90% of the original accuracy—which for many applications is an acceptable tradeoff.

Perhaps the biggest game-changer in SLM deployment has been *quantization*. Quantization reduces the precision of model weights from the usual 16- or 32-bit floating-point numbers down to 8-bit or even 4-bit integers. This can significantly reduce model size and memory usage: a 4-bit quantized model is one-quarter the size of a 16-bit model. The trick is to do this without wrecking performance. Basic 8-bit post-training quantization often works out of the box for transformers—you can quantize weights to `int8` and typically see minimal impact on accuracy (some specific layers, such as output projection, might need to be kept in higher precision, but tools handle that automatically).

Pushing to 4-bit is more challenging, but 2023 introduced new techniques, such as *QLoRA* (Quantized Low-Rank Adaptation), which made 4-bit training and fine-tuning feasible. QLoRA demonstrated that you could fine-tune a 65B-parameter LLM on a single GPU by keeping the model weights in 4-bit precision and only learning a small LoRA adapter on top. Astonishingly, this process preserved essentially 100% of the model's original performance. In their experiments, the 4-bit fine-tuned models achieved approximately 99% of the performance of a full 16-bit fine-tune on various instruction-following benchmarks. This means there was almost no penalty for quantizing, which is great news for practitioners. It implies you can take a powerful model and compress it to 1/4 of its memory footprint, with negligible impact on accuracy, which then lets you deploy that model in places you otherwise couldn't.

Building on this, there's active research into *sub-4-bit quantization*—using 3-bit or even 2-bit weights. That's challenging because at such low precision, information loss becomes severe. One idea is *quantization-aware training* (QAT), where you simulate low-precision during training, allowing the model to adjust to it. Another is hybrid schemes where most weights are 4-bit, but a few critical ones are higher precision.

CHAPTER 12 EMERGING TRENDS AND FUTURE DIRECTIONS IN NLP

Another approach is to quantize weights, while keeping activations in higher precision to mitigate errors. By combining these techniques, some experimental 3-bit models have been demonstrated to be effective, and 4-bit models are already quite mature.

The inclusion of *PEFT* (Parameter-Efficient Fine-Tuning) techniques with quantization, as QLoRA does by combining LoRA adapters with a quantized base, enables us to fine-tune locally on modest hardware. This democratizes the ability to adapt big models: a small organization can take a 30B model, quantize it, and fine-tune on a specific domain using just a consumer-grade GPU, yielding a specialized SLM that performs well and is deployable on CPUs or phones.

Let's paint a tangible picture of what these efficiency gains mean. Imagine a customer support chatbot for a large company: it needs to answer in real time, respect user privacy (ideally running on-device or at least without sending sensitive data), and handle potentially thousands of simultaneous users. A few years ago, the only way to get high-quality language understanding for such a bot was to use a cloud-based large model (like a BERT-based system or an API to GPT-3), which is expensive and could introduce latency. Today, thanks to SLM techniques, you can have a distilled and quantized model that's small enough to run in a mobile app or a web browser. For instance, a 100M-parameter model distilled from a big transformer, pruned down to 50M, and quantized to 8-bit would be just 50 million bytes (~50 MB)—easily loadable on a modern smartphone. That model might achieve, say, 90-95% of the accuracy of a state-of-the-art cloud model on the support domain, but with virtually zero latency (since it's on-device) and complete data privacy. This isn't science fiction; it's precisely the trend we're seeing with frameworks like ONNX and TensorFlow Lite enabling mobile BERT-like models, and with companies releasing "mobile GPT" variants. Apple, for example, has versions of its neural engine that run Transformer-based text models for Siri and on-device dictation.

Another example: the open-source community distilled the Alpaca/LLama models down to a 7B parameter variant that can run on a laptop CPU—and while it's not as fluent as GPT-4, it's surprisingly capable for many queries.

In the context of LLM agents, these efficiency gains are crucial. An LLM-based agent might need to run many LLM calls in a loop (planning, querying tools, reflecting, etc.). If each call is to a giant model, the agent will be slow and costly. By using a mix of model scales (for instance, a larger model only for the most complex steps and a smaller model for routine steps) or by compressing a model so that the agent's iterations are faster, we can approach real-time, affordable agents. In fact, some recent agent frameworks allow a form of *model cascading,* where the agent can use a less complex model to propose actions and occasionally verify or refine them with a stronger model. This extends the scaling tradeoff philosophy to interactive systems: not every action requires a 100B model; a 1B model might handle most of it, with a larger model stepping in only when needed.

To tie everything together, let's reflect on the leadership perspective and future. The industry is now awash with options on how to scale NLP solutions. The naive approach of "always use the biggest model available" is giving way to a more nuanced, optimal approach. Leaders in AI deployment (whether in tech companies or other sectors) are asking questions like these:

- Can we fine-tune a smaller model to get what we need instead of calling an expensive API?

- Do we really need real-time answers, or is batch processing okay (which might allow more complex reasoning)?

- Should we invest in training a single, large model, or several specialized ones?

- How do we serve millions of users without breaking the bank or the grid?

The topics in this chapter provide pieces of the answer. For example, if a company requires a multilingual assistant capable of covering 50 languages, a sparse MoE model may be the most efficient way to accommodate all languages within a single system. If a startup is building a mobile app that summarizes news, a distilled 6B-parameter model running on-device could offer a snappy user experience with privacy. Suppose a research lab is pushing the envelope on question answering. In that case, they might train a dense model to see how far pure scaling can take them, but also analyze whether specific capabilities plateau and might require new ideas beyond just scaling.

In conclusion, the tradeoffs between size, accuracy, and efficiency define the practical landscape of NLP today. Dense, large models provide raw power and are invaluable for general AI capabilities; sparse models, like MoEs, offer a clever shortcut to extreme scale; and small models with compression ensure that these advances are accessible and widely deployable. The beauty is that this is not a zero-sum game—we can use *all three approaches in harmony.* A real-world NLP system might utilize a compressed model for edge processing, a medium-sized model for routine server requests, and a large model for the most complex queries or research.

As the final takeaway from this chapter and the book: *modern NLP is about judiciously mixing innovation with optimization*—we have a spectrum of model sizes and techniques, and the frontier is about using the right tool for the job, often combining them, to create AI solutions that are powerful and practical. This balanced viewpoint sets the stage for the future directions of NLP, where progress will be measured not just by raw benchmark scores, but by how efficiently and inclusively we can deliver these AI capabilities to everyone.

12.5 Future Applications

The evolution of LLMs and SLMs is paving the way for a future where natural language becomes the primary user interface, enabling new ways to interact with technology:

- **Conversational interfaces and the agentic shift:** Beyond simple chatbots, the next wave of NLP applications will be AI agents that act autonomously to achieve user-defined goals. These agents, powered by an LLM "brain," will leverage memory (both short-term context and long-term knowledge in vector databases) and tools (APIs, calculators, search engines) to plan and execute multi-step tasks. For example, a legal agent could not only summarize a document but also query a legal database, draft a memo based on its findings, and even schedule a follow-up meeting by calling a calendar API. This enables LLMs to transition from being mere generators to orchestrating complex workflows. The architecture of such agents often includes a perception loop (sensing the environment), a planning module (breaking down goals), and actuators (executing actions via tools). The development of these agentic solutions is already underway, with companies exploring how to use them to streamline complex data access within large organizations, as seen in recent applications.

- **Domain-specialized AI agents:** While today's models are generalists, the future will likely see the rise of highly specialized AI agents in fields such as healthcare, law, and finance. These agents will be trained and aligned to specific compliance requirements and domain knowledge, enabling them to provide accurate, hallucination-free support for complex tasks. For example, a legal agent could summarize case law, or a medical agent could assist with electronic health record management.

- **Real-time multilingual and multimodal experiences:** Building on the trend of multimodality, future applications will seamlessly integrate text, voice, and vision to create a seamless user experience. We can expect real-time, live translation that captures emotional context and nuance, voice assistants that understand complex commands and context, and unified language-vision tools in consumer and professional settings.

- **Low-resource language inclusion:** A major ethical and business frontier is expanding NLP to support the world's over 7,000 languages, many of which are underserved by current models. Initiatives to create cross-lingual models (like mBERT, XLM-R, and NLLB) will continue, breaking down language barriers and allowing more people to participate in the digital world.

- **Quantum-augmented NLP (QNLP):** While still a nascent and experimental field, QNLP explores how quantum computing could fundamentally change NLP. By representing words and grammar as quantum circuits, QNLP aims to leverage quantum properties, such as superposition and entanglement, to process vast amounts of linguistic information simultaneously. While a commercially viable quantum computer is years away, this research suggests potential for unprecedented speed and efficiency in complex semantic tasks that are computationally prohibitive for classical computers.

12.6 Ethical Considerations

As NLP models become more powerful and ubiquitous, addressing their ethical implications is no longer an afterthought; it's a critical part of their development lifecycle. The rapid advancement of NLP necessitates a robust framework for ethical governance, focusing on bias mitigation, responsible development, and environmental sustainability.

12.6.1 Bias Mitigation

Bias is a significant challenge, as models can inherit and amplify societal prejudices present in their training data. Addressing this requires a multi-faceted approach:

- **Multi-stage mitigation:** Bias is addressed at every stage:
  - **Pretraining:** Curation of diverse and balanced datasets, along with the use of techniques such as pseudonymization to remove personally identifiable information, are crucial steps in preventing models from ingesting and perpetuating societal biases.
  - **Training:** Fairness-aware learning objectives can be introduced during fine-tuning to penalize outputs that exhibit bias against protected attributes.
  - **Post-processing:** The final output can be filtered or calibrated by a separate system to ensure fairness, acting as a final "safety net."
- **Practical techniques:** Developers are increasingly using red teaming (proactively and adversarially testing models for harmful outputs), counterfactual data augmentation (creating balanced datasets to teach models to be fair), and RLHF with specific safety policies to reduce harmful or biased responses.
- **Empathy-informed AI:** Beyond preventing harm, research is exploring how to instill positive traits in AI systems. Empathy-aware reinforcement learning uses human feedback that specifically rewards empathetic or supportive responses, teaching models to recognize and respond to emotional cues. This is a crucial step in building trustworthy and helpful AI assistants in high-stakes fields, such as mental health or customer support.

12.6.2 Responsible AI

Beyond technical fixes for bias, a comprehensive governance framework is essential for the responsible deployment of AI. Organizational and legal frameworks must support technical solutions to ensure accountability and transparency.

- **Governance frameworks:** A robust AI governance framework encompasses policies, processes, and tools that guide the responsible use of AI. Key pillars include:

    - **Transparency:** Documenting a model's training data, intended use cases, and known limitations (e.g., via Model Cards) helps stakeholders understand and trust the system.

    - **Accountability:** Clear lines of ownership and responsibility must be defined, ensuring that someone is accountable when a model produces a harmful output.

    - **Privacy:** Adhering to privacy-by-design principles, using data minimization, and implementing strict access controls are non-negotiable, especially for on-premise and sensitive data deployments.

    - **Human-in-the-loop:** For critical decisions, a human-in-the-loop protocol ensures human oversight, preventing autonomous AI from making unreviewed, high-stakes errors.

    - **Regulatory compliance:** Emerging regulations, such as the EU AI Act, are making responsible AI a legal requirement. These laws will likely mandate risk assessments, bias mitigation, and transparency, pushing organizations to adopt robust governance practices. Additionally, new research is exploring ways to address privacy concerns more efficiently, for example, by using a conditional generator to create privacy-preserving synthetic data without the burden of fine-tuning billion-parameter models.

12.6.3 Environmental Sustainability

The immense computational resources required to train large models have raised concerns about their environmental impact.

- **Efficiency-first designs:** A focus on efficiency is a core part of the solution. SLMs and compression techniques, such as quantization, pruning, and distillation, significantly reduce the compute, energy, and cost required to run NLP applications, leading to a greener footprint.

- **Edge/on-device inference:** Running models locally on devices instead of in data centers reduces energy consumption and latency. This approach, however, requires careful measurement of a device's energy use versus the cloud's, as a seemingly small local process can still be energy-intensive.

By addressing these ethical and sustainability considerations, the NLP community aims to ensure that the transformative power of language models is wielded responsibly, equitably, and sustainably for the benefit of all.

Index

A

Abstractive summarization
    BERT model, 287
    business document, 315
    constrained generation, 290
    fact-checking agents, 290
    fine-tuning, 304
    gap sentence prediction, 288
    GPT-3 and GPT-4 models, 289
    Hugging Face Transformers, 289
    large language models, 303
    open-source tools, 303
    PEGASUS, 288
    prominent models and approaches, 287
    SageMaker, 300–302
    text-to-text transfer transformer (T5), 288
ACT, *see* Active Custom Translation (ACT)
Active Custom Translation (ACT), 337
Amazon SageMaker
    Amazon Comprehend, 101, 102
    architecture workflow, 67
    autoscaling strategies, 70
    challenges, 60
    compilation process, 61
    Comprehend models
        advantages/models, 131
        GitHub link, 131
        handle complex documents, 132
        legal contracts/technical manuals, 132
        lightweight NER model, 133
        text analytics, 130
    compression, 71
    correction systems, 426–428
    cost-effective inference, 101
    deployment, 101
    distillation, 71
    edge deployments, 62
    endpoint, 69
    enterprise solutions, 62
    estimator interface, 58
    fine-tuning
        agents, 392–394
        capability and efficiency, 436
        conversation flows, 394
        deployment, 390
        dialogue system, 387
        domain, 391
        inference and scaling, 392
        JumpStart, 391
        larger models and distributed training, 434–436
        notebook/local environment, 432, 433
        Python environment, 389
        sequence-to-sequence models, 388
        serverless inference option, 392

INDEX

Amazon SageMaker (*cont.*)
    SLM/trade-offs, 435–438
    T5/BART, 431
    training data, 388
    training script, 431
  graph-level optimizations, 61
  high-quality content, 66
  inference/deployments, 104, 105
  instance selection, 70
  machine learning workflow, 57
  marketing automation tools, 67
  marketing workflows, 69
  memory constraints, 62
  model development, 467
    cost/scaling considerations, 473
    data preparation, 468
    managed endpoint, 471, 472
    NLP tasks, 467
    pretrained transformer, 468
    Python code, 469
    support ticket analysis, 474
    testing, 472, 473
    training script, 469
    workflow, 468
  multi-model endpoints, 71
  NER models
    BERT/RoBERTa, 134
    deployment, 136
    DistilBERT/TinyBERT, 138, 139
    fine-tuning and evaluation, 136
    labeled dataset, 134
    legal documents, 134
    parameters, 135
    pretrained model, 134
    Python implementation, 137
    real-time applications, 137
    SLM deployment, 138
    testing and iteration, 136
  on-device inference, 60
  pretrained models, 102–104
  PyTorch/TensorFlow model, 61
  real-time endpoints, 58, 67
  runtime execution, 61
  scalable inference
    A/B testing and continuous improvement, 441
    deployment model, 438, 442
    instance types, 439
    integration, 441
    latency considerations, 440
    LLM agent concept, 441
    real-time endpoint, 438
    subject-verb agreement, 442
    testing, 442, 443
  sentiment analysis, 199
    benefits, 205
    deployment, 201–204
    FinBERT, 199
    tradeoff, 205
    training job via Python code, 200, 201
  subsequent pages, 57
  training environment, 58
  translation system
    comparative view, 353–355
    compliance/proprietary reasons, 352
    evaluation, 356
    JumpStart, 355
    mBART-50, 352
    mT5 (multilingual T5), 352
    multilingual translation, 351
    multi-model endpoints, 355, 356
    quality triage, 353
    text-to-text transfer transformer (T5), 351

INDEX

    training large models, 355
    upcoming pages, 57
Amazon Web Services (AWS)
    classification tasks, 97–105
    cloud-based platforms, 18
    cloud platforms, 22
    computational demands, 18
    coreference/entailment
        models, 467
    correction systems, 426–443
    dialogue system, 379–404
    Fairseq, 21
    hugging face, 21, 22
    hybrid approach, 24
    key advantages, 18, 19
    Lambda
        classification tasks, 60
        high-level workflow, 59
        packages, 59
        real-time inference, 59
    LangChain/LlamaIndex, 21
    marketing content generation
        implementation, 68
        integration, 69
        marketing automation tools, 67
        problem statement, 66
        quality/consistency, 67
        real-time endpoints, 67
        resource constraints, 67
        SageMaker endpoint, 69
        scalable inference, 67
        triggering generation, 68
        workflow, 67
    NER implementation, 130–149
    Ollama, 21
    OpenNLP, 22
    open-source frameworks, 23
    QA systems, 243–258
    sentiment analysis, 189
    services/features, 19–23
    summarization techniques, 294–302
    technical support ticket triage
        API Gateway, 66
        architecture overview, 64
        automated classification, 64
        implementation steps, 65
        model and dependencies, 65
        problem statement, 63
        scalability, 64
        test/monitor, 66
    text generation techniques, 56
        cost optimization, 70
        hardware platforms, 60
        high-level architecture, 63
        inference requirement, 72
        Lambda, 59, 60
        local deployments, 56
        logging inference requests, 71
        marketing content generation, 66
        monitoring, 71
        on-device inference, 60
        performance profiling, 72
        practices/architectural
            considerations, 70
        prompt design, 72
        SageMaker, 57
        security measures, 72
        services, 56
        services working, 62
        technical support ticket
            triage, 63–66
        transition, 57, 58
        unclear model governance, 72
        X-Ray tracing, 71
    translation tasks, 335–356
AWS, *see* Amazon Web Services (AWS)

INDEX

B

BERT, *see* Bidirectional Encoder Representation from Transformers (BERT)
Bidirectional Encoder Representation from Transformers (BERT), 7, 90, 105
    abstractive summarization, 287
    extractive, 285
Bilingual evaluation understudy (BLEU), 447
BLEU, *see* Bilingual evaluation understudy (BLEU)

C

CI/CD, *see* Continuous integration and continuous deployment (CI/CD)
Classification tasks
    adaptive thresholding, 95
    AWS services, 97–105
        Amazon Comprehend, 97–99
        cost and computation comparison, 101, 102
        cost-effective inference, 101
        dataset preparation, 100
        deployment, 101
        fine-tune, 100
        inference/deployment, 104, 105
        Lambda function, 99
        pretrained models, 102–104
        sentiment analysis, 99
    Bayesian optimization, 96
    binary
        context awareness, 90
        DistilBERT/TinyBERT, 92
        handling ambiguity, 91
        tradeoffs, 93
        transfer learning, 90
        transformers library, 91
    considerations and tradeoffs, 97
    ensemble learning, 96
    fundamental task, 90
    industries, 113
        chatbots and virtual assistants, 115, 116
        customer reviews, 114
        social media sentiment, 115
        spam detection, 113, 114
    labels, 94–98
    multi-class tasks, 92, 93
    open-source frameworks, 105
        arguments, 106
        data preparation, 112, 113
        DistilBERT and TinyBERT, 105, 106
        Hugging Face's Transformers, 105
        inference methods, 108
        optimizations, 110
        padding and truncation, 112
        tokenization, 112
        training/inference, 106–112
    optimization techniques, 95
    pretraining strategies, 92
    sigmoid activation functions, 94
    temperature scaling, 95
Conditional Random Fields (CRFs), 121
Continuous integration and continuous deployment (CI/CD), 214
Coreference/entailment models, 457
    AWS services, 467–484
        LLMs models, 474–479
        model development, 467–474
        SLM approach, 479–484
    challenges, 461
    cloud-based/open-source solutions, 457

INDEX

coreference chain, 461–463
customer support teams
    analysis function, 492
    business value, 488
    ClauseFinder, 491
    deployment/scalability options, 490
    execution, 493
    hybrid QA pipeline, 489
    LawCorp, 490
    model loading, 492
    notebook, 489
    outcomes and benefits, 491
    workflow/features, 491
deep learning models, 462
entailment (*see* Entailment models)
fundamental concepts, 458
open-source models, 484–488
    coreference models, 485
    data augmentation, 487
    domain-specific fine-tuning, 488
    legal process, 487
    model accuracy, 486
    multi-faceted approach, 487
    textual entailment, 486
quick reference table, 463
semantic consistency, 459
span-prediction models, 462
understanding context/semantics, 459, 460
Correction systems
    AI-powered approaches, 423–426
    AWS services, 426–443
        Amazon SageMaker, 426–428
        augmentation, 429
        DistilGPT-2, 437
        grammar correction model, 428–431
        Hugging Face Transformers, 427
        MLOps/deployment pipelines, 427
        reliable/scalable, 438–443
        SageMaker, 431
        scalable infrastructure, 427
        secure/training environment, 427
        synthetic data generation, 429
    browsers, 420
    data augmentation, 423
    dictionary, 421
    EdgeFormer, 425
    email platforms, 420
    fine-tuning/prompting, 424
    fundamental application, 418
    grammatical errors/spelling mistakes, 418, 419
    heuristics, 422
    instruction-tuned model, 424
    knowledge and flexibility, 426
    lightweight/distilled versions, 425
    messaging/mobile keyboards, 420
    neural sequence-to-sequence models, 423
    open-source tools
        correctness, 445
        data curation, 448
        diversity/data quality, 444
        error analysis, 446
        evaluation metrics, 447, 448
        feedback collection, 448
        fine-tuning, 444
        Hugging Face ecosystem, 443
        human-in-the-loop, 446
        metrics/evaluation, 446
        retraining data, 448
        scale/computational requirements, 444
        selection model, 445
        user feedback loop, 447

INDEX

Correction systems (*cont.*)
  overview, 417
  proofreading application, 452
    backend microservice, 452
    continuous improvement, 454
    core text correction model, 453
    feedback pipeline, 454
    frontend integration, 452
    GlobalTech, 452
    intelligent suggestions, 453
    model retraining pipeline, 453
    outcomes/benefits, 454
    ProseGuard, 452
    real-time analysis, 453
    scalability, 454
    technical architecture, 452, 453
    user feedback and data pipeline, 453
    user interaction, 454
    workflow/features, 453
  rule-based grammar checking, 421
  sequence-to-sequence problem, 423
  SMT algorithms, 423
  traditional systems, 422, 423
  word processors, 419
  writing assistance tools, 420
    code integration, 451
    components, 451
    data pipeline, 450
    deployment options and scalability, 451
    error detection *vs.* correction, 449
    evaluation metrics, 450
    grammar correction, 450, 451
    grammatical acceptability, 450
    limited context and precision, 449
    minimal demonstration, 449
    overview, 449
    reference corrections, 449
CRFs, *see* Conditional Random Fields (CRFs)

D

Decomposed and Recomposed Attention (DoRA), 262, 305
Deep Learning Containers (DLCs), 469
Dialogue systems
  AWS services, 379–404
    hybrid approach, 379
    Lex, 380–387
    SageMaker, 387–394
    SLMs (DistilGPT-2/DistilT5), 394–404
  cloud and open-source tools, 367
  conversational model, 374
  design/implementation, 367
  fundamental paradigms, 368
  generative dialogue systems, 370–374
  handling structured dialogues, 369
  healthcare virtual assistants, 412–416
  implementation, 374, 375
  IVR system, 410–412
  key aspects, 372
  LLMs *vs.* SLMs
    agents, 379
    chatbot instances, 377
    comparison, 377
    gigantic models, 375
    lightweight conversational agent, 376
    LLM models, 375–379
    SLM models, 376
    smartphone, 377
  open-source community, 404–407

open-source tools
    fine-tuning, 407
    Hugging Face, 404
    intent and intent_ranking, 404
    project-Py file, 405
    Rasa robust, 404–407
rule-based system, 369, 370
travel requirements, 371, 372
virtual customer assistants, 407–410
DLCs, *see* Deep Learning Containers (DLCs)
DoRA, *see* Decomposed and Recomposed Attention (DoRA)

E

ECR, *see* Elastic Container Registry (ECR)
Elastic Container Registry (ECR), 196
Entailment models
    approaches, 465
    contradiction, 464
    GitHub, 465
    hypotheses, 464
    mention detection/linking, 461
    statements, 464
    textual entailment, 466
Extractive summarization techniques
    abstractive, 283
    agents/extraction, 287
    BERT model, 285
    differences, 283
    DistilBERT, 298–300
    fine-tuning, 305
    frequency/graph-based methods, 286
    fundamental approaches, 282–287
    hallucination, 283
    open-source tools, 303
    paragraph, 284

small language models, 304
strengths/limitations, 286
transformer-based models, 285

F

Facebook AI Research (FAIR), 21
FAIR, *see* Facebook AI Research (FAIR)
Fine-tuning
    abstractive summarization, 304
    BERT model, 103
    computational resources, 74
    Hugging Face's transformers, 149–152
    libraries, 76
    parameters, 74–76
    pretrained model, 76, 100
    pretraining phase, 73
    PyTorch or TensorFlow, 74
    save option, 77
    task-specific dataset, 74
    technical requirements, 74
    text-based dataset, 76
    text generation, 77
    tokenization, 76
    training loops/evaluation/logging, 77

G

GEC, *see* Grammatical error correction (GEC)
Grammatical error correction (GEC), 424

H

Healthcare virtual assistants
    business motivation, 412
    code implementation, 416
    deployment/scalability, 415

Healthcare virtual assistants (*cont.*)
   embedding model, 413
   generative dialogue, 413
   implementation, 413
   intent classification, 413
   symptom checker, 414
   technical architecture, 414
   technical issues, 415
Hidden Markov Models (HMMs), 6
HMMs, *see* Hidden Markov Models (HMMs)

I, J, K

ICL, *see* In-context learning (ICL)
In-context learning (ICL), 498
Interactive Voice Response (IVR) system
   chat interface, 412
   dialogue systems, 410
   GitHub, 411
   output view, 412
   web application, 411
IVR system, *see* Interactive Voice Response (IVR) system

L

Language translation, 325
   AWS services
      batching and streaming, 349
      casual style, 338
      compilation, 349
      content moderation system, 340
      cost-effective scaling, 338
      customization options, 337
      deployment, 347
      DistilBART, 346
      edge services, 350
      features, 336
      implementation, 335
      Inferentia chips, 345
      integration, 340
      knowledge distillation, 347
      language auto-detection, 337
      localization, 339
      MarianMT models, 341–346
      multilingual chatbot, 339
      optimization, 349
      profanity masking option, 337
      quality/speed, 349
      quantization, 349
      real-time/batch translation, 336
      real-world scenario, 348
      rule-based systems, 336
      SageMaker, 351–356
      security and privacy, 337
      streaming translation, 349
      terminology file, 337
      TinyBERT/MiniLM, 347
   content localization, 329
   data loading and indexing, 362
   definition, 326
   DistilBERT, 362
   Gemma/BitNet, 362
   global customer base, 361
   implementation components, 362
   key challenges, 326–328
   LLM models, 330
      context/nuance, 330
      cost/privacy, 332
      dynamic adaptation, 331
      limitations, 331
   machine translation, 328
   multilingual content/accessibility, 329
   multilingual content delivery, 363, 364
   NLP applications, 328

INDEX

open-source tools, 356–361
    cost-effective and responsive alternatives, 358
    deployment targets, 359
    DistilBART, 358
    evaluation metrics, 360
    high-resource languages, 357
    Hugging Face Transformers, 357
    inference optimization strategies, 360
    MarianMT, 358
    multilingual transfer, 357
    parameter-efficient fine-tuning, 358
    supervised translation, 357
    training time/memory consumption, 358
query interface, 363
SLM models
    approaches, 333
    BERT model, 333
    context-savvy model, 334
    DistilBART, 333
    efficiency and deployability, 332
    high-volume systems, 334
    machine translation systems, 335
    real-time translation, 334
    resource-constrained environments, 334
    scenarios, 334
    speed and resource efficiency, 332
standard metrics, 325
translation integration, 362
Large language models (LLMs), 1
    alignment strategies, 500
    approaches, 4, 5
    AWS (*see* Amazon Web Services (AWS))
    BERT/T5 pretraining, 496
    bias mitigation techniques, 500
    chain-of-thought (CoT) prompting, 499
    challenges, 8
    classification, 90–116
    cloud platform/open-source frameworks, 22, 23
    contextual comprehension, 6
    coreference/entailment models, 457–493
        architectures, 474
        distributed inference, 478
        driven agents, 479
        inference optimization, 478
        model training, 475
        question-answering style, 475
        span classification model, 475
        structured prediction, 475
        token representation, 476
        understanding context and semantics, 474
    cost/compute/effectiveness
        computational requirements, 34
        cost consideration, 34
        e-commerce platform, 36
        effectiveness, 36
        factors, 34
        infrastructure requirements, 36, 37
        performance and efficiency, 36
        scalability/maintenance, 37
        task suitability, 35
    data quality/bias, 9
    definition, 5
    dialogue (*see* Dialogue systems)
    dominant approaches, 499
    empathy-informed RL, 500
    energy consumption, 9
    evolution, 6, 7
    fairness-aware objectives, 500

INDEX

Large language models (LLMs) (*cont.*)
    foundation models, 496
    human-in-the-loop design, 500
    innovations, 9
    instruction-tuning/ICL, 498
    instruction-tuning/RLHF, 499
    interpretability, 9
    key clauses/flagging potential risks, 7, 8
    language translation, 325–365
    lifecycle activities
        augmentation, 25
        cloud platforms, 32
        cupcake flavors, 25
        data cleaning and augmentation, 25
        data sourcing, 25
        deployment strategies, 31–33
        distributed strategies, 32
        diverse sources, 25
        edge/browsers, 32
        ethical and legal compliance, 25
        evaluation/testing, 27–31
        filtering, 25
        fine-tuning, 26
        hardware, 26
        high-quality, 27
        hybrid approach, 24
        hyperparameter tuning, 27
        model training, 26
        monitoring/versioning, 33
        overfitting, 31
        quantitative metrics, 31
        scheduling updates, 27
    long-standing limitation, 499
    multimodal integration, 498
    NER (*see* Named Entity Recognition (NER))
    niche requirements, 9
    prompting paradigm, 496–498
    real-world impacts, 1
    red teaming, 499
    revolution, 6
    scaling models, 7
    scaling/power, 6
    sentiment analysis, 175–224
    SLMs (*see* Small language models (SLMs))
    summarization techniques, 281–323
    task handling paradigms, 497
    text generation (*see* Text generation techniques)
    transformer architecture, 2–4, 7
    word embeddings, 7
Lex (Amazon)
    augmented conversational agent, 383
    conversational interfaces, 380
    custom logic/backend systems, 381
    Electronic Travel Authorization (eTA), 382
    hybrid approach, 382
    intents/utterances/slots, 380
    interaction, 381
    Lambda function, 382
    limitations, 385
    LLM-based approach, 385–387
    performance, 385
    Python code, 383
    session attributes/active contexts, 381
    small language models, 397
    summarization, 385
    variations, 380
LLMs, *see* Large language models (LLMs)
LoRA, *see* Low-Rank Adaptation (LoRA)
Low-Rank Adaptation (LoRA), 501

M

Machine translation (MT) system, 328
Mixture-of-Experts (MoE), 506–510
    advantages, 508
    architectural factors, 508
    concepts, 506
    designing process, 507
    efficiency leverage (EL), 507
    expert granularity, 508
    multilingual model, 509
    refined techniques, 507
    sparse models, 510
    tradeoff option, 509
MT system, *see* Machine translation (MT) system
MoE, *see* Mixture-of-Experts (MoE)

N

Named Entity Recognition (NER), 117
    adaptive learning, 129
    agents/pipelines, 128, 129
    AWS services
        ACME tasks, 148
        approaches, 130
        auto scaling/load testing, 142
        Comprehend models, 130–133
        concurrency, 143
        decision-making/
           summarization, 146
        deployment
           considerations, 139–144
        end-to-end workflow/scaling, 147
        fallback strategies, 143
        inference optimization, 143
        key advantage, 148
        logging and monitoring, 144
        model size management, 143
        model storage and versioning, 143
        pipeline, 145
        post-processing and validation, 145
        real-world deployment, 144–147
        results/continuous
           improvement, 147
        SageMaker, 134–139
        security considerations, 143
        text extraction, 145
        unstructured texts, 144
    benefits, 129
    data processing, 120
    dynamic pipelines, 128
    entity categories, 119
    industries
        customer data extraction, 172
        legal documents, 171
        quantization benefits, 171
        resume parsing, 170, 171
    interactive debugging, 128
    key information, 118
    LangChain framework, 129
    low-level/small LLMs
        computational resources, 167
        cost-effective/efficient
           solution, 162–169
        cost saving/efficiency, 168
        DistilBERT model, 169
        fine-tuning, 169
        generalization, 168
        memory optimization, 167
        optimization techniques, 162
        real-time tasks, 169
        tradeoffs, 168
    open-source frameworks
        augmentation, 159
        AWS-based solutions, 149
        BERT model, 149, 151

INDEX

Named Entity Recognition (NER) (*cont.*)
    DistilBERT/TinyBERT, 153, 154
    domain text, 159
    evaluation metrics, 158–161
    financial reports, 160
    fine-tuning, 149–152
    Flair embedding, 156–158
    hyperparameter tuning, 159
    knowledge distillation, 159
    optimization techniques, 159
    pipeline optimization, 160
    pruning, 160
    quantization, 160
    quantization techniques, 161
    spaCy, 154–156
  real-world applications, 117, 120
  text classification, 118
  traditional models *vs.* LLMs, 121
    approaches, 121
    comparison, 126–128
    few-shot learning, 124
    Hugging Face's Transformers, 123
    key characteristics, 121, 124
    large language models, 124–126
    SLMs, 121, 122
    SpaCy, 122
    TensorFlow/PyTorch, 123
    zero-shot learning, 124
Natural Language Inference (NLI), 459
Natural language processing (NLP), 1, 496
  bidirectional models, 17
  conversational interfaces, 515
  digital systems, 14
  DistilBERT/ALBERT, 15
  domain-specialized AI agents, 515
  ethical considerations
    accountability/transparency, 518
    bias mitigation, 516
    edge/on-device inference, 519
    efficiency-first designs, 519
    environmental impact, 519
  extractive/abstractive summarization, 16
  key buzzwords and concepts, 495
  language models, 17
  language translation, 328–330
  LLM (*see* Large language models (LLMs))
  low-resource language inclusion, 516
  question-answering (QA) systems, 16–18
  real-time multilingual/multimodal experiences, 516
  sentiment analysis, 15
  SLMs (*see* Small language models (SLMs))
  text classification, 14
  text generation, 17
  unidirectional models, 17
  zero-shot and few-shot learning, 495
Natural language understanding (NLU)
  coreference (*see* Coreference/entailment models)
NER, *see* Named Entity Recognition (NER)
Neural machine translation (NMT), 336
NLI, *see* Natural Language Inference (NLI)
NLP, *see* Natural language processing (NLP)
NLU, *see* Natural language understanding (NLU)

O

Open-source frameworks
  AWS (*see* Amazon Web Services (AWS))
  benefits, 22

classification tasks, 105–113
cloud platforms, 23
correction systems, 443–448
dialogue systems, 404–407
Fairseq, 21
Hugging face, 21
hybrid approach, 24
LangChain/LlamaIndex, 21
named entity recognition (NER), 149–162
Ollama, 21
OpenNLP, 22
question-answering (QA) systems, 258–271
sentiment analysis, 210–220
summarization techniques, 303–308
text generation techniques, 73–87
translation system, 356–361
Opinion mining, *see* Sentiment analysis

P

Parameter-efficient fine-tuning (PEFT), 262, 305, 358, 513
PEFT, *see* Parameter-efficient fine-tuning (PEFT)

Q

QA systems, *see* Question-answering (QA) systems
QAT, *see* Quantization-aware training (QAT)
QLoRA, *see* Quantization-aware Low-Rank Adaptation (QLoRA)
QNLP, *see* Quantum-augmented NLP (QNLP)
Quantization-aware Low-Rank Adaptation (QLoRA), 262, 305, 512

Quantization-aware training (QAT), 512
Quantum-augmented NLP (QNLP), 516
Question-answering (QA) systems
  abstractive models
    approaches, 230
    architectures, 230, 231
    challenges, 232–234
    conversational assistants, 232
    hallucination and accuracy, 232
    Hugging Face Transformers, 231
    inference/synthesis, 231, 232
    resource intensity, 233
    sequence-to-sequence task, 230
    transparency, 233
  agents/multi-step reasoning
    chain-of-thought (CoT) prompting, 239
    concepts, 239
    engineering perspective, 241
    frameworks and libraries, 241
    overview, 240
    professional applications, 242
    ReAct (reasoning and acting), 240
    real-world implementation, 243
    retrieval-augmented generation, 241
    sentences, 242
    steps, 239
    user query, 241
  automated assistants, 276
    benefits, 278
    conversation history handling, 278
    deployment options, 278
    document ingestion and chunking, 277
    embedding and vector storage, 277
    implementation, 277
    query interpretation and routing, 277

INDEX

Question-answering (QA) systems (*cont.*)
    real-time interactions, 277
    response generation, 278
  AWS implementation
    benchmark candidate models, 252
    cold start mitigation, 253
    complexity, 245
    cost and performance, 253–257
    customization/domain adaptation, 247
    degradation, 252
    ensemble/cascaded approaches, 252
    extractive *vs.* abstractive, 243
    inference cost, 246
    JumpStart, 252
    Jupyter notebooks, 258
    latency and cost metrics, 254
    latency requirements, 246
    mapping models, 248
    memory and compute constraints, 247
    model selection strategies, 245–252
    monitor and iterate, 252
    pros/cons approaches, 243–245
    queries per second (QPS), 246
    simplest model, 252
    testing and chaos engineering, 253
    throughput and scalability, 246
    timeouts, 253
  contexts/general knowledge, 225
  extractive models, 226
    abstractive, 229
    approach/architecture, 227, 228
    characteristics, 229
    technical considerations, 228
    use cases, 228
  industries
    applications, 273
    chatbot implementation, 272
    customer support chatbots, 272
    deployment options, 274
    embeddings/vector storage, 273
    extractive, 272
    generative, 272
    knowledge documents, 273
    querying interface, 273
  knowledge base querying
    approaches, 274–276
    code integration, 275
    customer satisfaction insights, 276
    implementation, 275
  large *vs.* small languages
    adaptability, 235
    differences, 234, 236–238
    knowledge base, 234
    reasoning/multi-step inference, 235
    resource and deployment considerations, 236
    transformer models, 234
  methods, 226
  open-source tooling, 258–271
    BitNet models, 263
    cloud/edge deployments, 263
    comparative snapshot, 262
    computational demands, 262, 263
    DistilBERT, 261, 262
    dominant paradigms, 258
    extractive, 260
    generative, 259
    Hugging Face Transformers, 258–260
    inference optimization strategies, 264–267
    optimization strategies, 267–271
    performance metrics, 268

T5 fine-tuning, 260, 261
paradigms, 226

R

RAG, *see* Retrieval-augmented generation (RAG)
Recall-Oriented Understudy for Gisting Evaluation (ROUGE), 447
Recurrent neural networks (RNNs), 7
Reinforcement Learning from Human Feedback (RLHF), 495, 498
Retrieval-augmented generation (RAG), 408
Retrieval-augmented summarization, 292
RLHF, *see* Reinforcement Learning from Human Feedback (RLHF)
RNNs, *see* Recurrent neural networks (RNNs)
RoBERTa, *see* Robustly Optimized BERT Pre-Training Approach (RoBERTa)
Robustly Optimized BERT Pretraining Approach (RoBERTa), 90, 105
ROUGE, *see* Recall-Oriented Understudy for Gisting Evaluation (ROUGE)

S

Scaling models
    empirical relationships, 503
    guidance matrix, 505
    inference-time compute, 504
    knowledge-based tasks, 504
    mixture-of-experts (MoE), 502, 506–510
    multi-axis view, 506
    self-consistency, 504
    sparsity/compression techniques, 503
    striking differences, 504
    trade-offs, 502
SCoTD, *see* Symbolic Chain-of-Thought Distillation (SCoTD)
Sentiment analysis
    AWS services section, 189
    businesses/researchers, 177
    comparative overview
        accuracy, 208
        approaches, 208
        Comprehend *vs.* models *vs.* Lambda, 205
        cost-efficient, 207
        features, 208
        latency, 207
        SageMaker model, 209
    comparison, 184–187
    context and ambiguity, 178
    customer feedback analysis, 177
    customer reviews, 186–188
    data and multilingual challenges, 179
    definition, 176
    domains, 179
    financial text, 189
        comparative overview, 205–208
        Comprehend models, 191–193
        key components, 190, 191
        Lambda function, 190
        performance consideration, 199
        real-time inference, 193–199
        SageMaker, 200–206
        transformer models, 194
    healthcare and surveys, 177
    Hugging Face Transformers, 186
    industries
        applications, 224
        brand monitoring, 220
        customer satisfaction, 223

INDEX

Sentiment analysis (*cont.*)
    customer satisfaction insights, 221
    deployment options/scalability, 223
    implementation overview, 222
    social media analysis, 221
  intensity and nuance, 179
  key challenges, 178, 179
  Lambda function
    deployment, 196
    handler script, 195
    package model, 194–196
    terraform configuration, 196–199
    transformer models, 194
  LLMs approach, 180, 181
  negation handling, 178
  open-source tools, 210
    ambiguous, 219
    auto-scaling strategies, 214
    CI/CD pipelines, 214
    crowdsourcing, 219
    deployment, 215–218
    fine-tune TinyBERT, 212, 213
    Hugging Face transformers, 210–214
    inference optimization strategies, 217
    labeling data, 218–220
    lightweight deployment, 216
    quantization and pruning, 213
    retraining pipeline, 215
    synthetic data generation, 219
    text-to-text model, 211, 212
    versioning and experiment tracking, 214
  real-world applications, 175
  sarcasm, 178
  slang, 179
  SLMs models, 182, 183
  social monitoring, 177
  text analysis task, 176
  transfer learning, 188

SLMs, *see* Small language models (SLMs)

Small language models (SLMs), 1
  adaptive models, 13
  approaches/features, 4, 5
  challenges/limitations, 12
  classification, 90–116
  cloud-based large model, 513
  cloud platforms *vs.* open-source frameworks, 23
  compression techniques, 11, 510
  coreference/entailment models, 457–493
    base model, 481
    comparison, 482, 483
    DistilBERT, 480
    distilled/compressed versions, 479
    efficiency, 484
    knowledge distillation, 480
    MiniLM, 480
    multi-model endpoint, 483
    resource-constrained environment, 483
    TinyBERT, 480
  cost effective solution, 10
  cost-efficient serving, 396
  definition, 10
  deployment, 396
  deployment scenarios, 14, 401
  dialogue (*see* Dialogue systems)
  dialogues, 400–404
  DistilGPT-2 and DistilT5, 395
  DistilGPT-2 model, 396
  distillation, 511, 512
  efficiency/speed, 10, 500
  efficiency techniques, 502

INDEX

energy-efficient algorithms, 13
federated learning, 13
fine-tuning and optimization, 400
future developments, 39
hybrid pipelines, 502
IoT/Edge infrastructure, 401
knowledge distillation, 12, 394, 510
language translation, 325–365
lightweight adaptation methods, 501
lightweight version, 401
low-latency applications, 11
methods, 510
mobile phone/IoT device, 397
model cascading, 514
multilingual assistant, 514
narrow-domain tasks, 10
open-source community, 513
PEFT techniques, 513
performance comparison, 399
performance/cost/deployability, 396
privacy-focused deployments, 11
pruning/quantization, 12
pruning/retraining, 511, 512
QLoRA/LoRA, 512
quantization, 501, 512
quick chat response, 398–400
rationales/chain-of-thoughts, 511
real-time language processing, 11
real-world scenarios, 12
router systems, 502
rule-based systems, 11
sentiment analysis, 175–224
sparse transformers, 12
standard benchmarks, 501
summarization techniques, 281–323
task suitability, 14
techniques/innovations, 10, 13, 38
    knowledge distillation, 38
    pruning, 38
    quantization, 38
    sparse attention mechanisms, 39
text generation (*see* Text generation techniques)
tradeoffs, 514
SMT, *see* Statistical machine translation (SMT)
Statistical machine translation (SMT), 423
Summarization techniques
    abstractive, 287–290
    agents, 293
    aggregation
        deployment/scaling, 310
        extractive methods, 309
        sources, 309
    automated email summarization, 318
        approaches, 322
        batch, 320
        cleaning/preprocessing, 320
        email ingestion, 319
        general-purpose and performs, 322
        limitations, 322
        potential deployment
            architectures, 321
        real-world scenarios, 318
        results, 321
        security/compliance, 322
        solution overview, 319
        summarization model, 320
        traditional tools, 319
    AWS services, 294–302
        deployment pattterns, 295
        flexible/scalable tools, 295
        Lambda *vs.* SageMaker, 297
        reference architectures, 296
        stacks, 297
    business document

537

INDEX

Summarization techniques (*cont.*)
    abstractive, 315
    challenges and limitations, 317
    data preparation/fine-tuning, 312
    deployment pipeline, 313
    different models, 315
    DistilBERT, 311
    enterprise workflows, 316
    evaluation metrics, 314
    extensions, 317
    GitHub code, 318
    inference optimization, 313
    Jupyter notebook, 314
    model variants, 317
    multi-agent system, 316
    pipeline, 311
    post processing and storage, 312
    preprocessing and cleaning, 311
    scenarios, 310
    security/privacy concerns, 316
    time-consuming and error-prone, 310
  context distillation/state, 294
  DistilBERT, 298–300
  DistilGPT2, 291
  distilled and quantized models, 291
  extractive, 283–288
  multi-agent architectures, 294
  multi-document, 293
  open-source tools
    abstractive tasks, 303
    BitNet models, 306
    classical models, 303
    deployment, 306, 307
    evaluation metrics, 307
    extractive models, 304
    fine-tuning, 304
    Hugging Face Transformers, 303
    inference optimization strategies, 307
    methods, 305
  optimization, 302
  retrieval-augmented summarization, 292
  SageMaker, 300–302
  supervisor, 294
  TinyBERT, 291
  users/downstream systems, 282
Symbolic Chain-of-Thought Distillation (SCoTD), 511

T, U

Text generation techniques
  Amazon Web Services (AWS), 56–73
  chatbots and virtual assistants, 46, 47
  cloud environments, 55
  comparisons, 52, 53
  complexity, 53
  creative writing, 43
  deployment strategies, 55
  diverse genres/styles, 43
  edge devices, 47
  factors, 53
  fine-tuning/training, 53
  foundational aspects, 43
  generative/extractive, 47, 48
  industries
    chatbots/customer interaction, 84
    content generation/marketing, 81
    customization method, 79, 80
    execution process, 82
    on-device text generation, 86, 87
  knowledge assistant, 55

latency requirements, 53
mobile banking, 56
offering unique strengths, 42, 43
on-device deployment, 55
open-source frameworks, 73
    converting model, 78
    fine-tuning, 73–78
    local deployment, 78, 79
    mobile/edge devices, 79
    TensorFlow Lite, 78, 79
quantization techniques, 54
real-time applications, 41
reporting, 45
resource constraints, 53
structured/semi-structured content, 44, 45
technical considerations, 44, 48
training and deployment steps, 49–53
Transformer architecture, 2–4

V, W, X, Y, Z

VCAs, *see* Virtual Customer Assistants (VCAs)
Virtual Customer Assistants (VCAs)
    business, 409
    chat interfaces/internal portals, 410
    cloud deployment, 410
    containerization, 410
    core components, 408
    customers/internal employees, 407
    DistilBERT model, 408
    fallbacks, 409
    Gemma/BitNet, 408
    implementation, 408
    internal knowledge, 408
    monitoring/feedback, 410
    retrieval layer, 408
    technical issues, 409
    virtual assistant, 409

GPSR Compliance

The European Union's (EU) General Product Safety Regulation (GPSR) is a set of rules that requires consumer products to be safe and our obligations to ensure this.

If you have any concerns about our products, you can contact us on

ProductSafety@springernature.com

In case Publisher is established outside the EU, the EU authorized representative is:

Springer Nature Customer Service Center GmbH
Europaplatz 3
69115 Heidelberg, Germany

www.ingramcontent.com/pod-product-compliance
Lightning Source LLC
LaVergne TN
LVHW081344060526
838201LV00050B/1707